Research Methods in Occupational Health Psychology

Research Methods in Occupational Health Psychology

Measurement, Design, and Data Analysis

Edited by

Robert R. Sinclair, Ph.D.
Clemson University

Mo Wang, Ph.D.
University of Florida

Lois E. Tetrick, Ph.D.
George Mason University

Routledge
Taylor & Francis Group
NEW YORK AND LONDON

First published 2013
by Routledge
711 Third Avenue, New York, NY 10017

Simultaneously published in the UK
by Routledge
27 Church Road, Hove, East Sussex BN3 2FA

Routledge is an imprint of the Taylor & Francis Group, an informa business

Library of Congress Cataloging in Publication Data
Research methods in occupational health psychology : measurement, design, and data analysis / edited by Robert R. Sinclair, Mo Wang, Lois E. Tetrick.
p. cm.
Includes bibliographical references and index.
ISBN 978-0-415-87932-3 (hardback : alk. paper)
1. Industrial psychiatry. 2. Clinical health psychology—Research—Methodology. I. Sinclair, Robert R. II. Wang, Mo. III. Tetrick, Lois E.
RC967.5.R48
201 2616.890072–dc
232012012141

ISBN: 978-0-415-87932-3 (hbk)
ISBN: 978-0-203-09524-9 (ebk)

Typeset in Minion and Optima
by EvS Communication Networx, Inc.

Contents

About the Editors

Robert Sinclair is an Associate Professor of Industrial and Organizational Psychology at Clemson University, where he also serves as the Graduate Program Coordinator for the Department of Psychology. He completed his PhD in industrial/organizational psychology at Wayne State University in 1995. Prior to moving to Clemson University in 2008, he held faculty positions at the University of Tulsa and Portland State University. Dr. Sinclair is a founding member and Past-President of the Society for Occupational Health Psychology. He currently serves as an editorial board member for the *Journal of Occupational Health Psychology*, *Journal of Management*, and *Journal of Organizational Behavior* and as a panel member for the Occupational Safety and Health Study Section of the National Institute for Occupational Safety and Health. His recent work includes an edited volume (in press with Jonathan Houdmont and Stavroula Leka), *Contemporary Occupational Health Psychology: Global Perspectives on Research and Practice* (Vol. 2) and an edited volume (in press with Tom Britt) titled *Psychological Resilience in Military Personnel: Theory and Practice*. Dr. Sinclair's research focuses on individual (e.g., personality) and organizational (e.g., leadership) factors that contribute to occupational health concerns faced by military personnel, nurses, and entry-level hourly employees. His specific interests include economic stress, the employment relationship, work schedules, counterproductive workplace behavior, and psychological resilience.

Mo Wang is an Associate Professor of Management at University of Florida, where he also serves as the codirector of the Human Resource Research Center. He received his joint PhD in industrial-organizational psychology and developmental psychology at Bowling Green State University in 2005. Prior to moving to University of Florida, he held faculty positions at Portland State University (2005–2008) and the University of Maryland (2009–2011). Dr. Wang specializes in research and applications in the areas of retirement and older worker employment, occupational health psychology, cross-cultural HR management, leadership, and advanced quantitative methodologies. He has received Academy of Management HR Division Scholarly Achievement Award (2008), Careers Division Best Paper Award (2009), and Erasmus Mundus Scholarship for Work, Organizational, and Personnel Psychology (2009) for his research in these areas. He also received Early Career Achievement Awards from Society for Industrial-Organizational Psychology (2012), Academy of Management's HR Division (2011) and Research Methods Division (2011), and Society for Occupational Health Psychology (cosponsored by the APA and NIOSH, 2009). He currently serves as an Associate Editor for *Journal of Applied Psychology*. He also serves on the Editorial Boards of *Personnel Psychology, Journal of Management, Organizational Research Methods, Journal of Occupational Health Psychology*, and *Journal of Business and Psychology*.

Lois Tetrick received her doctorate in industrial and organizational psychology from Georgia Institute of Technology in 1983. Upon completion of her doctoral studies, she joined the faculty of the Department of Psychology at Wayne State University and remained there until 1995 when she moved to the Department of Psychology at the University of Houston. She joined the faculty at George Mason University as the Director of the Industrial and Organizational Psychology Program in 2003. Dr. Tetrick served as Editor of the *Journal of Occupational Health Psychology* (2006–2010), Associate Editor of the *Journal of Occupational Health Psychology* (2002–2006), and Associate Editor of the *Journal of Applied Psychology* (1996–2001). She currently serves on the editorial boards of *Journal of Organizational Behavior, Journal of Managerial Psychology, Journal of Applied Psychology*, and *Management and Organizational Review*. She coedited the first and second editions of the *Handbook of Occupational Health Psychology* with James C. Quick and *Health and Safety in Organizations* with David Hofmann. She also coedited *The Employment Relationship: Examining Psychological and Contextual Perspectives* with Jacqueline Coyle-Shapiro, Lynn Shore, and Susan Taylor. Dr. Tetrick is a founding member of the Society for Occupational Health Psychology and a fellow of the European Academy of Occupational Health Psychology, the American Psychological Association (APA), the Society for Industrial and Organizational Psychology (SIOP) and the Association for Psychological Science (APS).

She served as 2007–2008 President of SIOP, the Chair of the Human Resources Division of the Academy of Management (2001–2002), and has represented SIOP on the American Psychological Association (APA) Council of Representatives (2003–2005), and the APA Board of Scientific Affairs (2006–2009). Dr. Tetrick's research interests are in the areas of occupational health and safety, occupational stress, and the work–family interface. Her other area of research focuses on psychological contracts and the exchange relationship between employees and their organizations. A common underlying interest in both of these lines of research is incorporating a global perspective in understanding employees' experiences of the work environment.

About the Contributors

Amy B. Adler
U.S. Army Medical Research Unit-Europe,
Walter Reed Army Institute of Research,
Heidelberg, Germany.

Bengt B. Arnetz
Wayne State University,
Detroit, MI and
Uppsala University,
Uppsala, Sweden

Janet L. Barnes-Farrell
University of Connecticut
Storrs, CT

Björn Bergström
Pacific University,
Forest Grove, OR

Carmen Binnewies
University of Münster
Münster, Germany

Paul D. Bliese
Walter Reed Army Institute for Research
Silver Spring, MD

Emily Brede
University of Texas-Arlington
Arlington, TX

Michelle L. Burnett
Clemson University
Clemson, SC

Peter Y. Chen
University of South Australia, Adelaide
South Australia, Australia.

Yunhee Choi
University of Texas-Arlington
Arlington, TX

Konstantin P. Cigularov
Old Dominion University
Norfolk, VA

Lilia M. Cortina
University of Michigan
Ann Arbor, MI

Reeshad S. Dalal
George Mason University
Fairfax, VA

Erin M. Eatough
University of South Florida
Tampa, FL

Mark G. Ehrhart
San Diego State University
San Diego, CA

Gwenith G. Fisher
University of Michigan
Ann Arbor, MI

Lori Francis
St. Mary's University
Halifax, Nova Scotia, Canada

Robert J. Gatchel
University of Texas-Arlington
Arlington, TX

Kirby Gilliland
University of Oklahoma
Norman, Oklahoma

Ryan Ginter
George Mason University
Fairfax, VA

Nancy Goldstein
Sarpy and Associates, LLC.
Charlo Hesville, VA

M. Gloria González-Morales
University of Guelph
Guelph, Ontario, Canada

Krista Howard
Texas State University
San Marcos, TX

Seth Kaplan
George Mason University
Fairfax, VA

Lisa M. Kath
San Diego State University
San Diego, CA

Terence M. Keane
National Center for PTSD
Boston, MA

E. Kevin Kelloway
St. Mary's University
Halifax, Nova Scotia, Canada

Joseph N. Luchman
George Mason University
Fairfax, VA

Lisa A. Marchiondo
Wayne State University
Detroit, MI

Gerald Matthews
University of Cincinnati
Cincinnati, Ohio

Joseph J. Mazzola
Roosevelt University
Schaumburg, IL

James A. McCubbin
Clemson University
Clemson, SC

Lauren M. Menger
Colorado State University
Fort Collins, CO

Sandra Ohly
University of Kassel
Kassel, Germany

April Rose Panganiban
University of Cincinnati
Cincinnati, OH

June J. Pilcher
Clemson University
Clemson, SC

Felicia Rabito
Tulane University
New Orleans, LA

Scott C. Roesch
San Diego State University
San Diego, CA

Johan Rosqvist
Pacific University
Forest Grove, OR

Sue Ann Sarpy
Sarpy and Associates, LLC
Charlottesville, VA

Wilmar B. Schaufeli
Utrecht University
Utrecht, The Netherlands

Irvin Sam Schonfeld
The City College of the City
University of New York
New York, NY

Lindsay E. Sears
Healthways, Inc.
Franklin, TN

Sabine Sonnentag
University of Mannheim
Mannheim, Germany

Paul Spector
University of South Florida
Tampa, FL

Lois E. Tetrick
George Mason University
Fairfax, VA

Jay C. Thomas
Pacific University,
Forest Grove, OR

Matthew Ventimiglia
Wayne State University
Detroit, MI and
University of Detroit Mercy
Detroit, MI

Whitney E. Worzer
University of Texas-Arlington
Arlington, TX

Peter Warr
University of Sheffield
Sheffield, UK

Le Zhou
University of Florida
Gainesville, FL

Foreword

For the past 35 years, I have had the privilege of helping to foster the growth and development of the field of occupational health psychology (OHP). In looking back at my experiences, an inescapable conclusion for me is that the growth of OHP as a discipline has been commensurate with (and in part a result of) our attention to methodological rigor. In contrast to the early years of *Journal of Occupational Health Psychology* (*JOHP*) when both submission and rejection rates were low, we now receive over 225 manuscripts per year with a rejection rate (principally for methodological reasons) of over 93%. Over the years, scrupulous attention to study design, measurement, and analytic issues has increasingly allowed OHP researchers to both successfully compete for funding and publish study results. Needless to say, attention to methodological rigor continues to be an important key to successful OHP research funding, publishing, and the continued growth of our field. Indeed, given cutbacks by funding sources over recent years, this attention is perhaps more important than ever before. Therefore, this book on OHP research methodology is both significant and timely.

Attention to methodological rigor has also brought a modicum of much needed recognition of OHP from scholars (and funding sources) in other disciplines. Exceptionally high quality studies examining work related psychological factors as principal predictors, mediators, or outcomes are now published not only in *JOHP* and other well-respected psychology journals but increasingly in a variety of "mainstream" occupational medicine, public health, and occupational safety journals. While this is encouraging, it is quite clear that OHP has not yet achieved the level of recognition we are aiming for and until we gain full-fledged recognition within the broader contextual fields of psychology and, perhaps more importantly, occupational health and safety (OH&S), we cannot realize our full potential as a unique and independent discipline. To this end, it is important for us to fully recognize that our principal focus is on occupational health and safety, and while we may approach it from a psychological perspective, we share the occupational health and safety focus with researchers in the fields of occupational epidemiology, medicine, safety, and nursing; industrial hygiene; ergonomics; health physics; and others.

Each of these fields has rich (and long standing) research traditions and utilizes research methodologies that have applicability to OHP. Indeed, through the appreciation and use of these methodologies, we can attract greater research funding, make even greater contributions to the understanding of occupational health and safety, more easily publish in mainstream OH&S journals, and further increase our field's recognition. This book, while focusing primarily on the core methodological issues that arise in "traditional" OHP research, also provides the reader with an exposure to the benefits of some of the methodologies utilized in more mainstream OH&S research. This, in my view, is enormously important.

Finally, OHP is often described as an interdisciplinary field representing a broad range of backgrounds, interests and specializations. As noted above, we share with a variety of other fields the common interest of advancing understanding of OH&S. While I don't believe we are yet a fully interdisciplinary field, there has been a steady increase in interest in OHP and the methodologies of OHP among scholars in the more mainstream OH&S fields and others (e.g., sociology, industrial engineering and health, social and clinical psychology). This interest can only serve to improve the "health" of OHP. This book will provide such individuals with an excellent source of state of-the-art information and, by so doing, help continue our growth and allow us to become truly interdisciplinary in nature.

Joseph J. Hurrell, Jr., PhD,
Editor, *Journal of Occupational Health Psychology*

Preface

Occupational health psychology has arrived! Our discipline has established healthy professional organizations such as the Society for Occupational Health Psychology (SOHP) and the European Academy of Occupational Health Psychology (EA-OHP), highly cited journals, including *Work & Stress* and the *Journal of Occupational Health Psychology*, and thriving professional conferences such as the Work Stress and Health conference (sponsored by the National Institute for Occupational Safety and Health, the American Psychological Association, and SOHP) and the biannual conference of the EA-OHP. Many doctoral level training programs are producing graduates with OHP interests (including several coauthors and reviewers of the chapters in this volume). And, observers of the job market will note both a growing interest in the particular knowledge and skill set of OHP professionals and a steadily accumulating track record of success of OHP-trained graduates in competing for the best available jobs.

The empirical literature on OHP is growing at a similar pace; each year we see more and better research on occupational health—research that reflects contributions from many scientific disciplines. Similarly, there is been a steady growth of scholarly books on OHP including the *Handbook of Occupational Health Psychology* (Quick & Tetrick, 2011), the *Contemporary Occupational Health Psychology* series published by Wiley-Blackwell, and the *Research in Occupational Stress and Well-being* series published by Sage. In addition to these broad reviews, several handbooks have focused on OHP topics such as stress (Barling, Kelloway, & Frone, 2005), safety (Barling & Frone, 2004; Hofmann & Tetrick, 2003), and workplace violence (Kelloway, Barling, & Hurrell, 2006). Further, although calls for more and better OHP-focused interventions are a persistent refrain in the literature, many recent discussions of intervention-related topics suggest tremendous progress in this area in the last 5 to 10 years, both in terms of empirical support for certain kinds of interventions (cf. Lamontagne, Keegel, Louie, Ostry, & Landsbergis, 2007) and in terms of understanding best practices in how to conduct and evaluate inventions (Cox, Karanika, Griffiths, & Houdmont, 2007; Nielsen, Fredslund, Christensen, & Albertsen, 2006; Nielsen, Randall, Holten, & González, 2010; Scharf et al., 2008).

Although there are many reasons for excitement about the development of OHP to date, we have unfinished business. OHP scholarship deals with complex problems, including interactions between person and situation factors and potentially dynamic changes in some of these factors over time. Also the applied setting of OHP research often requires researchers to integrate multiple methods and/or levels of data analysis. In addition, researchers often face difficulties establishing causal sequences among variables, both because critical health outcomes are multiply determined, and because it is often difficult to conduct ecologically valid experimental research on occupational health problems. The difficulty of responding to these challenges is amplified by concerns about the appropriate conceptualization and operationalization of core OHP constructs.

Despite these concerns, few publications have discussed methodological issues associated with OHP research. This is an important gap in the literature and it is made even worse by the challenges of keeping up with relevant methodological developments in multiple OHP-related disciplines. Thus, with this volume, we set out to review, summarize, and make practical recommendations regarding methodological issues in different aspects of OHP research. We sought a balance between chapters focused on state-of-the-art summaries of issues on topics of long-standing concern in applied psychology and topics from researchers in other disciplines who brought their perspective to best practices in OHP research. Our goal was to help scholars by translating recent innovations in methodology into sets of useful concrete recommendations to improve their own research as well as their training of future researchers.

The book is organized into two sections. The first section includes brief chapters that review measurement issues in several OHP topics. We chose these topics with a couple of goals in mind. We wanted to include chapters on core OHP topics such as well-being, emotions, incivility, engagement, and work–family issues. These chapters are intended to provide overviews of recent developments and best practices in rapidly growing areas of scholarship. We also included several other chapters written by scholars who either (a) might not be regarded as traditional OHP scholars or (b) whose work is influenced by developments in related OHP disciplines. With these chapters, we sought to provide scholars whose training background is more in traditional applied psychology (e.g., industrial-organizational psychology) with overviews of topics that are less likely to be addressed in their training. These topics include fatigue, cognitive functioning, posttraumatic stress disorder, depression and anxiety, musculoskeletal functioning, and immune system functioning.

Because we wanted the chapters in the first section to be brief and focused on best-practices, we specifically asked the contributors not to include lengthy reviews of specific studies utilizing each method. In each case, we provided the authors with several questions related to their particular construct, process, or system, including: What are the common research questions/applications of this measurement strategy/ data source in the OHP context? What are the various kinds of available options for measurement in this area? What kinds of decisions/ issues should a researcher consider before embarking on study using this measurement strategy/data source? And, what differentiates outstanding studies that use this measurement strategy/data source from the rest of the research literature? The chapters place different amounts of emphasis on these questions, as appropriate to the topic; we expect readers to benefit tremendously from the guidance provided by the authors in these chapters.

The second section contains longer reviews addressing research design and statistical issues in OHP. Some chapters review widely used approaches, such as experimental, longitudinal, and multilevel design/ analyses, as well as survey sampling and self-report methods. Other chapters address data sources and research strategies that, in our view, have been underutilized by OHP scholars; these chapters include archival research and person-centered methods, event sampling, and qualitative research. As with the measurement chapters, we asked authors to address best practices with respect to their particular topic. We asked them to review, summarize, and integrate the literature on their chapter topic in relation to OHP research and to provide clear guidance and recommendations regarding practicing the research methods they discuss. As with the measurement chapters, we asked authors to consider some general questions about their topic, including: what are the implications of your chapter for constructing better theories in OHP? Do issues discussed in your chapter raise any particular concerns/requirements about samples? And, what are the common mistakes researchers make when dealing with issues covered by your chapter? We hope readers will find these chapters to be authoritative and informative reviews that will help them design better studies and more effectively and critically analyze research findings.

References

Barling, J., & Frone, M. R. (Eds.). (2004). *The psychology of workplace safety.* Washington, DC: American Psychological Association.

Barling, J., Kelloway, K. E., & Frone, M. R. (2005). *Handbook of work stress.* Thousand Oaks, CA: Sage.

Cox, T., Karanika, M., Griffiths, A., & Houdmont, J. (2007). Evaluating organizational-level work stress interventions: Beyond traditional methods. *Work & Stress, 21,* 348–362.

Hofmann, D. A., & Tetrick, L. E. (Eds.) (2003). *Health and safety in organizations: A multilevel perspective.* San Francisco, CA: Jossey-Bass.

Kelloway, E. K., Barling, J., & Hurrell, J. J, (Eds.). (2006). *Handbook of workplace violence.* Thousand Oaks, CA: Sage.

Lamontagne, A. D. Keegel, T., Louie, A.M., Ostry, A., & Landsbergis, P.A. (2007). A systematic review of the job-stress intervention evaluation literature, 1990–2005. *International Journal of Occupational & Environmental Health, 13,* 268–280.

Nielsen, K., Fredslund, H., Christensen, K. B., & Albertsen, K. (2006). Success or failure? Interpreting and understanding the impact of interventions. *Work & Stress, 20,* 272–287.

Nielsen, K., Randall, R., Holten, A., & González, E. R. (2010). Conducting organizational-level occupational health interventions: What works? *Work & Stress, 24,* 234–259

Quick, J. C., & Tetrick, L. E. (Eds.). (2011). *Handbook of occupational health psychology* (2nd ed.). Washington, DC: American Psychological Association.

Scharf, T., Chapman, L., Collins, J., Limanowski, J., Heaney, C., & Goldenhar, L. M. (2008). Intervention effectiveness evaluation criteria: Promoting competitions and raising the bar. *Journal of Occupational Health Psychology, 13,* 1–9.

Acknowledgments

We have many people to thank for their help with this book. All three of us are grateful for the love and support of our families, without whom our own work would be much less healthy. We also wish to express our appreciation to Anne Duffy and her staff for their support throughout the publication process. Of course, this would not be a book without the contributions of all of the chapter authors; one of the great things about doing work in Occupational Health Psychology is the people you meet—we are truly fortunate to have had the opportunity to work with an incredible group of scholars, and we thank them all for their work, as well as their patience with and responsiveness to the editorial process. Finally, we wish to think all of the graduate students who helped us with the review process. They include Laura Poms, Laura Wolkoff, Lindsay Sears, Crystal Burnette, Chad Peddie, Le Zhou, Melissa Waitsman, and Gary Giumetti. Their insightful comments definitely improved the book, and we look forward to seeing them as members of the next generation of leaders in the science and practice of occupational health psychology.

Acknowledgments

Part I

Measurement

1

Measurement of Immune System Functioning

Bengt B. Arnetz and Matthew Ventimiglia

Immune System Functioning and Occupational Health Psychology

Mind–Body Connections

The mind–body connection is a topic of great interest and the subject of many research studies. It is well documented that psychological processes have a significant impact on health. Within the occupational health psychology domain, numerous studies have reported associations among workplace stressors, unemployment, and adverse mental, somatic, and social health effects (Arnetz et al., 1987; Grawitch, Trares, & Kohler, 2007). In order to further develop the science and application of occupational health psychology, it is important to enhance our understanding of the causal biological links between workplace stressors, resiliency factors, and health. Such knowledge will inform theoretical models and offer a complimentary means to evaluate workplace interventions.

Psychological stress, including occupational stress, has been widely linked to altered immune functioning, and thus is an important factor to consider when assessing possible contributors to poor physical health (Meijman, von Dormolen, Herber, Rogen, & Kuioer, 1995; Zeier, Brauchi, & Joller-Jemelka, 1996). Occupational health psychologists should consider the bilateral relationship between the central nervous system and the immune system. Not only does the brain influence the immune system; the immune system influences the brain. For example, activation of proinflammatory parts of the immune system due to viral and bacterial infections or autoimmune disorders, such as rheumatoid disease, commonly induces "feeling sick" and "sickness behavior" (Rief, Hennings, Riemer, & Euteneuser, 2010). Sickness behavior is characterized by loss of energy, social withdrawal, and impaired cognitive functioning, which can ultimately affect performance at work.

The immune system is part of an intricate communication and feedback loop which includes the neuroendocrine, metabolic, and cardiovascular systems and involves cognitive, emotional, behavioral, and social processes (Howk & Bennett, 2010; McEwen, 1998). In this chapter, we briefly review the role and structure of the immune system and discuss some common, readily available, feasible, and reasonably priced measures of immune system functioning. Furthermore, the chapter will discuss sampling considerations and other methodological issues researchers need to consider prior to choosing specific immune markers.

The Role and Structure of the Immune System

The immune system protects us from pathogens with layered defenses of increasing sophistication. There are reliable and valid measures applicable to occupational health research to monitor the functionality of the respective layer of the immune system. Table 1.1 summarizes the various immune markers, what part of the immune system they represent, and means to sample bodily fluids to determine the concentration and function of these markers.

Mechanical Barrier

Surface barriers, such as skin, act to prevent pathogens from entering the body. The lungs, gastrointestinal, and genitourinary tracts have additional means to defend against pathogens. Coughing and sneezing mechanically eject foreign particles from the respiratory tract. Mucus secreted by the respiratory and gastrointestinal tracts traps and ejects pathogens. Flushing of tears and urine also facilitates the expulsion of pathogens.

Table 1.1 Markers of Immune System Functions in Occupational Health Psychology Research

Marker	Tissue to sample	Stress Type		Stress Effects*	Related Disorders#
		Acute	*Sustained*		
Wound healing	Skin	.		Stress delays healing	Diabetes, Immunodeficiency
Lysozyme/Phospholipase A2	Saliva/Tears/Breast milk	.		Stress decreases levels of markers	Alzheimer's, Epilepsy (?)#
Immunoglobulin A	Saliva	.	.	Anti-microbial. Stress hampers secretion	Rheumatoid Arthritis, Lupus, Eczema
Alpha-Amylase	Saliva, Blood		.	Sympathetic n.s. stimulates production	Immunodeficiency, Pancreatitis
Intradermal application of irritant	Skin		.	Stress decreases inflammatory response	Eczema
Cytotoxic T-lymphocytes	Blood		.	Decreased function with prolonged stress	HIV, Immunodeficiency
Natural Killer (NK) Cells	Blood		.	Decreased function with prolonged stress	Infection, Cancer
Neutrophils and Macrophages	Blood		.	Decreased function with prolonged stress	Aplastic anemia, Cancer
Cytokines (proinflammatory)	Blood	.	.	Interleukin 6, 8, 1, 2. Increase with stress	Fibromyalgia, Rheumatoid Arthritis, Cancer
Eicosanoids	Blood	.		Prostaglandins & Leukotrienes	Alzheimer's, Arthritis, Cancer, Depression
Interferon	Blood		.	Promotes inflammation & antimicrobial	Viruses, Immunodeficiency

Table 1.1 Continued

Marker	Tissue to sample	Stress Type		Stress Effects*	Related Disorders#
		Acute	*Sustained*		
C-reactive protein (CRP)	Blood , Saliva	.	.	Increases with stress.	Bacterial/viral infections, Cardio. Disease
Complement system	Blood	.		Peptides that promote inflammation	Lupus, Autoimmune disorders
Insulinlike growth factor-1 (IGF-1)	Blood		.	Anti-inflammatory. Decreases with stress	Cancer

*Most typical stress response described under Comments. In reality, stress responses depend on numerous factors, including genetics, general health status of participant, intensity and duration of stress, and levels of other factors of relevance for immune system functioning. For further discussions, please see text.

Marker	Tissue to sample	Stress Type		Comments	Related Disorders#
		Acute	Sustained		
White blood cells	Blood		.	Includes granulocytes/monocytes/lymphocytes	
Granulocytic cells	Blood		.	4 different cells; Neutrophils (destroy bacteria)	Infections
				Eosinophils (engulf antigen-antibody) Allergies, Asthma	
				Basphils (stimulates inflammation)	Allergies
				Mast cells (promote inflammation)	Allergies, Asthma

Monocytes	Blood	.	Divide into phagocytes that engulf antigens	Autoimmune diseases, Cushing's disease
Lymphocytes	Blood	.	Most important cells in the immune defense	
NK cells	Blood	.	Part of the innate system; exists since birth	Infections, Cancer
T cells	Blood	.	Cell-mediated immunity; recognizes "non-self"	HIV, Immunodeficiency
			Many subtypes of T-cells, including CD4+ T	
			Stress reduces funct. cap. to react to antigen	
B cells	Blood	.	Produces specific antibodies that target antigens	Lupus, Rheumatoid Arthritis

In many cases, the relationship to specific disease is complex. Some of the marker–disease relationships listed in the table are suggestive or speculative.

Chemical Barrier

There are also chemical barriers that protect against infections. The skin and respiratory tract secrete antimicrobial proteins (e.g., ß-defensins). Additionally, enzymes such as lysozyme and phospholipase A2 in saliva, tears, and breast milk are also part of the chemical defense system against pathogens.

Further, salivary alpha-amylase is a salivary enzyme, which is secreted by the salivary glands. Similar to salivary IgA, salivary amylase is a part of the oral mucosal defense system, the major function of which is to thwart the growth of antigens. High levels of amylase have been reported to be linked to activation of the sympathetic nervous system. Thus, high levels of amylase may be indirectly related to stress, which decreases immune functioning.

The Innate Immune System

The innate, or nonspecific, part of the immune system is activated once microorganisms or toxins have successfully entered the body. Cell-mediated immunity is an important part of this defense system. It depends on the activation of cytotoxic T-lymphocytes, natural killer cells (NK), macrophages, and cytokines to combat pathogens. The innate response is triggered when pathogens are recognized by the body, or by damaged cells that send out chemical stress signals.

The innate immune system is nonspecific and induces inflammation to defend the body against a vast array of different pathogens. Redness and swelling and signs of inflammation are caused by increased blood flow to infected and damaged tissue. Increased blood flow and increased temperature (fever) in the affected body region enhances the efficacy of the pathogen defense system. The inflammation response is caused by specific immune factors released from the damaged cells. Inflammation is further promoted by the release of eicosanoids, such as prostaglandins which produce fever and dilation of blood vessels, and leukotrienes that attract certain white blood cells. Damaged cells also release cytokines, including interleukins, such as interleukin 6 (IL-6), IL-8, and tumor necrosis factor-α, that are responsible for communication between white blood cells; chemokines that promote the migration of defensive white blood cells to affected tissue, and interferons with antiviral properties. Interleukins also stimulate the production and release of C-reactive protein (CRP) from the liver; an important component of the innate immune system. CRP, popular in psychoneuroimmunologic research, is a reliable prognostic marker for future cardiovascular disease (Ridker, Hennekens, Buring, & Rifai, 2000).

Concomitantly with the activation of the inflammatory process, the repair system is activated to accelerate recovery from damage caused by the infection. Thus, cytokines, for example IL-1 and IL-2, play an important role in regulating the initial immune response, but also restore the system to the status quo following an infection. Insulinlike growth factor (IGF)-1 is also an important anti-inflammatory peptide, which might decrease during severe stress (Cankaya, Chapman, Talbot, Moynihan, & Duberstein, 2009).

The complement system is another important component of the innate immune system. It contains more than 20 different proteins, which "complement" other systems in the killing of pathogens to protect the body against infection. The complement system binds to antibodies that are already attached to pathogens that the body wants to eliminate. The complement system also binds directly to proteins or carbohydrates on the surface of pathogens. It contains enzymes, proteases that destroy the structure of pathogens and also attract additional complement proteases. This way, the complement system amplifies an important cascade system promoting the migration of immune cells to damaged tissue, increasing vascular permeability to ensure cellular defense mechanisms get access to pathogens. The complement system also marks the surface of pathogens for destruction by other immune system components.

Cellular Barriers

White blood cells, or leukocytes, are an additional aspect of the innate immune system. White blood cells are part of the cellular (or cell-mediated) immune response system. This system consists of cell-engulfing phagocytes, which include macrophages and neutrophils; mast cells that promote inflammation; and natural killer cells. The ability for phagocytosis (consumption) of pathogens is an important feature of the innate immune system. Phagocytes constantly circulate in the body to scan for pathogens. They can also be called to specific locations by cytokines released from infected cells.

Adaptive Immune System

The adaptive immune system allows for a stronger and more targeted response to pathogens and also generates immunologic memory. Specific "memory cells" maintain a memory of the pathogen, which allows for an effective and speedy removal of the pathogen should it reappear in the future. The adaptive immune system consists of special types of white blood cells. There are three types: granulocytic cells, monocytes, and lymphocytes. Granulocytic cells consist of neutrophils which destroy bacteria; eosinophils which engulf antigen-antibodies and also defend against some parasites; and basophils which aid in inflammation and reaction to allergens.

Monocytes originate in the bone marrow and are able to move relatively quickly (within 8 to 12 hours) to sites of infection. They possess the ability to divide into macrophages and dendritic cells. Dendritic cells are phagocytes that are located in tissue that are in contact with the external environment, such as the skin and nose. They also present antigens to T cells, part of the adaptive immune system.

Lymphocytes, one of the most important cells in the immune defense system, consist of natural killer cells (NK), T cells, and B cells. NK cells are a part of the innate immune system, which is present at birth, and are crucial in defending the body from both tumors and virally infected cells. T cells are involved in cell-mediated immunity whereas B cells are primarily responsible for humoral immunity, producing specific antibodies). The function of T cells and B cells is to recognize specific "nonself" or foreign antigens. The cells generate specific responses that are tailored to maximally eliminate specific pathogens or pathogen infected cells. B cells respond to pathogens by dividing into plasma cells that secrete a large number of identical antigen-specific antibodies that circulate in the blood and the lymph. These antibodies bind to the specific antigen (pathogen) and mark them either for destruction by phagocytes or for destruction by the complement system. In response to pathogens, some T cells (i.e., T helper cells) produce cytokines that direct the immune response while other T cells (cytotoxic T cells) produce toxic granules that kill off pathogen-infected cells. There are numerous subtypes of T lymphocytes. One type, CD4+ T lymphocyte is of special interest in psychoneuroimmunologic research since it has been shown to have prognostic validity in HIV positive patients.

When B and T cells have been activated and begin to divide, some of the offspring will become long-lasting memory cells. These cells will "remember" specific antigens (pathogens) and react promptly should they reenter the body. Similarly, this same response is achieved by vaccination or immunization. The response to immunization, measured as the concentration of specific antibodies, is weakened in response to intense and sustained stress.

Inflammation

Inflammatory markers are strongly associated with cardiovascular disease, the leading cause of death in most developed countries, as well as Type 2 diabetes, arthritis, osteoporosis, Alzheimer's disease, periodontal disease, some cancers, frailty, and functional decline (Kiecolt-Glaser, 2009). Markers of inflammation include serum levels of interleukin (IL)-6 and other proinflammatory cytokines (e.g., IL-8). Stress and negative emotions such as anxiety and depression enhance the production of proinflammatory cytokines and ultimately increase inflammation (Sergerstrom & Miller, 2004).

Assessment of Humoral and Cellular Immune Reactions

Humoral and cellular immunity are the two main types of immune responses and can be measured with separate types of assays. For humoral immune responses, available assays measure production of antibodies. Cellular immune responses can be assessed through assays that measure functions of the responding T cells that occur from relatively early time points (major histocompatibility complex [MHC]-peptide binding, cytokine production, cytotoxicity) to later time points (proliferation) in the T cell activation process.

Enumerative Immune Assays

Immunoassays can also be differentiated by two classes: enumerative and functional. Enumerative assays consist of measures which assess the quantity or percentage of white blood cells (leukocytes) in the peripheral blood as well as those which measure the quantity of antibodies (e.g., immunoglobulins) or the antibody to pathogen ratio (Herbert & Cohen, 1993).

Measuring the quantity of specific types of white blood cells in the peripheral blood is a popular method researchers use to assess immune functioning. Such methods are commonly used because they are relatively simple to perform. On the other hand, the quantification of white blood cells as a means to test immune functioning can be difficult to interpret. For instance, a greater number of different cell types does not necessarily correspond with greater immune system functioning (Stites & Terr, 1991). Additionally, there exist alternative explanations such as cell migration outside the vascular system, which may explain a change in the number of cell types featured in blood samples (Herbert & Cohen, 1993).

Functional Immune Assays

Functional immunoassays, like enumerative assays, are performed in vitro but are different in that they measure not the quantity of certain immune cells, but instead the cells' abilities to combat antigens. Researchers using functional immunoassays can measure immune functioning by, for example, exposing lymphocytes to mitogens and then assessing the efficacy of lymphocyte response to the exposure. This process is referred to as *lymphocyte proliferation*. Functional assays assess how effectively lymphocytes divide when stimulated by mitogens (Arnetz et al., 1987; Davis, Kozinetz, & Noroski, 2006).

Another common activity performed by functional assays is to assess for NK cell cytotoxic activity. This approach determines how effectively NK cells kill damaged or altered cells. The process is done by incubating immune cells with tumor cells, which are natural targets for NK cells. The outcome is measured by ratio of immune to tumor cells with higher levels of immune cells indicating better immune response. Several studies have reported a link between stress, including occupational stress, and reduced NK activity and immune functioning in general (Boscolo, Youinou, Theoharides, Cerulli, & Conti, 2008; Di Dinato et al., 2006; Nakata et al., 2000).

Neuroendocrine Markers

The focus of the current chapter is on immune markers. However, as discussed earlier, the immune system is closely integrated with a number of biological systems in order to maximize the body's ability to withstand internal and external stressors. The sympathetic-adrenomedullary (SAM) system and the hypothalamic-pituitary-adrenal (HPA) axis directly influence important aspects of the immune system, including modulating the expression of cytokines and chemokines, with implications for the inflammatory response (Glaser, Rabkin, Chesney, Cohen, & Natelson, 1999; Ziemssen & Kern, 2007). Neuroendocrine markers include cortisol, catecholamines, and dehydroepiandrosterone.

Cortisol

Cortisol, the main stress hormone released from the HPA axis, is critical in the acute stress response (Ekman & Arnetz, 2006). Sustained HPA activation and thus higher prolonged concentration of cortisol and other glucocorticoids result in a number of pathological changes, however, including suppressed immune response. Glucocorticoids can enhance amygdala activity, heighten fear conditioning, and damage neurons, especially in the prefrontal cortex and the hippocampus (Charney, 2004; Makino, Gold, & Shulkin, 1994; Shepard, Barron, & Myers, 2000). In particular, elevated cortisol levels have been associated with several health problems including Type 2 diabetes, cardiovascular disease, cognitive impairment, and overall suppressed immune functioning. Cortisol can be measured in blood, saliva, and urine. Increasingly, assessment of free cortisol in saliva is used in occupational stress research (e.g., Groer et al., 2010; Rystedt, Cropley, Devereux, & Michalianou, 2008). However, glucocorticoids show substantial circadian variations; they respond to sleep deprivation, alcohol and other commonly used drugs, as well as lifestyle factors, including exercise and nutritional habits. Circadian variations can be controlled by restricting all sampling to a set time interval. This applies to cross-sectional studies, and it is especially important when individuals are followed over time. Switching from daylight saving time to standard time might also impact the validity of glucocorticoids in longitudinal studies.

Catecholamines

Catecholamines such as epinephrine (adrenaline) and norepinephrine (noradrenaline) are secreted by the SAM system in response to stress. Higher levels of catecholamines have been linked to health problems, such as cardiovascular disorders including higher blood lipid levels, increased blood clotting, atherosclerosis, hypertension, and myocardial infarction. In occupational health research, catecholamines are typically measured in the urine (Lundberg & Cooper, 2011). The alternative, though much more complex, is to sample catecholamines from arterial blood. Concentration of catecholamines in arterial blood is a valid representation of the actual peripheral tissue exposure to stress hormones. However, blood levels of catecholamines typically vary considerably over a short period of time.

Dehydroepiandrosterone

Dehydroepiandrosterone (DHEA) counteracts the effects of glucocorticoids in general, and cortisol in particular. DHEA is also neuroprotective. During short-term stress, DHEA secretion is increased along with cortisol. Sustained stress results in decreased DHEA levels, which hampers restorative processes and accelerates aging. Dehydroepiandrosterone-sulphate can be easily measured either in blood or saliva.

Considerations in the Selection of Immune System Markers

The main consideration in selecting which immune markers to use in occupational health research is, naturally, the overall objective of the study. Is the topic of interest short-term stress or sustained stressor exposure? For short-term stress, CRP, immunoglobulins, and interleukins are excellent indicators of altered immune responses. For longer term stress, changes in cytoxic T-lymphocytes and cell-mediated immunity are excellent additions. When the focus is on specific disease processes (e.g., occupational stress and the risk for cardiovascular disease), inflammatory markers such as CRP are more relevant. Is the design cross-sectional or longitudinal? All of these questions are important since they have an impact on the choice of immune markers. As discussed above, certain immune markers respond quickly to stress, while others could take days or weeks to respond. Another issue of great concern is the clinical validity of changes in immune parameters. Apart from CRP and, in the case of HIV positive persons, CD4+ T lymphocytes, there are few immune systems changes observed during everyday stress that are truly linked to future morbidity or mortality risks. Thus, immune markers need to undergo the same scrutiny as occupational health psychology researchers devote to the psychometric properties of psychological scales.

The venue of a study will also determine the "smorgasbord" of immune markers from which to choose. In controlled laboratory settings, or with infrequent sampling in naturalistic settings, blood samples are feasible. In workplace settings, saliva or collecting blood using blood spots (small sample of whole blood dried on filter paper) is more functional and acceptable. Saliva sampling for salivary IgA, lysozyme, CRP, or other biomarkers is a relatively convenient, effective, and easy method. However, saliva sampling does not allow for the assessment of other important biomarkers (e.g., white blood cell functioning) whereas blood samples are able to provide data regarding leukocytes, the complement system, leukotrienes, and ultimately a bigger picture of the immune system and its functioning as a whole.

Saliva or Blood

Many of the measures needed to assess the immune system require the collection of blood. Venipuncture is invasive, potentially anxiety provoking, and relatively costly since it must be performed by a trained phlebotomist. Samples also need to be promptly processed and stored.

There have been recent improvements in salivary immune assays and today we can measure CRP and Immunoglobulin A in saliva. We can also determine concentrations of a range of pathogen defense factors in saliva and tears; for example lysozyme and phospholipase A2. Saliva can also be used to measure DHEA and cortisol. Saliva is collected by having the participant place a cotton salivette dental swab in their mouth for a specific time period, typically 2 minutes. The swab is subsequently returned to a tube and frozen until further analysis. Alternatively, tubes with saliva swabs can be shipped to the lab by overnight express without being frozen. Thus, saliva is an ideal method for long-term and workplace based studies.

Another important consideration is cost. Saliva sampling and analysis is very cost-effective, particularly when compared to blood analysis. Unfortunately, saliva-based sampling does not allow for the measurements of more advanced immune functions. Increasingly the functional capacity of the immune system, for example, the ability of lymphocytes to react to pathogens, is of importance to assess the possible health impact of sustained stress, or exposure to pathogens.

Conclusion

The field of psychoneuroimmunology is rapidly expanding with clear implications for occupational health psychology research. Assessing immune functions in occupational health research will contribute to enhancing our understanding of causal biological mechanisms linking workplace psychosocial exposures to disease outcomes. Collecting immune markers in the workplace setting is also becoming more feasible with the rapid development of saliva-based immunological analysis, as well as the use of blood spots. However, many of the immune markers used in contemporary occupational health research still need further evaluation in terms of their prognostic validity and stability when sampled in the field. With the increased use of biological markers, we can begin to compare across studies to begin to quantify the physiological outcomes of workplace stress.

References

Arnetz, B. B., Wasserman, J., Petrini, B., Brenner, S. O., Levi, L., Eneroth, P., ... Theorell, T. (1987). Immune function in unemployed women. *Psychosomatic Medicine, 49*, 3–12.

Boscolo, P., Youinou, P., Theoharides, T. C., Cerulli, G., & Conti, P. (2008). Environmental and occupational stress and autoimmunity. *Autoimmunity Reviews, 7*(4), 340-343.

Cankaya, B., Chapman, B. P., Talbot, N. L., Moynihan, J., & Duberstein, P. R. (2009). History of sudden unexpected loss is associated with elevated interleukin-6 and decreased insulin-like growth factor-1 in women in an urban primary care setting. *Psychosomatic Medicine, 71*, 914–919.

Charney, D. (2004). Psychobiological mechanisms of resilience and vulnerability: Implications for successful adaptation to extreme stress. *Focus, 2*(3), 368–391.

Davis, C. M., Kozinetz, C. A., & Noroski, L. M. (2006). Decreased lymphocyte mitogen responses at diagnosis are associated with increased morbidity in common variable immunodeficiency patients. *The Journal of Allergy and Clinical Immunology, 117*(2), 106.

Di Donato, A., Di Giampaolo, L., Reale, M., Dadorante, V., Alparone, F., Fattorini, E., ... Boscolo, P. (2006). Effect of occupational stress and anxiety on natural killer lymphocyte activity of men and women employed in a university. *International Journal of Immunopathology and Pharmacology, 4*, 79-84.

Ekman, R., & Arnetz, B. B. (2006). The brain in stress—Influence of environmental and lifestyle stress-related disorders. In B. B. Arnetz & R. Ekman (Eds.), *Stress in health and disease* (pp. 196–213). Weinheim, Germany: Wiley-VCH.

Glaser, R., Rabkin, B., Chesney, M., Cohen, S., & Natelson, B. (1999). Stress-induced immunomodulation: Implications for infectious diseases? *Journal of the American Medical Association, 281*, 2268–2270.

Grawitch, M. J., Trares, S., & Kohler, J. M. (2007). Healthy workplace practices and employee outcomes. *International Journal of Stress Management, 14*, 275–293.

Groer, M., Murphy, R., Bunnell, W., Salomon, K., Van Eeopel, J., Rankin, B., ... Bykowski, C. (2008). Salivary measures of stress and immunity in police officers engaged in simulated critical incident scenarios. *Journal of Occupational and Environmental Medicine, 52*(6), 595–602.

Herbert, T., & Cohen, S. (1993). Stress and immunity in humans: A meta-analytic review. *Psychosomatic Medicine, 55*(4), 364–379.

Howk, C., & Bennett, M. (2010). Immune function and health outcomes in women with depression. *BioPsychoSocial Medicine, 4*(3), 1–9.

Kiecolt-Glaser, J. (2009). Psychoneuroimmunology: Psychology's gateway to the biomedical future. *Perspectives on Psychological Science, 4*(4), 367–369.

Lundberg, U., & Cooper, C. (2011). *The science of occupational health: Stress, psychobiology, and the new world of work.* Hoboken, NJ: Wiley-Blackwell.

Makino, S., Gold, P., & Shulkin, J. (1994). Corticosterone effects on corticotropin-releasing hormone mRNA in the central nucleus of the amygdala and the parvocellular region of the paraventricular nucleus of the hypothalamus. *Brain Research, 640*(1–2), 105–112.

McEwen, B. S. (1998). Protective and damaging effects of stress mediators. *New England Journal of Medicine, 338*, 171–179.

Meijman, T. F., von Dormolen, M., Herber, R. F. M., Rogen, H., & Kuioer, S. (1995). Job strain, neuroendocrine activation, and immune status. In S. L. Sauter & L. R. Murphy (Eds.), *Organizational risk factors for job stress* (pp. 113–126). Washington, DC: American Psychological Association.

Nakata, A., Araki, S., Tanigawa, T., Miki, A., Sakurai, S., & Kawakami, N. (2000). Decrease of suppressor-induced (CD4+CD45RA) T lymphocytes and increase of serum immunoglobin G due to perceived job stress in Japanese nuclear electric power plant workers. *Journal of Occupational and Environmental Medicine, 42*(2), 143-150.

Ridker, P. M., Hennekens, C. H., Buring, J. E., & Rifai, N. (2000). C-reactive protein and other markers of inflammation in the prediction of cardiovascular disease in women. *The New England Journal of Medicine, 342*, 836–843.

Rief, W., Hennings, A., Riemer, S., & Euteneuser, F. (2010). Psychobiological differences between depression and somatization. *Journal of Psychosomatic Research, 68*, 495–502.

Rystedt, L., Cropley, M., Devereux, J., & Michilianou, G. (2008). The relationship between long-term job strain and morning and evening saliva cortisol secretion among white-collar workers. *Journal of Occupational Health Psychology, 13*(2), 105–113.

Segerstrom, S., & Miller, G. (2004). Psychological stress and the human immune system: A meta-analytic study of 30 years of inquiry. *Psychological Bulletin, 130*(4), 601–630.

Shepard, J., K. Barron, K. W., & Myers, D. A. (2000). Corticosterone delivery to the amygdala increases corticotropin-releasing factor mRNA in the central amygdaloid nucleus and anxiety-like behavior. *Brain Research, 861*(2), 288–295.

Stites, D., & Terr, A. (1991). *Basic human immunology.* New York: McGraw-Hill/Appleton & Lange.

Zeier, H., Brauchi, P., & Joller-Jemelka, H. I. (1996). Effects of work demands on immunoglobulin A and cortisol in air traffic controllers. *Biological Psychiatry, 42*, 413–414.

Ziemssen, T., & Kern, S. (2007). Psychoneuroimmunology—Cross talk between the immune and nervous systems. *Journal of Neurology, 254*(2), 108–111.

2

Measurement of Musculoskeletal Functioning

Robert J. Gatchel, Emily Brede, Yunhee Choi, Krista Howard, and Whitney E. Worzer

Measurement of musculoskeletal functioning is important to the determination of physical disability, which can be defined as the impaired performance of functional tasks. Although there is no completely valid "gold standard" of measurement for disability, using physical measurements to define disability is more objective, valid, and reliable than other methods such as patient self-report. This chapter will discuss the reasons for quantifying musculoskeletal function with objective measures and the criteria that determine appropriate tests of musculoskeletal function. The process of complete functional testing will be illustrated with an example of functional capacity testing for injury to the lumbar spine.

Occupational musculoskeletal injuries are highly prevalent and costly. The prevalence of nonfatal occupational injury or illness in 2007 was 4.2% (U.S. Bureau of Labor and Statistics, 2007). Musculoskeletal injuries can either be localized to a specific body region, or can be widespread, depending on the nature of the injury. The U.S. Bureau of Labor Statistics (Courtney & Webster, 2001) reported that the majority of workplace injuries involve damage to the musculoskeletal system. Most injuries are related to sprains, strains, or tears of muscles; and overexertion, falls, and bodily reactions were the most commonly reported events causing such injuries. In 2007, 11% of all reported nonfatal occupational injuries were classified as a sprain, strain, or tear (U.S. Bureau of Labor Statistics, 2007). Upper-extremity injuries of the shoulder and wrist are more likely to be caused by overexertion or repetitive motions, while falls and overexertion are the most commonly reported causes of lower-extremity injuries.

While most individuals with musculoskeletal injuries are easily treated during the acute phase, about 10% will develop a chronic pain condition (Mayer & Polatin, 2000). It is estimated that $70 billion are spent annually on healthcare utilization and work productivity losses due to patients with chronic pain (Gatchel, 2004). However, despite major technological advances in medical care, the outcomes of many chronic pain treatments remain unsatisfactory (Flores, Gatchel, & Polatin, 1997). This has led to the development of a new theoretical framework for treating chronic pain, the biopsychosocial model, which takes into account the complexities of the disability process. The biopsychosocial approach is not aimed at the *disease*, a biological event that can be "cured," but rather is directed at the individual's *illness*, the complex interaction of biological, psychological, and social factors that shape the patient's perceptions of and responses to pain (Turk & Flor, 1999). The biopsychosocial approach is now a valuable model for assessing and treating patients with chronic musculoskeletal pain conditions. The most cost-effective and successful rehabilitation programs involve a comprehensive assessment of the physical, psychological, and socioeconomic factors contributing to disability in an individual. Objective data relating to physical function can be used to develop individualized treatment plans for patients and to evaluate the effectiveness of those treatments.

Disability and Function

Disability often arises as a result of chronic pain, and can be most easily defined in behavioral context. When assessed in socioeconomic terms, *decreased productivity* is used to measure disability and accounts for major societal losses (Gatchel, Mayer, Capra, Diamond, & Barnett, 1986; Levenstein & Kaplan, 1998). *Impairment in performance of functional tasks* defines the physical aspects of disability. While psychosocial symptoms including depression, anxiety, somatization, and substance abuse often accompany chronic pain, these symptoms also account for significant disability by themselves (Dersh, Mayer, Gatchel, Towns, Theodore, & Polatin, 2007; Fishbain, Cutler, Rosomoff, & Rosomoff, 1997; Kinney, Gatchel, Polatin, Fogarty, & Mayer, 1993; Polatin, Kinney, Gatchel, Lillo, & Mayer, 1993; Rush, Polatin, & Gatchel, 2000). Defining disability in legal terms involves *financial compensation* claims, which attempt to place blame for an injury that allegedly resulted in one losing the ability to work. Using *physical function* to define disability is far less subjective than self-report and can be advantageous as long as certain criteria are met. Performance measures that address disability in a vocational context as it relates to work appeal to agencies involved in defining and treating disability.

Why Quantify Function?

When documenting pain and human suffering, the individual's subjective perception of pain/disability has been used as a primary indicator. However, the inherently subjective nature of pain/disability perception is influenced by many factors, and, while a strong correlation exists between measures of physical pain/disability and treatment outcomes (Anagnostis, Mayer, Gatchel, & Proctor, 2003; Beals, 1984; Becker, Sjogren, Beck, Olsen, & Eriksen, 2000; Gatchel, Mayer, & Theodore, 2006; McGeary, Mayer, & Gatchel, 2006; Sullivan, Feuerstein, Gatchel, Linton, & Pransky, 2005), the actual quantification of individual human perceptions has proven to be an elusive task. Individuals with similar physical injuries can have vastly different perceptions of pain/disability, reinforcing the need for an objective way to quantify function. While standard self-report measures of pain/disability are clinically useful, they do not provide an objective assessment and limit the types of comparisons that can be made. Therefore the accurate evaluation of individuals' physical function in the context of performing safe tasks is increasingly important. Gatchel (2005) stressed that a physical function measure will always be more objective than a self-reported psychosocial measure. As he indicates:

> no matter what the level of accuracy of sophistication of a mechanical device used in collecting physiologic measures, it's always the case that human interpretation ultimately must be used in the understanding of the resulting findings. In addition, it must be remembered that a patient's performance during a physical assessment protocol can be greatly influenced by fear of pain or injury, motivation, instructional set, etc. (p. vi)

Quantification Testing Criteria

Certain requirements must be met when quantifying physical functioning (Gatchel, Polatin, Mayer, Robinson, & Dersh, 1998; King, Tuckwell, &

Barrett 1998; Mayer et al., 1988; Polatin, Gatchel, Barnes, Mayer, Arens, & Mayer, 1989). Meaningful clinical interpretations can be made only when tests measuring physical functioning are relevant, valid, reproducible, reliable, capable of identifying suboptimal effort, and have a normative database. For a measure to have *physiological relevance*, a specific and defined capacity must be assessed without reflecting additional unrelated information. For example, using a whole body lifting task to assess a specific muscle group would not be a physiologically relevant task. Additionally, devices used to measure human performance must be *accurate* in their measurement. Tests must also be *valid*, that is, they must actually provide a measurement of the specific musculoskeletal function they are intended to measure. Precise measurement of clinical variables must be *reproducible* by the same tester (intratester reliability) and by different testers (intertester reliability). It is possible to have a test that is reproducible but is invalid. Validity of a test is paramount because it is possible to address issues of reproducibility by adjusting the testing protocol. Once a functional measure of human performance meets these criteria, it is necessary to identify suboptimal effort. Fear avoidance, secondary gain, emotional distress, and other psychological factors can interfere with accurate assessment of function (Leeman, Polatin, Gatchel, & Kishino, 2000; Rainville, Sobel, Hartigan, & Wright, 1997). Therefore, it is essential for each test of function to have a means of objectively assessing effort. Finally, for clinically meaningful interpretations to be made, normative data are necessary. Access to a large, relevant *normative database* allows for extraction of meaningful information related to specific variables such as age, gender, occupation, and type of injury (Keeley, 1997). Problems with obtaining large normative databases have increased as technological advances have provided an abundance of new devices and variations in testing protocol. Therefore, clinicians must remain diligent in their efforts to provide relevant treatment, while normative data are being collected.

Functional Capacity Testing in the Lumbar Spine

At the onset, it should be pointed out that assessment of lumbar function clearly cannot be accomplished with any single test or measure of performance. However, combining individual physical evaluations provides valuable objective information regarding functional capacity (Flores et al., 1997). Computerized devices are available to assist in objectively assessing the patient's function with superior reliability in accounting for effort level. A more comprehensive review of the various tests can be found in Gatchel (2005).

Assessing range of motion provides information regarding the function of the intervertebral discs and facet joints. *Range of motion* is the achievable distance to which a bone joint can be flexed or extended from the reference position. Several standardized methods have been developed to measure range of motion in the spine, but the two-inclinometer technique has been most widely accepted (Nattrass, Nitschke, Disler, Chou, & Ooi, 1999). The handheld or computerized inclinometers are placed at the sacrum and at the level of T12-L1. The patient is asked to perform several movements, such as flexion (bend forward), extension (bend backward), and lateral flexion (bend to the right and left sides). The inclinometer at T12-L1 gives the gross range of motion measurement, and the inclinometer at the sacrum identifies the pelvic component of the motion.

Spinal stability can be assessed under radiography using the evaluation of alteration of motion segment integrity (AOMSI) protocol. Segments of the lumbar spine are observed for abnormally high levels of *translational motion*, the displacement of the vertebral body in the anterior-posterior direction during flexion and extension, and *angular motion*, the shifting of the vertebral end plates during rotation (Alam, 2002). This method is recommended by the American Medical Association (Rondinelli et al., 2008).

Trunk strength can be measured in a variety of ways, using devices that range in expense and sophistication. Multiple muscle groups are involved in trunk strength, and function together to move the lumbar spine in various ways (abduction/adduction, flexion/extension, and rotation). Due to the involvement of the intrinsics (erector spinae, multifidus, quadratus lumborum, psoas, and deep interspinalis and intertransversalis) and extrinsics (abdominals, glutealis, latissimys dorsi, and posterior thigh muscles), specific measures of trunk strength require the lumbopelvic unit to be isolated. New medical devices using isometric, isokinetic, or isodynamic technology now provide far more accurate measurement.

Measurement of *lifting capacity* usually involves performance tasks and does not isolate the trunk. Computerized lifting devices allow for assessment of a wide variety of body positions and lifting styles and are able to measure torque, work, power, and changes in curve shape. Lifting tests can also provide information about maximum capacity, as well as endurance, based on the frequency with which lifting may be required. One particularly useful measure of lifting capacity is the Progressive Isoinertial Lifting Evaluation (PILE; Flores et al., 1997). This test requires minimal special equipment and replicates the natural conditions of everyday lifting activities. The protocol involves lifting progressively heavier weights from floor to waist height and then from waist to shoulder height. The weights may be masked so the patient does not know how much weight he or she is lifting. Measurements can be taken of maximum weight lifted, final heart rate, and endurance.

General physical condition is frequently assessed by measuring *aerobic capacity*. This provides an overall estimation of cardiovascular endurance and indicates the level of activity in which the body is able to safely participate. Bicycle ergometry or treadmill walking estimates cardiovascular capacity by recording physiologic data on work performance and oxygen consumption. Bicycle ergometers for the upper or lower body may be used. In the assessment of chronic pain patients, submaximal exercise testing is most appropriate. Workload and speed are gradually increased at predetermined intervals until the patient reaches 85% of the age-related maximum heart rate. Oxygen consumption can be calculated, and time to complete test should be recorded as well (Noonan & Dean, 2000).

The use of *surface electromyography* (SEMG) to assess lumbar spine function, although still controversial, is becoming more widely used. SEMG provides a noninvasive measurement of the superficial electrical activity of the erector spinae muscles in the lower back. Different evaluation methods have been developed, using resting muscle activity, dynamic activity during movement, or combinations of the two. One of the issues hindering the widespread use of SEMG is that the accuracy of the measurement is highly dependent on the skill of the examiner, thus strict adherence to testing protocols is critical. The Comprehensive Muscular Activity Profile (CMAP) protocol is an FDA approved method for assessing range of motion and lifting capacity using SEMG (Gatchel, Ricard, Choksi, Mayank, & Howard, 2009). In addition to providing accurate measurements of physical parameters, the CMAP protocol can be used to identify maximal or submaximal effort on the test.

Functional Capacity Evaluation (FCE)

Work capacity is less directly related to low back pain. Work capacity tests do not specifically assess a particular lumbar function, but rather address functional tasks related to job demands and include tasks such as bending, stooping, and crawling. However, the measurement of functional tasks is frequently used when assessing work-related back pain (Feuerstein & Zastowny, 1996). Functional Capacity Evaluations (FCEs) attempt to measure the ability to perform the physical demands required on the job in a systematic and comprehensive manner (Strong, 2002; Tuckwell, Straker, & Barrett, 2002; Vasudevan, 1996). FCEs have become widely used by many agencies, including workers compensation authorities, insurance companies, welfare systems, government entities such as the Occupational Safety and Health Administration and the Social Security Administration, as well as other regulating agencies interested in determining precise levels of disability (Harten, 1998; Innes & Straker, 2002).

There has been some confusion regarding operational definitions for functional capacity. In a Delphi survey study, Soer and colleagues (Soer, van der Schans, Groothoff, Geertzen, & Reneman, 2008) collected worldwide FCE experts' opinions to determine operational definitions that could be agreed upon. Even though no consensus was reached for any single definition of FCE, two definitions with reasonable levels of consensus were identified. The two definitions with the highest agreement were (p. 395):

- An FCE is an evaluation designed to document and describe a person's current safe work ability from a physical ability and motivational perspective with consideration given to any existing medical, impairment,or syndromes.
- An FCE is an evaluation of capacity of activities that is used to make recommendations for participation in work while considering the person's body functions and structures, environmental factors, personal factors, and health status.

The most common application of FCE is to determine an injured worker's ability to return to work, and while evaluations are primarily based on assessing impairment of physical abilities, psychosocial factors have been proven to greatly influence performance (Rudy, Lieber, Boston, Gourley, & Baysal, 2003). FCEs typically contain components and questionnaires to assess psychosocial factors, such as fear of pain, but still have significant limitations. In addition, the specific battery of tests and testing protocols can vary between different FCEs. Reliability and validity have not been thoughtfully measured in these various versions of the FCE. Ideally, a single standardized battery of tests would be used to assess all functional capacity. However, due to various disagreements and controversies, this has not yet been achieved.

International Classification of Functioning (ICF)

In response to the desire for a standardized model to consistently measure the components of disability and function, The World Health Organization (WHO) created a new paradigm called the "International Classification of Functioning, Disability and Health" (ICF; WHO, 2001). ICF does not specifically measure disability, but rather classifies functional abilities in various domains. Functioning and disability are assessed in the context of environmental and personal factors, taking into account the social aspects of disability. Individual abilities and limitations are viewed as interactive and dynamic rather than linear or static, and an emphasis is placed on function rather than condition or disease (Wind, Gouttebarge, Kuijer, & Frings-Dresen, 2005). The framework of this classification system measures health and disability at individual levels, as well as population levels. All health conditions are viewed in the same context, which allows a common metric to be used when making comparisons of health and disability. ICF *activities* are defined as the actions people accomplish without assistance or barriers, and they range from basic (eating, bathing, dressing) to complex (work, school, civic activity). *Participation* refers to functioning and accounts for the impact of barriers in the environment (WHO, 2001).

The ICF has a specific classification system for activities and participation which includes: learning and applying knowledge; general tasks and demands; communication; movement; self-care; domestic life areas; interpersonal interactions; major life areas, community and social/civic life. Empirical evidence supports a distinction between Activity and Participation dimensions within the ICF measuring factors of Mobility Activities, Daily Activities, and Social/Participation (Jette, Haley, & Kooyoomjian, 2003).

The ICF was developed to provide a scientific basis for assessing the impact of health conditions. It also establishes the capability to communicate and compare data between countries and health care disciplines. Finally, the ICF provides the conceptual framework for FCE.

Summary and Conclusions

The biopsychosocial model is now viewed as the most useful heuristic approach in the assessment and treatment of chronic pain disorders (Gatchel, 2005; Gatchel & Bruga, 2005; Turk & Monarch, 2002). There are three broad categories of measures—physical, psychological, and socioeconomic. However, these three major measurement categories may not always display high concordance with one another when measuring a construct as multidimensional as chronic pain. In terms of quantification of function in chronic pain patients, one must be aware that there is no totally valid "gold standard" measurement. Clinicians and researchers must be mindful of clearly operationally defining how they are measuring function, as well as using the most reliable measurement devices. The fact that a device is mechanical does not automatically rule out the influence of psychosocial factors (e.g., fear-avoidance, patient motivation, etc.) on final performance. A comprehensive biopsychosocial approach needs to be employed, in which physical function is just one component of the overall evaluation process (Gatchel, 2005).

References

Alam, A. (2002). Radiological evaluation of lumbar intervertebral instability. *Methods in Aerospace Medicine, 46*(2), 48–52.

Anagnostis, C., Mayer, T. G., Gatchel, R. J., & Proctor, T. (2003). The Million Visual Analog Scale: Its utility for predicting tertiary rehabilitation outcomes. *Spine, 28*, 1–10.

Beals, R. (1984). Compensation and recovery from injury. *Western Journal of Medicine, 140*, 233–237.

Becker, N., Sjogren, P., Beck, P., Olsen, A. K., & Eriksen, J. (2000). Treatment outcome of chronic non-malignant pain patients managed in a Danish multidisciplinary pain centre compared to general practice: A randomised controlled trial. *Pain, 84*(2–3), 203–211.

Courtney, T. K., & Webster, B. S. (2001). Antecedent factors and disabling occupational morbidity—Insights from the new BLS data. *American Industrial Hygiene Association Journal, 62*, 622–632.

Dersh, J., Mayer, T., Gatchel, R., Towns, B., Theodore, B., & Polatin, P. B. (2007). Do psychiatric disorders affect functional restoration outcomes in chronic disabling occupational spinal disorders? *Spine, 32*, 1045–1051.

Feuerstein, M., & Zastowny, T. R. (1996). Occupational rehabilitation: Multidisciplinary management of work related musculoskeletal pain and disability. In R.

J. Gatchel & D. C. Turk (Eds.), *Psychological approaches to pain management: A practitioner's handbook* (pp. 458–485). New York: Guilford.

Fishbain, D. A., Cutler, R., Rosomoff, H. L., & Rosomoff, R. S. (1997). Chronic pain-associated depression: Antecedent or consequence of chronic pain? A review. *Clinical Journal of Pain, 13*, 116–137.

Flores, L., Gatchel, R. J., & Polatin, P. B. (1997). Objectification of functional improvement after nonoperative care. *Spine, 22*(14), 1622–1633.

Gatchel, R. J. (2004). Comorbidity of chronic mental and physical health disorders: The biopsychosocial perspective. *American Psychologist, 59*, 792–805.

Gatchel, R. J. (2005). *Clinical essentials of pain management*. Washington, DC: American Psychological Association.

Gatchel, R. J., & Bruga, D. (2005, September/October). Multidisciplinary intervention for injured workers with chronic low back pain. *SpineLine*, 8–13.

Gatchel, R. J., Mayer, T. G., Capra, P., Diamond, P., & Barnett, J. (1986). Quantification of lumbar function. Part VI: The use of psychological measure in guiding physical functional restoration. *Spine, 11*, 36–42.

Gatchel, R. J., Mayer, T., & Theodore, R. (2006). The Pain Disability Questionnaire: Relationship to one-year functional and psychosocial rehabilitation outcomes. *Journal of Occupational Rehabilitation, 16*, 75–94.

Gatchel, R. J., Polatin, P. B., Mayer, T. G., Robinson, R., & Dersh, J. (1998). Use of the SF-36 health status survey with a chronically disabled back pain population: Strengths and limitations. *Journal of Occupational Rehabilitation, 8*, 237–246.

Gatchel, R. J., Ricard, M. D., Choksi, D. N., Mayank, J., & Howard, K. H. (2009). The Comprehensive Muscular Activity Profile (CMAP): Its high sensitivity, specificity and overall classification rate for detecting submaximal effort on functional capacity testing. *Journal of Occupational Rehabilitation, 19*(1), 49–55.

Harten, J. A. (1998). Functional capacity evaluation. *Occupational Medicine, 13*, 209–212.

Innes, E., & Straker, L. (2002). Workplace assessments and functional capacity evaluations: Current practices for therapists in Australia. *Work, 18*(1), 51–66

Jette, A. M., Haley, S. M., & Kooyoomjian J. T. (2003). Are the ICF Activity and Participation dimensions distinct? *Journal of Rehabilitation Medicine, 35*(3), 145–149.

Keeley, J. (1997). Quantification of function. In T. G. Mayer, V. Mooney, & R. J. Gatchel (Eds.), *Contemporary conservative care for painful spinal disorders* (pp. 290–307). Philadelphia, PA: Lea & Febiger.

King, P. M., Tuckwell, N., & Barrett, T. E. (1998). A critical review of functional restoration. *Physical Therapy, 78*(8), 852–866.

Kinney, R. K., Gatchel, R. J., Polatin, P. B., Fogarty, W. J., & Mayer, T. G. (1993). Prevalence of psychopathology in acute and chronic low back pain patients. *Journal of Occupational Rehabilitation, 3*(2), 95–103.

Leeman, G., Polatin, P., Gatchel, R., & Kishino, N. (2000). Managing secondary gain in patients with pain-associated disability: A clinical perspective. *Journal of Workers Compensation, 9*, 25–44.

Levenstein, S. & Kaplan, G. A. (1998). Socioeconomic status and ulcer. A prospective study of contributory risk factors. *Journal of Clinical Gastroenterology, 26*(1), 14–17.

Mayer, T. G., Barnes, D., Kishino, N. D., Nichols, G., Gatchel, R. J., Mayer, H., … Mooney, V. (1988). Progressive isoinertial lifting evaluation. Part I. A standardized protocol and normative database. *Spine, 13*, 993–997.

Mayer, T. G., & Polatin, P. B. (2000). Tertiary non-operative interdisciplinary programs: The functional restoration variant of the outpatient chronic pain management program. In T. G. Mayer, R. J. Gatchel, & P. B. Polatin (Eds.), *Occupational musculoskeletal disorders: Function, outcomes, and evidence* (pp. 639–649). Philadelphia, PA: Lippincott, Williams, & Wilkins.

McGeary, D. D., Mayer, T. G., & Gatchel, R. J. (2006). High pain ratings predict treatment failure in chronic occupational musculoskeletal disorders. *Journal of Bone and Joint Surgery, 88*(2), 317–325.

Nattrass, C. L., Nitschke, J. E., Disler, P. B., Chou, M. J., & Ooi, K. T. (1999). Lumbar spine range of motion as a measure of physical and functional impairment: An investigation of validity. *Clinical Rehabilitation, 13*, 211–218.

Noonan, V., & Dean, E. (2000). Submaximal exercise testing: Clinical application and interpretation. *Physical Therapy, 80*(8), 782–807.

Polatin, P. B., Gatchel, R. J., Barnes, D., Mayer, H., Arens, C., & Mayer, T. G. (1989). A psychosociomedical prediction model of response to treatment by chronically disabled workers with low back pain. *Spine, 14*, 956–961.

Polatin, P. B., Kinney, R., Gatchel, R. J., Lillo, E., & Mayer, T. G. (1993). Psychiatric Illness and chronic low back pain: The mind and the spine—which goes first? *Spine, 18*, 66–71.

Rainville, J., Sobel, J., Hartigan, C., & Wright, A. (1997). The effect of compensation involvement of the reporting of pain and disability by patients referred for rehabilitation of chronic low back pain. *Spine, 22*(17), 2016–2024.

Rondinelli, R. D., Genovese, E., Katz, R. T., Mayer, T. G., Mueller, K., Ranavaya, M., … Brigham, C. R. (2008). *Guides to the evaluation of permanent impairment* (6th ed.). Chicago, IL: American Medical Association.

Rudy, T. E., Lieber, S. J., Boston, J. R., Gourley, L. M., & Baysal, E. (2003). Psychosocial predictors of physical performance in disabled indibiduals with chronic pain. *The Clinical Journal of Pain, 19*, 18–30.

Rush, A., Polatin, P., & Gatchel, R. J. (2000). Depression and chronic low back pain: Establishing priorities in treatment. *Spine, 25*, 2566–2571.

Soer, R., van der Schans, C. P., Groothoff, J. W., Geertzen, J. H. B., & Reneman, M. F. (2008). Towards a consensus in operational definitions in functional restoration capacity evaluation: A Delphi survey. *Journal of Occupational Rehabilitation, 18*, 389–400.

Strong, S. (2002). Functional capacity evaluation: The good, the bad, and the ugly. *Occupational Therapy Now*, 5–9.

Sullivan, M. J. L., Feuerstein, M., Gatchel, R. J., Linton, S. J., & Pransky, G. (2005). Integrating psychosocial and behavioral interventions to achieve optimal rehabilitation outcomes. *Journal of Occupational Rehabilitation, 15*(4), 475–489.

Tuckwell, N. L., Straker, L., & Barrett, T. E. (2002). Test–retest reliability on nine tasks of the physical work performance evaluation. *Work, 19*, 243–253.

Turk, D. C., & Flor, H. (1999). Chronic pain: A biobehavioral perspective. In R. J. Gatchel & D. C. Turk (Eds.), *Psychosocial factors in pain: Clinical perspectives* (pp. 18–34). New York: Guilford.

Turk, D. C., & Monarch, E. S. (2002). Biopsychosocial perspective on chronic pain. In D. C. Turk & R. J. Gatchel (Eds.), *Psychological approaches to pain management: A practitioner's handbook* (2nd ed., pp. 159-186). New York: Guilford.

U.S. Bureau of Labor and Statistics. (2007). *Occupational injury and illness report.* Retrieved from http://www.bls.gov/IIF

Vasudevan, S. V. (1996). Role of functional capacity assessment in disability evaluation. *Journal of Back Musculoskeletal Rehabilitation, 6*, 237–248.

Wind, H., Gouttebarge, V., Kuijer, P. F. M., & Frings-Dresen, M. H. W. (2005). Assessment of functional capacity of the musculoskeletal system in the context of work, daily living, and sport: A systematic review. *Journal of Occupational Rehabilitation, 15*(2), 253–272.

World Health Organization. (2001). *International classification of functioning, disability and health* (*ICF*). Paper presented at the World Health Organization, Geneva, Switzerland.

3

Measurement Issues in Work–Family Research[1]

M. Gloria González-Morales, Lois E. Tetrick, and Ryan Ginter

The relationships between work and family, or work and nonwork domains, have become one of the focal interests of occupational health psychologists. However, along with the growth of the field comes the proliferation of different constructs and measures. In this scenario, the use of different measures creates potential concerns for generalizability. In this chapter we aim to help by identifying the key considerations in work–family measurement. We begin with a brief description and definition of the main constructs in work–personal life research: work–family conflict and the work–family positive relationship (enrichment, facilitation, and positive spillover). We suggest various measures for each construct, describing their major strengths, discussing their weaknesses, and indicating how frequently they are used in scholarly research and have been published in academic journals. Finally, we discuss important issues relative to item construction and response scales that researchers should take into account when choosing a measure or adapting it.

Work–Family Conflict

In 1985, Greenhaus and Beutell defined work–family conflict (WFC) as "a form of interrole conflict in which the role pressures from the

1. The tables were created by searching published papers in PsycINFO that included the corresponding complete measures in their method section. Search terms included *work–family/life conflict, balance, interference, spillover, facilitation, enrichment,* and their permutations.

work and family domains are mutually incompatible in some respect. That is, participation in the work (family) role is made more difficult by virtue of participation in the family (work) role" (p. 77). This definition implies that work–family conflict can be bidirectional: work can interfere with family, and family can interfere with work. Although early studies and measures were not designed to explore these two directions (they were only focused on work to family issues), the most recent WFC measures include several, and mostly parallel, items in each direction. In addition to direction, Greenhaus and Beutell (1985) suggested that in the work–family literature there were three major forms of conflict: time-based (time devoted to one role interferes with the enactment of the other role), energy-based (strain that arises from one role affects participation in the other), and behavior-based (behaviors that are useful in one role are incompatible with behaviors needed in the other).

Table 3.1 presents descriptions of the WFC measures used in the last 10 years sorted by frequency of use. The counts shown in the last two columns indicate that during the first half of this decade, the Gutek, Searle, and Klepa (1991) and Netemeyer, Boles, and McMurrian's (1996) measures were the most frequently used measures in published papers. However, after 2006, Carlson et al.'s (2000) measure started to be more frequently used, along with Netemeyer et al.'s measure. The number of items and measured dimensions vary among measures. Carlson and colleagues' (2000) measure is the most frequently used in recent empirical research (18 papers in the last 10 years).

As it can be observed in the content of Table 3.1, two major issues in WFC assessment are parallelism (the equivalence of work-to-family and family-to-work conflict directions) and form (time, strain, or behavior-based conflict). The existence of work-to-family conflict and family-to-work conflict raises the question of parallelism of the items in terms of the nature of the conflict in each direction. For instance, Bellavia and Frone (2005) noted:

> the items developed by Carlson et al. (2000), the three items assessing strain-based work-to-family interference collectively assess the extent to which work causes someone to be frazzled, emotionally drained, and stressed at home. In contrast, the three items assessing strain-based family-to-work interference collectively assess the extent to which family causes someone to be preoccupied with family, unable to concentrate, and tense at work. It seems that the items assessing work-to-family conflict primarily refer to affective or emotional reactions whereas the items assessing family-to-work conflict primarily refer to cognitive reactions. (p. 132)

Given their research questions, researchers should assess if it is necessary to build new measures (items) that are fully parallel, or whether this lack of parallelism is an essential characteristic of the different types of interference that occur from one domain to the other. Bellavia and Frone advise researchers to "assess separately and analyze

Table 3.1 Work–Family Conflict Measures

Authors (year)	Construct (number of items): *example*	Response scale	2005	2010
Netemeyer et al. (1996)	**WtF, time-based (4), strain-based (1):** *The demands of my work interfere with my home and family life.* **FtW, time-based (4), strain-based (1):** *Family-related strain interferes with my ability to perform job-related duties.*	7-point agreement	11	23
Carlson et al. (2000)	**WtF, time-based (3), strain-based(3), behavior-based(3):** *My work keeps me from my family activities more than I would like.* **FtW, time-based (3), strain-based (3), behavior-based(3):** *Tension and anxiety from my family life often weakens my ability to do my job.*	5-point agreement	3	18
Gutek et al. (1991)	**WtF, time-based (2), strain-based (2):** *After work, I come home too tired to do some of the things I'd like to do.* **FtW, time-based (2), strain-based(2):** *My personal life takes up time that I'd like to spend at work.*	5-point agreement	8	6
Frone et al. (1992)	**WtF, time-based (2):** *How often does your job or career keep you from spending the amount of time you would like to spend with your family?* **FtW, time-based (2):** *How often does your home life keep you from spending the amount of time you would like to spend on job or career-related activities?*	5-point frequency	4	6
Kopelman et al. (1983)	**WtF, time-based (2), strain-based (3), global (1):** *My family dislikes how often I am preoccupied with my work while I am home.*	5-point agreement	4	5

(*continued*)

Table 3.1 Continued

Authors (year)	Construct (number of items): *example*	Response scale	2005	2010
Bohen-Viveros & Long (1981)	**WtF, time-based (9), strain-based (2):** *Sometimes I have difficulties in balancing my time between work and family activities.* **WtF, time-based (1), strain-based (2):** *Sometimes I feel torn between my work and my family.* **Bidirectional, time based (3):** *I have a good balance between my job and my family time.* **Other (2):** *I have as much patience with my children as I would like.*	5-point frequency	2	2
Carlson & Frone (2003)	**WtF, external (3), internal (3):** *How often does your job or career interfere with your home life?* **FtW, external (3), internal (3):** *When you are at work, how often do you think about family related problems?*	5-point frequency	0	3
Frone & Yardley (1996)	**FtW (6):** *I'm too tired at work because of the things I have to do at home.* **WtF (6):** *My work takes up time that I'd like to spend with family/ friends.*	5-point frequency	1	1
Grzywacz et al. (2007)	**WtF (3):** *In the last 6 months how often did your job or career Interfere with your home life?* **FtW(3)** *In the last 6 months how often did your home life interfere with your job or career?"*	6-point frequency, objective anchor (days per week)	0	1
Dolcos & Daley (2009)	**WtF, time-based (2), strain-based (3):** *In the past 3 months, how often have you not had enough time for yourself because of your job?*	5-point frequency	0	1
Stephens & Sommer (1996)	**WtF, time-based (4), strain-based (4), behavior-based (6):** *My work keeps me from my family more than I would like.*	7-point agreement	0	1

Notes. 2005: Number of articles that used the measure between 2000 and 2005; 2010: Number of articles that used the measure between 2006 and 2010 simultaneously both directional dimensions" (2005, p. 132) and to take into consideration the difficulty of comparing both directions when the items in each direction are not parallel. In any instance, Carlson and Grzywacz (2008) claim that there is solid evidence for the distinction between work-to-family and family-to-work conflict, and there is a need to include both directions in studies of work–family conflict (Byron, 2005; Mesmer-Magnus & Viswesvaran, 2005). These concerns indicate the need to carefully examine measurement options in order to decide which measure, in terms of direction, is appropriate for the particular research question.

The second measurement issue to consider is related to the type of conflict. For example, the Carlson, Kacmar, and Williams (2000) measure covers the three types of conflict (time, strain, behavior) in each direction, yielding six different work–family conflict scores. Even though there is theoretical value in distinguishing the different types of conflict, some authors have pointed out an interesting limitation about this type of measurement. Bellavia and Frone (2005) suggest that time, strain, and behavior are types of antecedents or sources of work–family conflict. Therefore, measures that ask for these three "types of conflict" are building in causal attributions (e.g., "Due to all the pressures at work, sometimes when I come home I am too stressed to do the things I enjoy"). In this case, in order to agree with this item, respondents need to attribute the cause of their feelings of stress and the inability to do the things they enjoy to the pressures they experience at work.

Research on attribution (Heider, 1958) indicates that people are not very good at making correct causal attributions. In fact, MacDermid (2005) discussed that given the inherent complexity of the phenomenon, respondents may not have the cognitive capacity to respond to typical work–family conflict items. Consequently, according to Bellavia and Frone (2005) the wording of the actual conflict items should be as general as possible without asking respondents to think too much (e.g., "How often did your home life interfere with your job or career?"; Grzywacz, Frone, Brewer, & Kovner, 2006). This approach thus allows examination of how the three types of predictors (lack of time, lack of energy, or incongruent behaviors) relate to a very general measure of conflict. However, this suggestion has not been overwhelmingly integrated in work–family research.

As we will describe below, the attributional process is a clear characteristic, and from our perspective, an important limitation of the instruments that measure the positive side of the work–family interface. For an interesting discussion on this issue and other related issues and limitations such as the work and nonwork conceptualization (e.g., Fisher, Bulger, & Smith, 2009), response scales or the difference between conflict and interference, previous chapters on work–family measurement and conceptualization can be reviewed (Bellavia & Frone, 2005; Carlson & Grzywacz, 2008; Tetrick & Buffardi, 2006).

The Positive Relationship between Family and Work

The study of the positive relationship between family and work has been proposed using three different constructs. We present them here from the least to the most broad and inclusive. First, work–family *positive spillover* is the most commonly used and refers to the transfer of positively valenced elements (affect, skills, behaviors, and values) from one domain to the other resulting in beneficial effects in the receiving domain (Hanson, Hammer, & Colton, 2006). *Enrichment* refers to "the extent to which experiences in one role improve the quality of life in the other role" (Greenhaus & Powell, 2006, p. 73). Finally, *facilitation* is the broadest concept and tries to capture a superordinate level of analysis that examines the extent to which an individual's engagement in one social system or domain contributes to growth and enhanced functioning in another system or life domain (Grzywacz, Carlson, Kacmar, & Wayne, 2007; Wayne, Grzywacz, Carlson, & Kacmar, 2007).

Table 3.2 shows the most frequently used measures of the work–family positive constructs. Hanson et al.'s measure (2006) is a validated measure of *positive spillover*; however, Carlson and Gryzwacz (2008) indicate that it also contains items assessing enrichment. Carlson, Kacmar, Wayne, and Grzywacz's (2006) measure of *enrichment* is designed to capture the three elements described in the construct of enrichment: individuals' activity in one domain, type of benefit derived from the activity, and better performance or quality of life in the other role (Carlson & Gryzwacz, 2008). Although there is evidence of solid reliability and validity, it should be taken into account that these items, as discussed below in more detail, are double barreled.

The interest of researchers in examining the positive side of the family–work interface has led several scholars to propose omnibus scales that measure both negative and positive aspects. Among the omnibus measures presented in Table 3.3, there are two that measure *facilitation*: Gryzwacz and Bass (2003) and Wayne, Musisca, and Fleeson (2004). However, according to Carlson and Gryzwacz (2008), who present a

Table 3.2 Positive Work–Family Measures

Authors (year)	Construct (number of items): *example*	Response scale	2005	2010
Carlson et al. (2006). Enrichment	**WtF Development (3):** *Helps me to gain knowledge and this helps me be a better family member.* **WtF Affect (3):** *Puts me in a good mood and this helps me be a better family member* **WtF Capital (3):** *Helps me feel personally fulfilled and this helps me be a better family member.* **FtW Development (3):** *Helps me acquire skills and this helps me be a better worker.* **FtW Affect (3):** *Puts me in a good mood and this helps me be a better worker.* **FtW Efficiency (3):** *Causes me to be more focused at work and this helps me be a better worker.*	5-point agreement	0	4
Hanson et al. (2006). Positive Spillover	**WtF Affective (4):** *Having a good day at work allows me to be optimistic with my family.* **WtF Behavior (4):** *Skills developed at work help me in my family life.* **WtF Value (3):** *I apply the principles my workplace values in family situations.* **FtW Affective (4):** *Being happy at home improves my spirits at work.* **FtW Behavior (4):** *Skills developed in my family life help me in my job* **FtW Value (3):** *Values developed in my family make me a better employee.*	5-point agreement	0	3

Notes. 2005: Number of articles that used the measure between 2000 and 2005; 2010: Number of articles that used the measure between 2006 and 2010

deep discussion in both theoretical and practical terms around these three constructs, the items actually reflect enrichment or positive spill-over because facilitation implies a system-level construct not captured in their items. Therefore, even though each concept has been carefully conceptualized from a theoretical perspective, their operational definitions and measurement are confounded in practice. For instance, we can find academic papers that label the positive relationship between work and family as facilitation and describe in the method section a measure of positive spillover.

In order to clarify the operational definition issue, we can think of spillover, enrichment, and facilitation as three different inclusive levels of analysis. Therefore, their operationalization becomes more complex as we advance through the levels. Positive spillover items are relatively simple given that they just have to reflect how a particular experience, element, or its effects transfers domains (e.g., "When something positive happens at work, I am in a good mood at home"). However, enrichment gets a little more complicated because it requires wording that suggests a transfer of elements, or in the case of enrichment, experiences, and that this transfer improves the general quality of life in the other domain. Based on Greenhaus and Powell's (2006) theory of enrichment, the items should include the individual's activity in one domain, the type of benefit derived from that activity, and an appraisal of improved performance or quality of life for the individual in the

Table 3.3 Omnibus Work–Family Measures

Authors (year)	Construct (number of items): *example*	Response scale	2005	2010
Geurts et al. (2005)	**Negative Work–Home Interaction (8):** *You are irritable at home because your work is demanding* **Negative Home–Work Interaction (4):** *Problems with your spouse/family/friends affect your job performance.* **Positive Work–Home Interaction (5):** *You manage your time at home more efficiently as a result of the way you do your job* **Positive Home–Work Interaction (5):** *You have greater self-confidence at work because you have your home life well organized.*	4-point frequency	0	6
Gryzwacz & Bass (2003)*	*Items selected from the MIDUS scale.* **WtF Conflict (4):** *Stress at work makes you irritable at home.* **FtW Conflict (4):** *Responsibilities at home reduce the effort you can devote to your job.* **WtF Facilitation (3):** *The things you do at work make you a more interesting person at home.* **FtW Facilitation (3):** *The love and respect you get at home makes you feel confident about yourself at work.*	5-point frequency	0	4

Authors (year)	Construct (number of items): *example*	Response scale	2005	2010
MacDermid et al. (2000)	**WtF interference self/spouse:** energy (2/2), strain (2/2), time (2/2), behavior (2/2), representativeness item (1), severity item (2) **FtW interference self**: energy (2), strain (2), time (2), behavior (1), representativeness item (1), severity item (2) **WtF enhancement self/spouse**: energy (1/1), strain (2/1), time (2/2), behavior (2/1), support (2/1), representativeness item (1), severity item (1) **FtW enhancement self**: energy (1), strain (2), time (2), behavior (2), support (2), representativeness item (1), severity item (1) **Impact of Work–Life Tension on Work (17), Impact of Work–Life Tension on Personal Life (18):** *Items available from http://wfnetwork.bc.edu/pdfs/measure_tension.pdf*	Multiple Response Scales	0	3
Wayne et al. (2004)	**WtF Conflict (4):** *Your job reduces the effort you can give to activities at home.* **WtF Facilitation (4):** *The skills you use on your job are useful for things you have to do at home.* **FtW Conflict (4):** *Responsibilities at home reduce the effort you can devote to your job.* **FtW Faciliation (4):** *Talking with someone at home helps you deal with problems at work.*	5-point frequency	0	2
Kirchmeyer (1992)	**Positive Spillover from nonwork to work: privileges gained (3), status security (4), status enhancement (4), personality enrichment (4):** *Provides me with contacts that are helpful for my work.* **Negative Spillover from nonwork to work: time-based (1), strain-based (4), behavior-based (3):** *Makes me behave in ways that are unacceptable at work.*	6-point agreement	1	0

Notes. 2005: Number of articles that used the measure between 2000 and 2005; 2010: Number of articles that used the measure between 2006 and 2010

* Same items are used in Gryzwacz and Marks (2000) but with positive and negative spillover label.

other domain. In this case, we find double-barreled items: questions that measure more than one statement in which it is possible that the respondent might agree with one part of a question but not another. With this type of item, it may be unclear, for example, whether people agree that involvement at work puts them in a good mood or that the good mood results in them being a better family member. Even though researchers should follow the theory to design measures, it is important to take into account that trying to measure various ideas in one item is not only cognitively challenging for some respondents, but a threat to measurement validity.

This issue becomes more critical as we progress to the next level: facilitation. Following the definition of the construct, Carlson and Grzywacz (2008) suggest the following as potential items for measuring facilitation:

> My engagement in my family/work provides me with [*insert spillover-like effects* such as "new skills" or "a positive attitude"] which improves functioning in my workplace/family. (p. 64)

> My engagement in work *provides my family members* with unique opportunities, and this improves overall functioning in my family. (p. 64)

In this case, facilitation items would be triple-barreled because the question needs to assess that the person is engaged at work, that this engagement provides their family members with unique opportunities, and that these opportunities improve overall functioning.

The simplest way to resolve problems with double or triple barreled items is to split the items into separate assessments. For example, facilitation items can be divided into three sets of elements: I am engaged at work; My job provides me with new skills/positive attitudes; My family functioning is good. Moreover, longitudinal designs would enable researchers to study the causal structure of these attributions, rather than relying on participants' attributional processes. In the specific case of the positive interface, diary or experience-sampling methods (Ohly, Sonnentag, Niessen, & Zapf, 2010) are more appropriate for the dynamic nature of the research object and the involved time intervals. For example, the elements of facilitation can be measured at different times: today/week/month I am engaged at work; today/week/month my job has provided me with new skills/positive attitudes; today/week/month my family functioning has been good. This way we can study if there is a positive relationship between the day/week/month in which the individual is engaged at work, the acquisition of skills/positive attitudes during that day/week/month, and the quality of the individual's family during that time frame. This approach would not require participants to make causal attributions and would allow researchers to explore the positive (and negative) relations between work and family as they unfold over time as well as examining boundary conditions (e.g., work centrality) that explain how and when these processes take place. In sum, as suggested by Tetrick and Buffardi (2006), longitudinal approaches, both theoretical and methodological, should be considered in work–family research.

On Response Scales

A final issue with respect to work–family measurement concerns the nature of the response scale used by most measures from both the negative and positive sides of the work–family (non/work) interrelations. In the 1980s, the publication of the Daily Hassles Scale (Kanner, Coyne, Schaefer, & Lazarus, 1981) generated a debate about the proper manner of measuring hassles without confounding them with distress/strain indicators. Authors like Dohrenwend, Dohrenwend, Dodson, and Shrout (1984) argued that only "objective measures" such as event frequency were appropriate for measuring the impact of hassles on the individual. On the other hand, Lazarus, DeLongis, Folkman, and Gruen (1985) argued that only measures of appraisal of subjective features of experience such as intensity would capture the stress experience. In 1988, Reich, Parrella, and Filstead suggested that both ideas could be reconciled by measuring both aspects in order to have a complete understanding of the impact of the environment on individuals. The frequency measure that accounts for the number of times the event has occurred and the intensity of the experience that reflects the appraisal of the stressor may be two conceptually distinct aspects of the stress experience: the first would be the more "objective" experience and the latter would be the more "subjective" experience, although Lazarus et al. (1985) point out that even in the recall and report of events there is certain degree of subjectivity. The subjectivity–objectivity debate was brought back 10 years later in a written debate among Perrewe and Zellars (1999), Schaubroeck (1999), and Frese and Zapf (1999), from which Spector (1999) concluded that job stress researchers agreed that the job stress process involves both objective and subjective aspects.

Frequency and intensity measures have been employed in the stress literature when measuring daily hassles (e.g., Jose & Ratcliffe, 2004;

Maybery, Neale, Arentz, & Jones-Ellis, 2007; Reich et al., 1988) but they have not been the most widespread method for rating work stressors. In the same line, although WFC is considered a stressor and could be conceptualized as a hassle rather than a major life event, we have not found any measure that assesses the frequency and severity/intensity of WF events, either negative or positive, at the same time.

In fact, most work–family interface measures use agreement scales (e.g., Carlson, Kacmar, Wayne, & Grzywacz, 2006; Carlson, Kacmar, & Williams, 2000; Gutek et al., 19991; Kirchmeyer, 1992; Kopelman, Greenhaus, & Connolly, 1983; Netemeyer et al., 1996; Stephens & Sommer, 1996; Sumer & Knight, 2001). Agreement response scales cannot distinguish between the occurrence of the event and its appraisal; they assume the occurrence of the event. Or alternatively, agreement response scales may indicate if the event of work/family interfering with family/work occurred, but they do not give any information about the frequency of that event or the extent it bothered or affected the individual depending on how the individual interprets the scale. In addition, they do not measure the intensity of the appraisal: does more agreement mean more severity, more frequency of the event, or more certainty about its occurrence?

Other WFC interface measures use frequency scales (e.g., Bohe & Viveros-Long, 1981; Frone, Russell, & Cooper, 1992; Frone & Yardley, 1996; Grzywacz & Marks, 2000; Wayne et al., 2004). Frequency alone cannot explain how stressful the event is, it only measures how many times an event has occurred in a given time frame (Maybery et al., 2007). In addition, the frequency scales that we usually see use response options that are themselves subjective (e.g., never, sometimes, often). Bellavia and Frone (2005) suggested that these scales should have more specific anchors (e.g., daily, weekly, monthly).

The use of intensity or severity scales is rare; we only found one instrument that measured how much internal conflict each situation posed for the participant relative to interrole conflict between job and parent roles and job and spouse roles items (Holahan & Gilbert, 1979). Assessments of severity or intensity alone do not provide a complete idea of the different levels of potential harmfulness of the stressor (it is not the same to experience a severe stressor once a week rather than once a day); it is necessary to assess the frequency as well as the severity (Peiró, 2000).

The *Measurement of Work/Life Tension* report (MacDermid et al., 2000) sponsored by the Work and Family Research Network Sloan

Foundation, in which many of the prominent researchers in the field participated, recommends including both frequency and severity in the measurement of tension between work and family. However, the way they propose to do so entails specific items related to time, energy, strain, and behavior for frequency and general questions on WF interference for severity. Similarly, González-Morales and Tetrick (2009) proposed a measurement approach that allowed the participants to rate the work–family conflict items in terms of both frequency and severity in order to: (a) measure the "objective" occurrence of the event separately from the "subjective" appraisal; and (b) capture a complete picture of the harmfulness of the event.

Researchers may want to consider the option of asking about how frequent and severe/important are the events in which work and family domains interact rather than using agreement response scales. However, using measures of both frequency and intensity doubles the number of items required, which may be a practical consideration to take into account in terms of the length of a survey. In addition, González-Morales and Tetrick (2009) found that the correlations between frequency and severity ranged between .52 and .84, suggesting that they may be interchangeable to some extent but not necessarily in all cases. In addition, we do not have data on the correlation of frequency or severity scales with agreement scales. This may be a venue for future research that could clarify empirically the importance of choosing an appropriate response scale.

Summary and Concluding Comments

The goal of this chapter was to provide a simple guide to work–family measurement and to offer some insight into important issues that should be taken into account when selecting or adapting a measure. We organized the first part of the chapter around the measures of negative (work–family conflict) and positive (enrichment, facilitation, and positive spillover) constructs that are used to explain the relationships between the two domains.

Work–family conflict is a multidimensional and bidirectional construct that can be assessed with different measures. The Carlson et al. (2000) and Netemeyer et al. (1996) measures were the most frequently used in the last decade. We discussed issues in relation to the parallelism of the work-to-family and family-to-work directions and the use of causal attributions when answering the items related to the different types of conflict.

The positive relationship between work and family has been conceptualized into three main constructs that try to capture different levels of complexity and this complexity is reflected in the items of the measures. We suggest longitudinal approaches like experience-sampling as an alternative for capturing the complex and dynamic nature of the relationships between family and work, especially in its positive conceptualization.

To close the chapter, we consider the limitations associated with agreement and frequency response scales. We review previous stress research on the issue of subjectivity-objectivity appraisal and describe the response scales of extant work–family measures. This discussion leads us to propose that both frequency and severity scales are used in combination in order to capture the different levels of harmfulness of the work–family conflict situation.

Finally, we would like to point out other constructs of interest for work–personal life research such as work–family balance: for a review on its conceptualization and measurement we refer the reader to Tetrick and Buffardi (2006) and Greenhaus and Allen (2010); work–family culture (perceptions of the extent to which organizations facilitate balancing work and family responsibilities; Thompson, Beauvis, & Lyness, 1999); perceived organizational family support (perceptions of tangible and intangible support provided by an organization; Jahn, Thompson, & Kopelman, 2003); and work–family centrality (the relative importance of work versus family in one's life; Carr, Boyar, & Gregory, 2008).

We would like to close this chapter by addressing new trends in measurement of work–family issues. In a recent review of the impact of work–family research on organizations, Kossek, Baltes, and Mathews (2011) mention multisource data (from family members, colleagues, supervisors, and customers), triangulation of quantitative and qualitative methods and longitudinal designs as improvements in measurement. However, they call for creative measurement approaches such as non-same-source scale development that incorporates objective indicators of interference (e.g., number of family-related activities missed because of work) or outcomes (e.g., health of the family members) that enable us to understand dyadic and multilevel relationships. In addition Agars and French (2011) note the need to adequately incorporate context in work–family research, especially by looking closer into different populations and how their unique needs shape work and family domains.

References

Agars, M. D., & French, K. A. (2011). What if work and family research actually considered workers and their families? *Industrial and Organizational Psychology, 4,* 375–378. doi: 10.1111/j.1754-9434.2011.01355.x

Bellavia, G., & Frone, M. R. (2005). Work–family conflict. In J. Barling, E. K. Kelloway, & M. R. Frone (Eds.), *Handbook of work stress* (pp. 113–147). Thousand Oaks, CA: Sage.

Bohen, H. H., & Viveros-Long, A. (1981). *Balancing jobs and family life: Do flexible schedules really help?* Philadelphia, PA: Temple University Press.

Byron, D. (2005). A meta-analytic review of work–family conflict and its antecedents. *Journal of Vocational Behavior, 67,* 169–198. doi:10.1016/j.jvb.2004.08.009

Carlson, D. S., & Frone, M. R. (2003). Relation of behavioral and psychological involvement to a new four-factor conceptualization of work–family interference. *Journal of Business and Psychology, 17,* 515–535. doi:10.1023/A:1023404302295

Carlson, D. S., & Grzywacz, J. G. (2008). Reflections and future directions on measurement in work–family research. In K. Korabik, D. S. Lero, & D. L. Whitehead (Eds.), *Handbook of work–family integration: Research, theory, and best practices* (pp. 57–74). London: Academic Press.

Carlson, D. S., Kacmar, K. M., Wayne, J. H., & Grzywacz, J. G. (2006). Measuring the positive side of the work–family interface: Development and validation of a work–family enrichment scale. *Journal of Vocational Behavior, 68,* 131–164. doi:10.1016/j.jvb.2005.02.002

Carlson, D. S., Kacmar, K. M., & Williams, L. J. (2000). Construction and initial validation of a multidimensional measure of work–family conflict. *Journal of Vocational Behavior, 56,* 249–276. doi:10.1006/jvbe.1999.1713

Carr, J. C., Boyar, S. L., & Gregory, B. T. (2008). The moderating effect of work–family centrality on work–family conflict, organizational attitudes, and turnover behavior. *Journal of Management, 34,* 244–246. doi: 10.1177/0149206307309262

Dohrenwend, B. S., Dohrenwend, B. P., Dodson, M., & Shrout, P. E. (1984). Symptoms, hassles, social supports, and life events: Problem of confounded measures. *Journal of Abnormal Psychology, 93,* 222–230. doi: 10.1037/0021-843X.93.2.222

Dolcos, S. M., & Daley, D. (2009). Work pressure, workplace social resources, and work–family conflict: The tale of two sectors. *International Journal of Stress Management, 16,* 291–311. doi:10.1037/a0017319

Fisher, G. G., Bulger, C. A., & Smith, C. S. (2009). Beyond work and family: A measure of work/non-work interference and enhancement. *Journal of Occupational Health Psychology, 14,* 441–456. doi: 10.1037/a0016737

Frese, M., & Zapf, D. (1999). On the importance of the objective environment in stress and attribution theory. Counterpoint to Perrewé and Zellars. *Journal of Organizational Behavior, 20,* 761-765. doi: 10.1002/(SICI)1099-1379(199909)20:5<761::AID-JOB951>3.0.CO;2-Y

Frone, M. R., Russell, M., & Cooper, M. L. (1992). Antecedents and outcomes of work–family conflict: Testing a model of the work–family interface. *Journal of Applied Psychology, 77,* 65–78. doi:10.1037/0021-9010.77.1.65

Frone, M. R., & Yardley, J. K. (1996). Workplace family-supportive programmes: Predictors of employed parents' importance ratings. *Journal of Occupational and Organizational Psychology, 69,* 351–366. doi: 10.1111/j.2044-8325.1996.tb00621.x

Geurts, S. A. E., Taris, T. W., Kompier, M. A. J., Dikkers, J. S. E., van Hooff, M. L. M., & Kinnunen, U. M. (2005). Work–home interaction from a work psychological perspective: Development and validation of a new questionnaire, the SWING. *Work & Stress, 19,* 319–339. doi:10.1080/02678370500410208

González-Morales, M. G., & Tetrick, L. E. (2009). *Frequency and intensity of work–family conflict: An enriched measurement strategy.* Paper presented at the SIOP Annual Conference. New Orleans, LA.

Greenhaus, G. H., & Allen, T. D. (2010). Work–family balance: A review and extension of the literature. In L. Tetrick & J. C. Quick (Eds.), *Handbook of occupational health psychology* (2nd ed., pp. 165–183). Washington, DC: American Psychological Association.

Greenhaus, J. H., & Beutell, N. J. (1985). Sources and conflict between work and family roles. *The Academy of Management Review, 10,* 76–88. doi: 10.2307/258214

Greenhaus, J. H., & Powell, G. (2006). When work and family are allies: A theory of work–family enrichment. *Academy of Management Review, 31,* 72–92. doi: 10.2307/20159186

Grzywacz, J. G., & Bass, B. L. (2003). Work, family, and mental health: Testing different models of work–family fit. *Journal of Marriage and Family, 65,* 248–261. doi: 10.1111/j.1741-3737.2003.00248.x

Grzywacz, J. G., Carlson, D., Kacmar, K. M., & Wayne, J. (2007). A multi-level perspective on the synergies between work and family. *Journal of Occupational and Organizational Psychology, 80,* 559–574. doi: 10.1348/096317906X163081

Grzywacz, J. G., Frone, M. R., Brewer, C. S., & Kovner, C. T. (2006). Quantifying work–family conflict among registered nurses. *Research in Nursing & Health, 29,* 414–426. doi:10.1002/nur.20133

Grzywacz, J. G., & Marks, N. F. (2000). Reconceptualizing the work–family interface: An ecological perspective on the correlates of positive and negative spillover between work and family. *Journal of Occupational Health Psychology, 5,* 111–126. doi:10.1037/1076-8998.5.1.111

Gutek, B. A., Searle, S., & Klepa, L. (1991). Rational versus gender role explanations for work–family conflict. *Journal of Applied Psychology, 76,* 560–568. doi:10.1037/0021-9010.76.4.560

Hanson, G. C., Hammer, L. B., & Colton, C. L. (2006). Development and validation of a multidimensional scale of perceived work–family positive spillover. *Journal of Occupational Health Psychology, 11,* 249–265. doi:10.1037/1076-8998.11.3.249

Heider, F. (1958). *The psychology of interpersonal relations.* New York: Wiley.

Holahan, C. K., & Gilbert, L. A. (1979). Interrole conflict for working women: Careers versus jobs. *Journal of Applied Psychology, 64,* 86–90. doi: 10.1037/0021-9010.64.1.86

Jahn, E., Thompson, C. A., & Kopelman, R. E. (2003). Rationale and construct validity evidence for a measure of perceived organizational family support (POFS): Because purported practices may not reflect reality. *Community, Work and Family, 6,* 123–140. doi: 10.1080/13668800302545

Jose, P. E., & Ratcliffe, V. (2004). Stressor frequency and perceived intensity as predictors of internalizing symptoms: Gender and age differences in adolescence. *New Zealand Journal of Psychology, 33,* 145–154. Retrieved from http://www.psychology.org.nz/ cms_show_download.php?id=632.

Kanner, A., Coyne, J., Schaefer, C., & Lazarus, R. (1981). Comparison of two modes of stress measurement: Daily hassles and uplifts versus major life events. *Journal of Behavioral Medicine, 4,* 1–39. doi: 10.1007/BF00844845

Kirchmeyer, C. (1992). Perceptions of nonwork-to-work spillover: Challenging the common view of conflict-ridden domain relationships. *Basic and Applied Social Psychology, 13,* 231–249. doi: 10.1207/s15324834basp1302_7

Kopelman, R. E., Greenhaus, J. H., & Connolly, T. F. (1983). A model of work, family, and interrole conflict: A construct validation study. *Organizational Behavior & Human Performance, 32,* 198–215. doi:10.1016/0030-5073(83)90147-2

Kossek, E. E., Baltes, B. B., & Matthews, R. A. (2011). How work–family research can finally have an impact in the workplace. *Industrial and Organizational Psychology: Perspectives on Science and Practice, 4,* 352–369. doi: 10.1111/j.1754-9434.2011.01353.x

Lazarus, R. S., DeLongis, A., Folkman, S., & Gruen, R. (1985). Stress and adaptational outcomes: The problem of confounded measures. *American Psychologist, 40,* 770–779. doi: 10.1037/0003-066X.40.7.770

MacDermid, S. M. (2005). (Re)considering conflict between work and family. In E. E. Kossek & S. J. Lambert (Eds.), *Work and life integration: Organizational, cultural, and individual perspectives* (pp. 19–40). Mahwah, NJ: Erlbaum.

MacDermid, S. M., Barnett, R., Crosby, F., Greenhaus, J., Koblenz, M., Marks, S. S., … Sabbatini-Bunch, L. (2000). *The measurement of work/life tension: Recommendations of a virtual think tank.* Boston, MA: Sloan Work and Family Research Network. Retrieved from http://wfnetwork.bc. edu/pdfs/measure_tension.pdf

Maybery D. J., Neale, J., Arentz, A., & Jones-Ellis, J. (2007). The Negative Event Scale: Measuring frequency and intensity of adult hassles. *Anxiety, Stress and Coping, 20,* 163–176. doi: 10.1080/10615800701217654

Mesmer-Magnus, J. R., & Viswesvaran, C. (2005). Convergence between measures of work-to-family and family-to-work conflict: A meta-analytic examination. *Journal of Vocational Behavior, 67,* 215–232. doi:10.1016/j.jvb.2004.05.004

Netemeyer, R. G., Boles, J. S., & McMurrian, R. (1996). Development and validation of work–family conflict and family–work conflict scales. *Journal of Applied Psychology, 81,* 400–410. doi:10.1037/0021-9010.81.4.400

Ohly, S., Sonnentag, S., Niessen, C., & Zapf, D. (2010). Diary studies in organizational research: A practical examination and some recommendations. *Journal of Personnel Psychology, 9*, 78–93. doi: 10.1027/1866-5888/a000009

Perrewe, P. L., & Zellars, K. L. (1999). An examination of attributions and emotions in the transactional approach to the organizational stress process. *Journal of Organizational Behavior, 20*, 739-752. doi: 10.1002/(SICI)1099-1379(199909)20:5<739::AID-JOB1949>3.0.CO;2-C

Peiró, J. M. (2000). Assessment of psychosocial risks and prevention strategies: The AMIGO model as the basis of the prevenlab/psicosocial methodology. *Psychology in Spain, 4*, 139–166. Retrieved from http://www.psychologyinspain.com/content/reprints/2000/12.pdf

Reich, W. P., Parrella, D. P., & Filstead, W. J. (1988). Unconfounding the hassles scale: External sources versus internal responses to stress. *Journal of Behavioral Medicine, 11*, 239–249. doi: 10.1007/BF00844430

Schaubroeck, J. (1999). Should the subjective be the objective? On studying mental processes, coping behavior, and actual exposures in organizational stress research. *Journal of Organizational Behavior, 20*, 753-760. doi: 10.1002/(SICI) 1099-1379(199909)20:5<753::AID-JOB950>3.0.CO;2-W

Spector, P. (1999). You have full text access to this content objective versus subjective approaches to the study of job stress. *Journal of Organizational Behavior, 20*, 737–738. doi: 10.1002/(SICI)1099-1379(199909)20:5<737::AID-JOB949>3.0.CO;2-X

Stephens, G. K., & Sommer, S. M. (1996). The measurement of work to family conflict. *Educational and Psychological Measurement, 56*, 475–486. doi: 10.1177/0013164496056003009

Sumer, H. C., & Knight, P. A. (2001). How do people with different attachment styles balance work and family? A personality perspective on work–family linkage. *Journal of Applied Psychology, 86*, 653–663. doi: 10.1037/0021-9010.86.4.653

Tetrick, L. E., & Buffardi, L. C. (2006). Measurement issues in research on the work-home interface. In F. Jones, R. J. Burke, & M. Westman (Eds.), *Work–life balance: A psychological perspective* (pp. 90–114). New York: Psychology Press.

Thompson, C. A., Beauvais, L. L., & Lyness, K. S. (1999). When work–family benefits are not enough: The influence of work–family climate on benefit utilization, organizational attachment, and work–family conflict. *Journal of Vocational Behavior, 54*, 392–415. doi:10.1006/jvbe.1998.1681

Wayne, J. H., Grzywacz, J. G., Carlson, D. S., & Kacmar, K. M. (2007). Work–family facilitation: A theoretical explanation and model of primary antecedents and consequences. *Human Resource Management Review, 17*, 63–76. doi:10.1016/j.hrmr.2007.01.002

Wayne, J. H., Musisca, N., & Fleeson, W. (2004). Considering the role of personality in the work–family experience: Relationships of the big five to work–family conflict and facilitation. *Journal of Vocational Behavior, 64*, 1081–1130. doi:10.1016/S0001-8791(03)00035-6

4

Measurement of Sleep and Sleepiness

June J. Pilcher, Michelle L. Burnett, and James A. McCubbin

The role of sleep in improving health and well-being is gradually gaining acceptance in modern industrialized societies. The Alameda County Study, which began in 1965, was one of the first studies that suggested sleeping 7 to 8 hours a night would result in better health and lower mortality rates (Belloc, 1973). Unfortunately, sleep loss continues to be a common occurrence in modern life. Short sleep duration of 6 hours or less a day has increased and occurs across a range of occupations (Luckhaupt, Tak, & Calvert, 2010). Furthermore, chronic levels of decreased sleep and increased sleepiness in modern society contribute to human error in work and nonwork related accidents (Åkerstedt & Nilsson, 2003; Kuhn, 2001) and result in detrimental effects on societies and their economies (Leger, 1994). Our focus in this review is to examine the methods used to document sleep and sleepiness as well as the current methodological issues surrounding study designs that examine simulated shift work under sleep deprivation conditions.

Circadian Rhythms, Sleep, and Shift Work

All animals are subject to endogenous rhythms that affect a variety of internal biological systems. Many of these rhythms, the circadian rhythms, cycle on a 24-hour basis in part due to the regulating influence of the sun. Numerous behavioral and physiological processes in humans are affected by circadian rhythms, including body temperature (Kräuchi & Wirz-Justice, 1994), cortisol levels (Chan & Debono, 2010), physical strength (Guette, Gondin, & Martin, 2005), and cognitive performance (Balkin et al., 2004). The timing of many of our biological rhythms is such that we naturally are at our optimal state during the day (Arendt, 2010).

A particularly powerful circadian rhythm is the sleep-wake cycle which encourages humans to be awake and alert during the day and to sleep at night (Franken & Dijk, 2009). Researchers have long recognized that a person's ability to remain alert and functioning varies across the 24-hour day and theorized that the change in performance was due in some way to a growing need for sleep (e.g., Thorndike, 1926). More recent research has suggested that two factors, time since awakening and circadian rhythms, are primarily responsible for the variations seen in sleepiness and performance across the wake period (e.g., Carrier & Monk, 2000).

The global economy created by modern industrial nations relies on a work force that is available and alert throughout the day and night, creating an environment where many people must work and sleep outside of their natural sleep-wake cycle. Work that starts in the early morning or involves working on shifts increases the likelihood that persons will have short sleep durations (Pilcher, Lambert, & Huffcutt, 2000). Furthermore, night workers (Åkerstedt, 2007; Pilcher et al., 2000) and on-call workers (Pilcher & Coplen, 2000) often experience more fragmented and less sleep than other workers with a concomitant impairment in alertness and performance (Åkerstedt, Fredlund, Gillberg, & Jansson, 2002). Problems adapting to working at night occur even if there is adequate daytime sleep because the circadian clock for most night workers does not realign with working at night and sleeping during the day (Sharkey, Fogg, & Eastman, 2001). This lack of entrainment to night work is due at least in part to exposure to sunlight during the day (Simon, Weibel, & Brandenberger, 2000). Difficulties with sleep and excessive sleepiness under shift-work conditions are diagnosed as shift-work sleep disorder (American Academy of Sleep Medicine, 2005), a disorder reported in approximately 30% of all shift workers (Drake, Roehrs, Richardson, Walsh, & Roth, 2004).

Shift work is often accompanied by a decrease in performance efficiency and an increase in accident and injury rate, particularly during the night shift (Folkard & Tucker, 2003). Furthermore, shift work and the resulting sleep loss results in a number of detrimental short-term and long-term health-related physiological changes (Knutsson, 2003). In addition to the relatively short-term effects of sleepiness and inattention, a number of longer term health issues are related to shift work and sleep loss including cardiovascular disease (Suwazono et al., 2008), gastrointestinal complaints (Scott & LaDou, 1994), obesity (Wright, 2006), diabetes (Gangwisch et al., 2006), and inflammatory processes (Simpson & Dinges, 2007). The underlying mechanism in many of the physiological side-effects of working shifts may be due in part to displaced sleep, sleep loss (McCubbin, Pilcher, & Moore, 2010), and the resultant periods of desynchronization with the endogenous circadian clock (Arendt, 2010).

Measuring Sleep and Sleep Loss

Polysomnography is the gold standard for measuring sleep. Polysomnography usually takes place in a sleep clinic or laboratory with a complete setup of electrophysiological equipment. The sleep technician prepares for the recordings by attaching electrodes to the scalp to provide a measure of the electrical activity in the brain—an electroencephalogram. The technician also attaches electrodes to the muscles under the chin to provide a measure of muscle activity (an electromyogram) and around the eyes to provide a measure of eye movement, an electrooculogram. The electroencephalogram, electromyogram, and electrooculogram provide the basic information necessary to identify sleep versus wakefulness and the individual sleep stages experienced during the night. Other physiological measures can be taken depending upon the needs of the researcher or to monitor for specific sleep disorders such as blood oxygen saturation, heart rate, and respiration. Polysomnography; however, is not convenient for measuring sleep habits in a larger population or sleep habits outside of the sleep clinic or laboratory. Other methods, including subjective and objective measures, can be used to provide information about the sleep-wake cycle across the 24-hour day for days or weeks at a time.

A commonly used method of assessing sleep-wake patterns outside of the clinic and laboratory is self-report instruments. Sleep logs are a widespread method used to evaluate sleep-wake habits and the occurrence of sleep loss in day-to-day life. They have face validity and can provide measures of both quantitative and qualitative aspects of sleep (Sack et al., 2007). Sleep logs often contain questions about sleep onset time and sleep offset time to provide a measure of sleep quantity. To better estimate actual sleep time, sleep logs can contain questions that differentiate between estimated sleep onset time versus the time going to bed with the intention of sleeping. Some individuals, especially those who self-identify as poor sleepers, overestimate how long it takes them to go to sleep. In those individuals, asking for the time they go to bed with the intention of sleeping can provide a more accurate estimate of sleep onset time. Furthermore, sleep quality can be assessed in a sleep log using several questions such as simply providing a rating of the quality of sleep, indicating how often the person awakened during the night, or rating how rested the person feels upon awakening. Sleep logs are typically completed immediately after awakening for a number of consecutive days.

Another frequently used method is to assess sleep habits in a one-time assessment. This can be accomplished with a one-time multi-item questionnaire or with a few sleep-related questions as part of an interview. A commonly used questionnaire is the Pittsburgh Sleep Quality Index (Buysse, Reynolds, Monk, Berman, & Kupfer, 1989). The Pittsburgh Sleep Quality Index is a 19-item survey that measures seven aspects of sleep including sleep quality, sleep latency, sleep duration, habitual sleep efficiency, sleep disturbances, use of sleep medications, and daytime dysfunction. These seven separate areas can also be combined and scored as a single overall sleep quality measure. Sleep habits can also be measured by one-time questionnaires that provide self-report estimates of time in bed and time spent asleep such as the Sleep Timing Questionnaire (Monk et al., 2003). Alternatively, some researchers and clinicians use single questions asking about sleep length or sleep quality as part of a larger interview, especially when completing epidemiological studies (e.g., Drake et al., 2004; Luckhaupt et al., 2010).

The Morningness-Eveningness Questionnaire is another commonly used survey to assess sleep; however, in this case the 19-item survey aims to assess the individual's morning and evening preferences (Horne & Ostberg, 1976). Sleep scientists and clinicians use the Morningness-Eveningness Questionnaire as a subjective estimate of how the human circadian system influences our sleep-wake cycle. Scores on the questionnaire that indicate a person naturally prefers staying up late (i.e., an "owl") have been used as a predictor of better tolerance of working the night shift (Hilliker, Muehlbach, Schweitzer, & Walsh, 1992).

An objective measure of sleep-wake activity that is often used in conjunction with sleep logs is actigraphy. Actigraphs are watch-sized devices that are typically worn on the wrist. They contain accelerometers to measure body movement and adequate memory to record movement for several weeks. Movement can be sampled at different intervals depending on the device and the precision needed and stored for download to a computer for analysis. Computer software is used to analyze the raw data to indicate times of activity and times of inactivity. Times of activity are generally interpreted as wakefulness and extended periods of inactivity are interpreted as sleep. The software programs use algorithms to estimate a number of sleep-related parameters including time asleep, sleep efficiency, number of awakenings, and time awake (Ancoli-Israel et al., 2003). Actigraphy provides a reliable measure of sleep habits and sleep patterns in normal, healthy adults and in clinical populations with circadian rhythm abnormalities, such as shift-work sleep disorder (Morgenthaler et al., 2007).

Best Practices

When gathering sleep-related data in day-to-day life, the best method is to use the daily sleep log in combination with actigraphy. This combination provides a subjective and objective estimate of daily sleep times and sleep quality. If it is necessary to examine sleep habits in a larger population it may become more practical to use one-time measures of sleep.

Measuring Circadian Rhythms

Chronobiology researchers have developed markers for the endogenous circadian phase that are suitable for use with humans. Core body temperature is a commonly used marker of the circadian system. In humans in phase with the sun, core body temperature fluctuates around its highest level from approximately 11 a.m. until 7 p.m. with a small early afternoon dip and at its nadir between about 4 and 5 am. However, because core body temperature can be altered by activity, food intake, and sleep times, it is best measured under settings where the individual is kept on a constant routine including bed rest, fed frequent, small meals at equal intervals, and kept awake (Sack et al., 2007).

Another commonly used measure is the timing of melatonin secretion from the pineal gland. Melatonin is released in low amounts during the day and in increasing amounts as ambient lighting decreases. Immunoassays are available that provide a sensitive measure of melatonin levels in blood plasma or saliva. There are also immunoassays that measure the metabolite of melatonin in urine. Melatonin release, however, can be suppressed by artificial lighting. Thus, melatonin samples should be obtained under dim light conditions, a procedure called dim light melatonin onset (Lewy & Sack, 1989).

When core body temperature and dim light melatonin onset are measured correctly, they can each provide a reasonably accurate indication of the endogenous circadian phase. When comparing the efficacy of the two measures, melatonin is considered to be the more stable marker of the circadian system (Sack et al., 2007). Both measures, however, require specific conditions for accurate measurement that are best accomplished in a clinical or laboratory setting. An alternative measure that can be used under normal daily living conditions is wake up time. If an individual maintains a regular sleep-wake pattern of sleeping at night and waking up in the morning, the wake up time can be used as a reasonable estimate of the internal circadian clock (Burgess & Eastman, 2005). Since the sun and the endogenous human circadian clock encourage us to sleep at night and be awake during the day, waking up at about the same time each morning without the regular use of an alarm clock is indicative of the circadian rhythm being in phase. It is important to note that using the wake-up time as an indicator of the internal circadian phase is only effective in individuals who sleep at about the same time each night. Sleep patterns that vary throughout the day and night suggests that the individual is not maintaining a sleep-wake cycle that is in phase with the natural circadian rhythm.

Best Practices

Dim light melatonin onset, when measured correctly following the dim light routine in a laboratory or clinical setting, is the best measure currently available for accurately assessing the endogenous circadian phase. If this measure is not feasible due to the large number of persons being assessed or due to cost, spontaneous wake up time can be a viable alternative. However, if using wake up time as an indication of circadian phase, it is important to ensure that the person maintains a regular sleep-wake cycle, which would include going to bed and waking up at approximately the same time each day and that the major sleep episode is at night and thus in congruence with the sun.

Measuring Sleepiness

Sleepiness and alertness are important components of an individual's response to the sleep-wake cycle. In a clinical or laboratory setting, polygraphs can be used to provide objective measures of sleepiness and the ability to remain awake. The Multiple Sleep Latency Test provides a measure of the physiological tendency to fall asleep (Carskadon & Dement, 1982). In the Multiple Sleep Latency Test, electrodes are applied to the scalp, chin muscles, and around the eyes to monitor for the onset of sleep. The individual is put in bed for 20 minutes and told to try to go to sleep. If polygraph-defined sleep occurs as defined by Stage 1 sleep, the individual is awakened after one minute and the time to sleep onset (sleep latency) is used as a measure of sleepiness with a lower sleep latency time indicating increased levels of sleepiness.

The Maintenance of Wakefulness Test can be used as a measure of the ability to stay awake (Mitler, Gujavarty, & Browman, 1982). After having electrodes attached to the scalp, chin muscles, and around the eyes, the individual is seated in a quiet, dimly lit room and told to remain awake. If the individual remains awake for 20 minutes the test is concluded. If polygraph-defined sleep occurs, the sleep latency is used as a measure of sleepiness with a lower sleep latency time indicating increased levels of sleepiness. Both the Multiple Sleep Latency Test and the Maintenance of Wakefulness Test are used as markers of the results of inadequate sleep quantity and sleep quality (Arand et al., 2005).

Although polygraph-based measures of sleepiness can be used in clinics and laboratories, they have limited usability in the larger population. An alternative is to use subjective sleepiness measures. The Epworth Sleepiness Scale and the Stanford Sleepiness Scale are commonly used self-report surveys. The Epworth Sleepiness Scale asks individuals to rate how likely they are to doze off in eight scenarios such as sitting and reading (Johns, 1991). The Stanford Sleepiness Scale uses a one item scale from 1 (feeling active and vital) to 7 (almost in reverie, sleep onset soon) as a measure of subjective sleepiness (Hoddes, Zarcone, Smythe, Phillips, & Dement, 1973). Visual Analogue Scales can also be used to measure subjective sleepiness by posing a range of questions on alertness, sleepiness, feeling tired, or likelihood of falling asleep and, as such, measure different dimensions of subjective sleepiness while using the same scale (McClelland & Pilcher, 2007).

Best Practices

When gathering measures of sleepiness outside of a laboratory or clinical setting, subjective sleepiness scales provide the best measures. The

Epworth Sleepiness Scale is frequently used in clinics and research settings. It is important to note, however, that subjective sleepiness appears to be composed of two primary factors: a basic feeling of sleepiness and a behavioral component where the likelihood of falling asleep is the better measure. Visual Analogue Scales can be useful to capture information from both major components of sleepiness by using the same metric to implement several sleepiness-related questions.

Methodological Issues with Simulated Shift Work

Studies simulating nighttime shift work have a number of limitations. These methodological considerations are impractical to address in a scientific manner largely due to the endogenous circadian system and the sun. One limitation is the lack of a true control group. A scientific study that includes a well-rested control group working a simulated night shift would allow one to specifically examine the effects of sleep deprivation versus the sustained performance aspect of working at night (Odle-Dusseau, Bradley, & Pilcher, 2009). Unfortunately, it is virtually impossible to create a scenario where there is a well-rested control group working at night. To do that, one would have to seclude the group from the sun for 3 to 5 days prior to the onset of the study where they would be awake at night and sleep during the day, since even short exposure to sunlight entrains the circadian system to the 24-hour day imposed by the sun. Another concern is that many tasks are often used when simulating a shift work scenario. Fully controlling all potential aspects of the tasks such as the time on task, the order of task administration, and number of tasks is not feasible. To attempt to control for all possible task(s) related issues would create a large number of experimental cells that would have to be counter-balanced (Balkin et al., 2004).

Finally, it is important to note that performance or subjective responses in studies using a simulated shift work paradigm cannot be attributed solely to sleep deprivation due to the strong circadian influence on our homeostatic processes. For individuals working at night, the effects of sleep loss per se cannot be fully separated from the natural phase of the individual's endogenous circadian rhythm (Groeger et al., 2008). The negative effect of sleep loss when working at night coincides with the natural change in our endogenous circadian rhythms. As such, sleep loss and circadian influences could have an interactive effect that negatively affects the individual's ability to cope with shift work (Williamson, Feyer, Mattick, Friswell, & Finlay-Brown, 2001). It is not feasible to complete a study using a simulated night shift with a well-rested control group without totally inverting the participants' circadian rhythms; something that would be difficult to accomplish and something that few potential participants would be willing to do.

Summary and Conclusion

Sleep is essential both for long-term good health and to maximize daily performance. Using methods that will allow us to reliably measure sleep and sleepiness is essential in modern society where loss of sleep and shift work is prevalent. In spite of the methodological concerns with simulated night shifts, these studies are an essential part of the scientific effort to better understand how humans can best adapt to shift work. Simulated night shift studies imitate the real world where the worker is increasingly sleep deprived within each night and across consecutive nights. At the same time, the worker experiences the natural circadian rhythm to be awake during the day and to sleep at night. Even a brief exposure to sun light in the early morning when driving home after a night shift maintains the endogenous circadian clock, thus guaranteeing that the night worker never fully adapts to being awake at night and sleeping during the day.

Best Practices

The best practice is to establish a sleep habit where sleep occurs primarily at night. More specifically, it is best to go to sleep and wake up at the same time every day. The body and brain will quickly adapt to a stable sleep-wake pattern, making sleep more efficient and restorative. Unfortunately, no one can do this all the time, but the best practice is to maintain stable sleep times as much as possible. Shift workers, especially night shift workers can experience more difficulty maintaining a regular sleep habit than nonshift workers. Even under shift work conditions, however, it is best to create a schedule where the person sleeps as close to the same time each day as possible. For night shifts, this could be two sleep episodes; once in the morning right after getting home from work and again later in the day prior to reporting back to work. Ultimately, placing priority on good sleep habits is the answer.

References

Åkerstedt, T. (2007). Altered sleep/wake patterns and mental performance. *Physiology and Behavior, 90*, 209–218.

Åkerstedt, T., Fredlund, P., Gillberg, M., & Jansson, B. (2002). Work load and work hours in relation to disturbed sleep and fatigue in a large representative sample. *Journal of Psychosomatic Research, 53*, 585–588.

Åkerstedt, T., & Nilsson, P. M. (2003). Sleep as restitution: An introduction. *Journal of Internal Medicine, 254*, 6–12.

American Academy of Sleep Medicine. (2005). *The international classification of sleep disorders: Diagnostic and coding manual* (2nd ed.). Darien, IL: Author.

Ancoli-Israel, S., Cole, R., Alessi, C., Chambers, M., Moorcroft, W., & Pollak, C. P. (2003). The role of actigraphy in the study of sleep and circadian rhythms. American Academy of Sleep Medicine Review Paper. *Sleep, 26*, 342–392.

Arand, D., Bonnet, M., Hurwitz, T., Mitler, M., Rosa, R., & Sangal, R. B. (2005). The clinical use of the MSLT and MWT. *Sleep, 28*, 123–144.

Arendt, J. (2010). Shift work: Coping with the biological clock. *Occupational Medicine, 60*, 10–20.

Balkin, T. J., Bliese, P. D., Belenky, G., Sing, H., Thorne, D. R., Thomas, M., … Wesenten, N. J. (2004). Comparative utility of instruments for monitoring sleepiness-related performance decrements in the operational environment. *Journal of Sleep Research, 13*, 219–227.

Belloc, N. (1973). Relationships of health practices and mortality. *Preventive Medicine, 2*, 67–81.

Burgess, H. J., & Eastman, C. I. (2005). The dim light melatonin onset following fixed and free sleep schedules. *Journal of Sleep Research, 14*, 229–237.

Buysse, D. J., Reynolds, C. F., Monk, T. H., Berman, S. R., & Kupfer, D. J. (1989). The Pittsburgh Sleep Quality Index: A new instrument for psychiatric practice and research. *Psychiatry Research, 28*, 193–213.

Carrier, J., & Monk, T. H. (2000). Circadian rhythms of performance: New trends. *Chronobiology International, 17*, 719–732.

Carskadon, M. A., & Dement, W. C. (1982). The multiple sleep latency test: What does it measure? *Sleep, 5*(Suppl. 2), S67–S72.

Chan, S., & Debono, M. (2010). Replication of cortisol circadian rhythm: New advances in hydrocortisone replacement therapy. *Therapeutic Advances in Endocrinology and Metabolism, 1*, 129–138.

Drake, C. L., Roehrs, T., Richardson, G., Walsh, J. K., & Roth, T. (2004). Shift work sleep disorder: Prevalence and consequences beyond that of symptomatic day workers. *Sleep, 27*, 1453-1462.

Folkard, S., & Tucker, P. (2003). Shift work, safety and productivity. *Occupational Medicine, 53*, 95–101.

Franken, P., & Dijk, D. (2009). Circadian clock genes and sleep homeostasis. *European Journal of Neuroscience, 29*, 1820–1829.

Gangwisch, J. E., Heymsfield, S. B., Boden-Albala, B., Buijs, R. M., Kreier, F., Pickering, T., … Malaspina, D. (2006). Short sleep duration as a risk factor for hypertension: Analyses of the first National Health and Nutrition Examination Survey. *Hypertension, 47*, 833–839.

Groeger, J. A., Viola, A. U., Lo, J. C. Y., von Schantz, M., Archer, S. N., & Dijk, D. J. (2008). Early morning executive functioning during sleep deprivation is compromised by a PERIOD3 polymorphism. *Sleep, 31*, 1159–1167.

Guette, M., Gondin, J., & Martin, A. (2005). Time-of-day effect on the torque and neuromuscular properties of dominant and non-dominant quadriceps femoris. *Chronobiology International, 22*, 541–558.

Hilliker, N. A., Muehlbach, M. J., Schweitzer, P. K., & Walsh, J. K. (1992). Sleepiness/alertness on a simulated night shift schedule and morningness-eveningness tendency. *Sleep, 15*, 430–433.

Hoddes, E., Zarcone, V., Smythe, H., Phillips, R., & Dement, W. C. (1973). Quantification of sleepiness: A new approach. *Psychophysiology, 10*, 431–436.

Horne, J. A., & Ostberg, O. (1976). A self-assessment questionnaire to determine morningness-eveningness in human circadian rhythms. *International Journal of Chronobiology, 4*, 97–110.

Johns, M. W. (1991). A new method for measuring daytime sleepiness: The Epworth sleepiness scale. *Sleep, 14*, 540–549.

Knutsson, A. (2003). Health disorders in shift workers. *Occupational Medicine, 53*, 103–108.

Kräuchi, K., & Wirz-Justice, A. (1994). Circadian rhythm of heat production, heart rate, and skin and core temperature under unmasking conditions in men. *American Journal of Physiology, 267*, R819–R829.

Kuhn, G. (2001). Circadian rhythm, shift work, and emergency medicine. *Annals of Emergency Medicine, 37*, 88–98.

Leger, D. (1994). The cost of sleep-related accidents: A report for the national commission on sleep disorders research. *Sleep, 17*, 84–93.

Lewy, A. J., & Sack, R. L. (1989). The dim light melatonin onset as a marker for circadian phase position. *Chronobiology International, 6*, 93–102.

Luckhaupt, S. E., Tak, S., & Calvert, G. M. (2010). The prevalence of short sleep duration by industry and occupation in the National Health Interview Survey. *Sleep, 33*, 149–159.

McClelland, L. E., & Pilcher, J. J. (2007). Assessing subjective sleepiness during a night of sleep deprivation: Examining the internal state and behavioral dimensions of sleepiness. *Behavioral Medicine, 33*, 17–26.

McCubbin, J. A., Pilcher, J. J., & Moore, D. D. (2010). Blood pressure increases during a simulated night shift in persons at risk for hypertension. *International Journal of Behavioral Medicine, 17*, 314–320.

Mitler, M. M., Gujavarty, K. S., & Browman, C. P. (1982). Maintenance of wakefulness test: A polysomnographic technique for evaluation treatment efficacy in patients with excessive somnolence. *Electroencephalography and Clinical Neurophysiology, 53*, 658–661.

Monk, T. H., Buysse, D. J., Kennedy, K. S., Potts, J. M., DeGrazia, J. M., & Miewald, J. M. (2003). Measuring sleep habits without using a diary: The sleep timing questionnaire (STQ). *Sleep, 26*, 208–212.

Morgenthaler, T., Alessi, C., Friedman, L., Owens, J., Kapur, V., Boehlecke, B., … Swick, T. J. (2007). Practice parameters for the use of actigraphy in the assessment of sleep and sleep disorders: An update for 2007. *Sleep, 30*, 519–529.

Odle-Dusseau, H. N., Bradley, J. L., & Pilcher, J. J. (2010). Subjective perceptions of the effects of sustained performance under sleep-deprivation conditions. *Chronobiology International, 27*, 318–333.

Pilcher, J. J., & Coplen, M. K. (2000). Work/rest cycles in railroad operations: Effects of shorter than 24-h shift work schedules and on-call schedules on sleep. *Ergonomics, 43*, 573–588.

Pilcher, J. J., Lambert, B. J., & Huffcutt, A. I. (2000). Differential effects of permanent and rotating shifts on self-report sleep length: A meta-analytic review. *Sleep, 23*, 155–163.

Sack, R. L., Auckley, D., Auger, R. R., Carskadon, M. A., Wright, K. P., Vitiello, M.V., & Zhdanova, I. V. (2007). Circadian rhythm sleep disorders: Part I, basic principles, shift work and jet lag disorders. *Sleep, 30*, 1460–1483.

Scott, A. J., & LaDou, J. (1994). Health and safety in shift workers. In C. Zenz, O. B. Dickerson, & E. P. Horvath (Eds.), *Occupational Medicine* (pp. 960–986). St. Louis, MO: Mosby.

Sharkey, K. M., Fogg, L. F., & Eastman, C. I. (2001). Effects of melatonin administration on daytime sleep after simulated night shift work. *Journal of Sleep Research, 10*, 181–192.

Simon, C., Weibel, L., & Brandenberger, G. (2000). Twenty-four-hour rhythms of plasma glucose and insulin secretion rate in regular night workers. *American Journal of Physiology—Endocrinology and Metabolism, 278*, E413–E420.

Simpson, N., & Dinges, D. F. (2007). Sleep and inflammation. *Nutrition Reviews, 65*, S244–S252.

Suwazono, Y., Dochi, M., Sakata, K., Okubo, Y., Oishi M, Tanaka K, … Nogawa, K. (2008). Shift work is a risk factor for increased blood pressure in Japanese men: A 14-year historical cohort study. *Hypertension, 52*, 581–586.

Thorndike, E. L., (1926). *Mental work and fatigue and individual differences and their causes.* New York: Columbia University Press.

Williamson, A. M., Feyer, A. M., Mattick, R. P., Friswell, R., & Finlay-Brown, S. (2001). Developing measures of fatigue using an alcohol comparison to validate the effects of fatigue on performance. *Accident Analysis and Prevention, 33*, 313–326.

Wright, K. (2006). Too little sleep: A risk factor for obesity? *Obesity Management, 2*, 140–145.

5

Measurement of Emotions

Seth Kaplan, Reeshad S. Dalal, and Joseph N. Luchman

Work is an emotional experience. Employees experience fear over losing their jobs, anger over being treated unfairly, and anxiety about impending deadlines. However, they also derive warmth and gratitude when interacting with coworkers, pride from working for a virtuous organization, and a sense of personal growth from job accomplishments. Mixed in with these experiences are the emotions that individuals bring with them to work, such as distress about a family member's illness, frustration over morning traffic, or underlying psychopathological issues that cannot simply "be left at the company's door." Collectively, these numerous and varied emotions have significant implications for employees' physical and psychological well-being and for their ability to function effectively at work (for reviews, see articles in Table 5.1).

Given the consequences of these workplace emotions, their assessment is an important endeavor, both for scholarly and applied purposes. Thus, the goal of this chapter is to provide an overview of the strategies available for assessing work-related emotions and to offer practical guidance for those engaged in such practice. Below, we first mention general considerations regarding the measurement of emotions and then discuss specific measurement strategies.

General Issues and Considerations

Emotions and Related Terms and Phenomena

Measuring emotions entails consideration of some fundamental issues. One important consideration is deciding what type of affective

Table 5.1 Overview of Emotion Measurement Strategies

Type of Measurement	Examples	Advantages	Drawbacks/ Considerations	Key References
Self-Report	-Paper-and-pencil or Internet surveys -Day-reconstruction surveys -Experience-sampling surveys -Moment-to-moment elicitations -Interviews	-High face validity -Well-suited to assess self-referential perceptions -Convenient and cheap to administer	-Rely on people's ability and willingness to provide true reports of their emotional experiences -Potential for inflation of relationship estimates due to common-method variance	-Gray & Watson (2007) -Sloan & Kring (2007) -Watson & Clark (1999)
Observational	-Behavioral observation -Facial coding -Coding written or verbal accounts	-May overcome people's inability or unwillingness to provide true reports of emotions -Provides rich data -Useful for examining emotion display/expression	-Expensive -Time consuming -May not correlate strongly with self-reports -Not a measure of emotion experience	-Coan & Allen (2007) -Paul Ekman's work -Reed & McIntosh (2008)
Psychophysiological	-Heart rate -Blood pressure -Respiratory output -Skin response -EEG -PET scans -fMRI	-May overcome people's inability or unwillingness to provide true reports of emotions experiences -Some of the measurements (e.g., blood pressure) also mediate health outcomes	-Expensive -Impractical -Time consuming -May not correlate strongly with self-reports	-*Biological Psychology* (2010) -Curtin et al. (2007) -Larsen et al. (2008)
Key References for Emotion at Work		-Barsade & Gibson (2007); Brief & Weiss (2002); Elfenbein (2007)		
Key References for Measuring Emotions		Coan & Allen (2007); Larsen & Prizmic-Larsen (2006); Mauss & Robinson (2009); Sloan & Kring (2007)		

experience is of interest. Important distinctions exist between emotions and related phenomena (see Barsade & Gibson, 2007). Emotions represent intense affective, valenced reactions and are directed at a specific cause (e.g., anger over interaction with a rude customer). A closely related concept is that of moods, which are less intense than emotions and not consciously linked to a specific target. Whereas moods are common, intense emotional reactions are rare, both at work and in general (Miner, Glomb, & Hulin, 2005). A final related concept is that of trait affect (i.e., affective disposition), which represents a stable tendency to experience certain affective states across time and situations. Someone regarded as an "angry person" for instance, would likely score high on a measure of trait anger. *Affect* generally is used an umbrella term, subsuming these various feeling states.

The context and the purpose of measurement should dictate the type of affect assessed. For example, interest in how customers feel after interacting with particular service agents would call for assessing the customers' moods or emotions, as these types of affect are ephemeral and linked to discrete events. In contrast, an organization interested in selecting service agents based on their emotional characteristics should assess trait affect which is more stable and predictive over the longer term (Kaplan, Bradley, Luchman, & Haynes, 2009).

Aspects of Emotional Responding

Another paramount decision is determining which aspect(s) of emotional responding to assess. Typically, interest centers on emotional *experience,* or one's subjective assessment of emotion. In some cases, though, other phenomena, such as emotional awareness (the ability to recognize emotions in oneself and others), emotional expression (outward display of emotions), emotional regulation (the process of influencing emotional experience and expression), and psychophysiological response are of interest (see Sloan & Kring, 2007).

Again, the issue of ultimate scientific or applied concern must drive the measurement strategy. This dictum is especially true for emotions because the various aspects of emotional response are loosely coupled and oftentimes weakly correlated with one another (Cacioppo, Berntson, Larsen, Poehlmann, & Ito, 2000). Thus, measurement of one aspect does not necessarily serve as a proxy for others. As an illustration of this point, consider that customer service agents' outward emotional displays need not correspond to their subjective emotional experiences, which in turn need not correspond to their physiological emotional responses.

Another maxim of measurement is that, as far as indicators (methods of assessment, sources, etc.) go, "more is better." This axiom holds special significance for emotions. The loose coupling of emotion components suggests that employing multiple assessment strategies will be most informative. Convergent results from these different measurements (e.g., observation of the agents' emotional displays and a self-report measure of agents' emotional experience) will yield more reliable and valid conclusions (Larsen & Prizmic-Larsen, 2006). In addition, the *pattern* of results across these components can also be quite informative. In sum, assessing multiple emotional phenomena, and using multiple measures to assess each, will be the most prudent and informative strategy.

Dimensional versus Categorical Perspectives on Emotions

A final consideration is whether to assess emotion as consistent from a dimensional or a categorical perspective. Proponents of the former view suggest that emotional experience can be captured by a set of underlying dimensions. According to one dimensional framework, valence (also known as "hedonic tone" or pleasantness-unpleasantness) and activation (also known as arousal) are the two fundamental dimensions organizing emotional experience (Russell & Barrett, 1999). In another popular model, the two primary factors (i.e., dimensions) of emotional experience are positive activation and negative activation (i.e., PA and NA; Watson, Wiese, Vaidya, & Tellegen, 1999). Researchers need to think carefully about which of these models is consistent with the purpose and theory of the investigation and need to be explicit in reporting which model they are following/measuring.

Considerable evidence supports the dimensional perspective. For instance, many emotions (e.g., guilt, anxiety) often co-occur, implying that they reflect the same underlying dimensions and processes (Watson et al., 1999). Also, many measures can distinguish among emotions at opposite poles of a dimension (e.g., positive vs. negative emotions) but cannot distinguish well between specific emotions closely located on that dimension (e.g., fear vs. shame along the negative dimension).

A well-known alternative view is that there are a certain number of discrete emotions (e.g., anger, fear, disgust), each of which corresponds to and represents a unique pattern of emotional experience, behavior, and physiology (e.g., Ekman, 1999). Adherents of this view argue that, by collapsing distinct emotions (e.g., guilt and anxiety) into dimensional emotional experiences (e.g., high negative activation), the uniqueness of each emotion is lost.

Measures differ in the degree to which they are indicative of dimensional versus discrete emotional experience. For instance, psychophysiological measures are especially useful for assessing negative activation (Cacioppo et al., 2000), whereas different self-report measures vary in how well they capture discrete emotions (see Mauss & Robinson, 2009). Again, the purpose and context of the measurement must be the focus when choosing a measurement strategy. With these general considerations in mind, we now turn to specific emotion measurement strategies.

Self-Report Measures

Types of Self-Report Measures

Self-report measures of emotions are those in which individuals are asked to describe their emotions, and they are the most common way of assessing emotions (Larsen & Prizmic-Larsen, 2006). Self-reports typically involve surveys (e.g., paper-and-pencil, online), although they could also take other forms (e.g., telephone or in-person interviews, or methods described below). Technical issues related to the construction of emotion surveys (e.g., the optimal numbers of items and response options, whether "reverse-scored" items should be used) are discussed in detail by Dalal and Credé (in press). In the current chapter, we focus on technical issues related to the measurement of emotion traits (i.e., dispositions) versus emotion states.

In the case of traits, people respond to cross-sectional (i.e., one-shot) measures in which they indicate how they *typically* feel or feel "*in general*." Here, the researcher's focus is generally on comparisons across people (e.g., "Harry is typically angrier than Sally"). In the case of states, people are asked indicate their feelings over a short time interval (e.g., during the past week, the past 24 hours, or even at the current moment). Researchers are occasionally interested in cross-sectional measures of states, where the focus continues to be on comparisons across people, albeit on a particular occasion (e.g., "After they both interacted with a rude customer yesterday, Harry was angrier than Sally"). More often, though, researchers are interested in multiple, repeated measures of states, where the focus is on within-person comparisons across time (e.g., "Harry was angrier yesterday afternoon than he was yesterday morning"). Importantly, assessing (and perhaps controlling for) trait affect is essential in these "repeated measures" scenarios, for both conceptual and statistical reasons.

Several techniques recently have been developed to measure emotional states. In the day reconstruction method (Kahneman, Krueger, Schkade, Schwarz, & Stone, 2004), respondents construct a sequence of episodes from the previous day and then describe the situational features of each episode (e.g., times, location, type of activity, extent of social interaction) and how they felt during each episode. In experience sampling methods (Hektner, Schmidt, & Csikszentmihalyi, 2007; for an example, see Dalal, Lam, Weiss, Welch, & Hulin, 2009), respondents complete repeated surveys in near real-time (e.g., several times per day for 3 weeks) in their natural setting (e.g., the workplace, home). Finally, a truly real-time technique involves moment-to-moment reports of how the respondent feels, using equipment such as a joystick, rating dial, slider, or computer mouse (for an example, see Schuldberg & Gottlieb, 2002). These sophisticated measurement strategies provide very rich data, but they potentially sacrifice "breadth" as they generally necessitate using brief surveys to minimize the burden on respondents.

Advantages of Self-Reports

Self-reports of emotions have high "face validity" in that they "look like" they measure what they are supposed to measure. As such, stakeholders (e.g., respondents, the organization's management) may more readily accept results from self-report measures than from less direct and transparent measures (e.g., using blood pressure monitors to assess emotion). Another advantage of self-reports stems from the fact that emotions are inherently perceptual and self-referential in nature. Self-report measures have an advantage over observational measures because, in order for observational measures to be valid, each of the following requirements must be met: (a) the person's emotional state must translate into potentially observable behavior, (b) the behavior must in fact be observed, and (c) the observer must be able to accurately infer the person's emotional state from the observed behavior (Chan, 2009).

Finally, self-report measures rarely require expensive equipment. Although some of the more intricate techniques (see above) may require equipment such as hand-held computers or joysticks, self-report measures are usually more convenient and cheaper than alternatives like psychophysiological measures.

Disadvantages of Self-Reports

Despite these advantages, self-report measures also have some drawbacks. One concern with them is that respondents may be unable to provide accurate reports of their emotional experiences. Certain clinical populations, although able to react to emotional stimuli, are unable to identify and label their emotional experiences in a manner conducive to self-report. Of perhaps greater relevance to OHP, inaccurate responding is likely when respondents are asked to describe their "characteristic" level of emotionality (as is the case when assessing emotion *traits*). In such cases, responses are based partly on beliefs about emotions (e.g., sex-role stereotypes) rather than on actual recall of experienced emotions (Robinson & Clore, 2002).

One potential solution to this issue is to use near real-time self-report techniques, such as those mentioned above. The characteristic (trait) level of emotionality can then be operationalized as the average of the state levels of emotion across measurement occasions. Moreover, another trait score, level of emotional *variability* (Eid & Diener, 1999), can be operationalized as the standard deviation across measurement occasions.

Even when respondents are able to provide accurate reports of their emotional experiences, they may be unwilling to do so. Due to the transparency of self-reports, respondents may distort their responses in order to present themselves in a more socially desirable light. However, not all domains of emotional content are equally susceptible to socially desirable responding (Chan, 2009). For example, respondents may be reluctant to report high levels of negative emotional states (e.g., anger) but less reluctant to report low levels of positive emotional states (e.g., joy). In addition, research suggests that respondents typically do not distort their responses to the degree to which they actually are able to do so (Chan, 2009).

Another common concern is that, when only one measurement method is used to measure all variables of interest, relationships between variables may be inflated due to common-method variance (Podsakoff, MacKenzie, Lee, & Podsakoff, 2003). A relevant example would be when a person with a high score on negative affectivity reports experiencing more negative events than is accurate. Depending on one's purpose, such a phenomenon could be of theoretical interest or could be a "method effect." Common-method variance often, but not always, inflates observed relationships: its effects must be considered in the context of the specific study (Chan, 2009). Both methodological and statistical remedies for common-method variance have been described (see Podsakoff et al., 2003), though statistical remedies are unlikely to be a panacea.

Observational Measures

Behavioral Observation

One way to supplement self-report methods is to observe and record behaviors such as laughing or yelling, which are linked to specific emotions (e.g., Tracy, Robins, & Schriber, 2009). Structuring the rating process, by using behaviorally anchored rating scales, for example, can significantly increase interrater reliability (Maurer, 2002). After coding these behaviors, one then can examine frequencies, patterns, and correlates of displayed emotions.

More recently, emotion researchers have begun to systematically evaluate body postures and other "whole body movements" for emotional content (e.g., Reed & McIntosh, 2008). Using this approach, one might assess work engagement, for instance, by focusing on body postures such as the degree to which one has an upward pointed chin and a broad chest (Reed & McIntosh, 2008). Evaluating entire body postures is important because many emotional experiences (e.g., pride) cannot be decoded from vocal behavior or other observations but can be identified by body movements and posture (Reed & McIntosh, 2008).

Facial Behavior

Other observation methods focus on facial movements, such as "nonsocial" (i.e., Duchenne or genuine) smiles, discernible by observing "wrinkling" near the eyes, which developed in humans, at least in part, to convey positive emotion (Ekman, 1999). Several procedures exist to assess emotion-encoded facial behavior. One procedure, called the Facial Action Coding System (FACS; Ekman & Freisen, 1978) is a coding system that focuses on 44 different facial movements linked to the experience of emotions. A second method involves using electromyography (EMG). EMGs assess electrical potentials created by facial movements and, like the FACS, can reliably measure facial indicators of emotion. This latter technique is especially useful for assessing "the startle response"—an evolutionarily adaptive reaction that is related to fear and anxiety (Mauss & Robinson, 2009).

Narrative Accounts

Emotional experiences also can be investigated using qualitative methods such as the coding of written or oral narratives. These methods are especially useful in assessing particular, discrete emotional experiences (e.g., a bad mood at work on a particular morning) and in identifying the specific conditions that give rise to them (e.g., a child's illness). Also, writing narratives about one's own work and the emotions it generates can improve respondents' well-being (e.g., Barclay & Skarlicki, 2005).

Advantages of Observational Measures

Observational methods offer several notable benefits and can overcome many of the drawbacks associated with self-reports. For instance, observational methods can provide insights into emotional experiences that individuals may not recognize or be willing to divulge (Tracy & Robins, 2007). Also, some of these methods (e.g., narrative accounts) provide richer information than do self-reported numerical ratings and can offer insights into causal processes and intrapersonal motivation.

Observational methods generally are most useful when research interest is in the display (versus the experience) of emotion. Assessing emotional displays can provide important information. Recurring instances of inappropriate emotional displays (e.g., expressing too much or too little emotion) can indicate poor emotional functioning at work. Also, given that interaction partners respond to the actor's emotional expressions (not necessarily to the actor's emotional experiences;

Keltner & Haidt, 1999), these expressions are the proximal predictor of workplace social outcomes (e.g., quality of interpersonal relationships, reputation).

Disadvantages of Observational Measures

Although observational measures can be quite informative, they also have some potential drawbacks. First, as noted above, some of them are more indicative of emotional expression/displays than emotional experience per se. Thus, when primary interest centers around emotional experience, observational measures should supplement, not replace, self-reports. Another consideration with observational measures is their practicality. Most of them require having raters observe or code emotional displays. The time and effort that goes into this work, as well as the training for it, is substantial. Another practical factor is the potential expense involved. Materials such as audio and video equipment as well as qualitative and observational coding software can be quite costly, as can the required training for some of these methods (e.g., FACS).

Psychophysiological Measures

Measures of Autonomic Nervous System Functioning

The final measurement type considered here is psychophysiological measurement. The most frequently used physophysiological measures of emotion are indicators of the autonomic nervous system (ANS), including indicators of respiratory functioning (e.g., oxygen uptake), cardiac functioning (e.g., blood pressure, heart rate), and electrodermal measures (e.g., skin conductance response; see Mauss & Robinson, 2009). While some studies suggest that specific patterns of ANS functioning correspond to specific emotions (e.g., finger temperature decreases less with anger than with fear), others suggest that ANS measures are most useful for assessing the general affective dimensions of valence (i.e., positive/negative) and arousal/activation (e.g., Cacioppo et al., 2000). Although debate about this specificity issue persists (see Stephens, Christie, & Friedman, 2010), the use of these measures to assess negative arousal is uncontested (see Larsen, Bernston, Poehlman, Ito, Cacioppo,

2008). Importantly, the incidence of negative, high emotional arousal has been linked to various health problems (cf. Smith & MacKenzie, 2006). A particularly consistent finding is the link between increased ambulatory blood pressure and both job strain and cardiovascular disease (e.g., Schnall, Schwartz, Landsbergis, Warren, Pickering, 1992).

Measures of Central Nervous System Functioning

Other methods provide emotion-relevant information by assessing central nervous system (CNS) functioning. One such method—electroencephalography (EEG)—is a noninvasive technique in which electrical brain activity is measured through electrodes placed on the scalp. Emotion studies employing EEG primarily have focused on hemispheric activation and, more specifically, frontal asymmetry in such activation (see Harmon-Jones, Gable, & Peterson, 2010). Researchers also have begun to assess emotional functioning using neuroimaging technology (e.g., functional magnetic resonance imaging [fMRI]; positron emission tomography [PET]). These techniques provide far more precise assessment of brain activity location than do EEGs, and they therefore hold the promise of identifying which brain structures and circuits are connected to specific emotions (see Mauss & Robinson, 2009).

Advantages of Psychophysiological Measures

The main advantage of psychophysiological measures is that they have the potential to overcome potential unwillingness or inability to self-report emotions. With respect to willingness, these measures appear likely to be less susceptible to intentional distortion (e.g., malingering) than are self-report measures. With regard to ability, there are some emotional states about which individuals are less aware and therefore less able to self-report, but that are potentially accessible to physiological indicators (Bradley, 2000). Also, these measures may be especially useful for people who are especially poor at identifying or verbally expressing their emotions. Another practical benefit of physiological indicators is that some of these measures (e.g., heart rate and blood pressure) are not only emotion markers but also represent mechanisms through which emotion impacts health outcomes, such as cardiovascular disease and mortality (e.g., Schnall et al., 1992).

Disadvantages of Psychophysiological Measures

There are also several considerations/drawbacks associated with the use of these measures. First, all of these measurements are multiply determined, in that they reflect multiple bodily systems as well as factors such as appetite and bodily movement. Thus, attempting to isolate the degree to which a particular physiological indicator is reflective of emotional experience can be challenging (see Myrtek, 2004). At a minimum, one is advised to aggregate multiple, repeated measurements to obtain more reliable and valid results.

The other major consideration regarding these measures is their practicality. As a rule, administration and monitoring of these measures is labor intensive for both participants and researchers. In addition, these measures are costly and require significant researcher training. For these reasons, considerable deliberation should go into the decision to use these measures and into their actual implementation (see Curtin, Lozano, & Allen, 2007). As with observational measures, the most useful strategy generally is to use these measures in conjunction with self-report measures because doing so provides both convergent and nonredundant information.

Summary and Conclusions

In this chapter, we have provided an overview of different approaches for measuring work-related emotions, focusing mainly on the practical considerations of each approach. In the hopes of providing an easily accessible synopsis of these various approaches, we provide in Table 5.1 a summary of the main points discussed above. This table provides examples, advantages, and drawbacks/considerations associated with each type of measurement and also lists some key references for each measurement approach. In addition, the table lists citations to important papers about emotions at work and about emotion measurement in general. As this table and the chapter in general make clear, measuring emotions is not a simple or formulaic undertaking. As with all measurement, there is no substitute for thoughtfulness in assessing workplace emotion. We hope this chapter provides a useful resource for those involved in this challenging, but very important endeavor.

References

Barclay L. J., & Skarlicki, D. P. (2005). Healing the wounds of organizational injustice: Examining the benefits of expressive writing. *Journal of Applied Psychology, 94,* 511–523.

Barsade, S., & Gibson, D. E. (2007). Why does affect matter in organizations? *Academy of Management Perspectives, 21,* 36–59.

Brief, A. P., & Weiss, H. M. (2002). Organizational behavior: Affect in the workplace. *Annual Review of Psychology, 53,* 279–307.

Cacioppo, J. T., Berntson, G. G., Larsen, J. T., Poehlmann, K. M., & Ito, T. A. (2000). The psychophysiology of emotion. In M. Lewis & J. M. Haviland-Jones (Eds.), *The handbook of emotion, 2nd Edition* (pp. 173-191). New York: Guildford.

Chan, D. (2009). So why ask me? Are self-report data really that bad? In C. E. Lance & R. J. Vandenberg (Eds.), *Statistical and methodological myths and urban legends: Doctrine, verity and fable in the organizational and social sciences* (pp. 309–336). New York: Routledge.

Coan, J. A., & Allen, J. J. B. (Eds.). (2007). *Handbook of emotion elicitation and assessment.* Oxford, England: Oxford University Press.

Curtin, J. L., Lozano, D. L., & Allen, J. J. B. (2007). The psychophysiological laboratory. In J. A. Coan & J. J. B. Allen (Eds.), *Handbook of emotion elicitation and assessment* (pp. 398–425). Oxford, England: Oxford University Press.

Dalal, R. S., & Credé, M. (in press). Job satisfaction. In K. F. Geisinger (Ed.), *American Psychological Association handbook of testing and assessment in psychology.* Washington, DC: American Psychological Association.

Dalal, R. S., Lam, H., Weiss, H. M., Welch, E., & Hulin, C. L. (2009). A within-person approach to work behavior and performance: Concurrent and lagged citizenship—Counterproductivity associations, and dynamic relationships with affect and overall job performance. *Academy of Management Journal, 52,* 1051–1066.

Eid, M., & Diener, E. (1999). Intraindividual variability in affect: Reliability, validity, and personality correlates. *Journal of Personality and Social Psychology, 76,* 662–676.

Ekman, P. (1999). Basic emotions. In T. Dalgleish & M. J. Power (Eds.), *Handbook of cognition and emotion* (pp. 45–60). New York: Wiley.

Ekman, P., & Friesen, W. V. (1978). *The facial action coding system.* Palo Alto, CA: Consulting Psychologists Press.

Elfenbein, H. A. (2007). Emotion in organizations: A review and theoretical integration. *Academy of Management Annals, 1,* 371–457.

Gray, E. K., & Watson, D. (2007). Assessing positive and negative affect via self-report. In J. A. Coan & J. J. B. Allen (Eds.), *Handbook of emotion elicitation and assessment* (pp. 171–184). Oxford, England: Oxford University Press.

Harmon-Jones, E., Gable, P. A., & Peterson, C. K. (2010). The role of asymmetric frontal cortical activity in emotion-related phenomena: A review and update. *Biological Psychology, 84,* 451–462.

Hektner, J., Schmidt, J., & Csikszentmihalyi, M. (2007). *Experience sampling method: Measuring the quality of everyday life*. Thousand Oaks, CA: Sage.

Kahneman, D., Krueger, A. B., Schkade, D. A., Schwarz, N., & Stone, A. A. (2004). A survey method for characterizing daily life experience: The day reconstruction method. *Science, 306,* 1776–1780.

Kaplan, S., Bradley, J. C., Luchman, J. N., & Haynes, D. (2009). On the role of positive and negative affectivity in job performance: A meta-analytic investigation. *Journal of Applied Psychology, 94,* 162–176.

Keltner, D., & Haidt, J. (1999). Social functions of emotions at four levels of analysis. *Cognition and Emotion, 13,* 505–521.

Larsen, J. T., Berntson, G. G., Poehlmann, K. M., Ito, T. A., & Cacioppo, J. T. (2008). The psychophysiology of emotion. In R. Lewis, J. M. Haviland-Jones, & L. F. Barrett (Eds.), *The handbook of emotions* (3rd ed., pp. 180–195). New York: Guilford.

Larsen, R. J., & Prizmic-Larsen, Z. (2006). Measuring emotions: Implications of a multimethod perspective. In M. Eid & E. Diener (Eds.), *Handbook of multimethod measurement in psychology* (pp. 337–351). Washington, DC: American Psychological Association.

Maurer, S. (2002). A practitioner-based analysis of interviewer job expertise and scale format as contextual factors in situational interviews. *Personnel Psychology, 55*(2), 307–327. doi:10.1111/j.1744-6570.2002.tb00112.x.

Mauss, I. B., & Robinson, M. D. (2009). Measures of emotion: A review. *Cognition and Emotion, 23,* 209–237.

Miner, A. G., Glomb, T. M., & Hulin, C. L. (2005). Experience sampling mood and its correlates at work. *Journal of Organizational and Occupational Psychology, 78,* 171–193.

Myrtek, M. (2004). *Heart and emotion. Ambulatory monitoring studies in everyday life.* Gottingen, Germany: Hogrefe & Huber.

Podsakoff, P. M., MacKenzie, S. B., Lee, J-Y., & Podsakoff, N. P. (2003). Common method biases in behavioral research: A critical review of the literature and recommended remedies. *Journal of Applied Psychology, 88,* 879–903.

Reed, C., & McIntosh, D. (2008). The social dance: On-line body perception in the context of others. In R. L. Klatzky, M. Behrmann, & B. MacWhinney (Eds.), *Embodiment, ego-space, and action* (pp. 79–111). Hillsdale, NJ: Erlbaum.

Robinson, M. D., & Clore, G. L. (2002). Episodic and semantic knowledge in emotional self-report: Evidence for two judgment processes. *Journal of Personality and Social Psychology, 83,* 198–215.

Russell, J. A., & Barrett, L. F. (1999). Core affect, prototypical emotional episodes, and other things called emotion: Dissecting the elephant. *Journal of Personality and Social Psychology, 76,* 805–819.

Schnall, P. L., Schwartz J. E., Landsbergis, P. A., Warren K., & Pickering, T. G. (1992). Relation between job strain, alcohol and ambulatory blood pressure. *Hypertension, 19,* 488–494.

Schuldberg, D., & Gottlieb, J. (2002). Dynamics and correlates of microscopic changes in affect. *Nonlinear Dynamics, Psychology, and Life Sciences, 6*, 231–257.

Sloan, D. M., & Kring, A. M. (2007). Measuring changes in emotion during psychotherapy: Conceptual and methodological issues. *Clinical Psychology: Science and Practice, 14*, 307–322.

Smith, T. W., & MacKenzie, J. (2006). Personality and risk of physical illness. *Annual Review of Clinical Psychology, 2*, 435–467.

Stephens, C. L., Christie, I. C., & Friedman, B. H. (2010). Autonomic specificity of basic emotions: Evidence from pattern classification and cluster analysis. *Biological Psychology, 82*, 463–473.

Tracy, J., & Robins, R. (2007). The prototypical pride expression: Development of a nonverbal behavior coding system. *Emotion, 7*, 789–801. doi:10.1037/1528-3542.7.4.789.

Tracy, J., Robins, R., & Schriber, R. (2009). Development of a FACS-verified set of basic and self-conscious emotion expressions. *Emotion, 9*, 554–559. doi:10.1037/a0015766.

Watson, D., & Clark, L. A. (1999). *The PANAS-X: Manual for the positive and negative affect schedule—Expanded form.* Retrieved from http://www.psychology.uiowa.edu/Faculty/Watson/PANAS-X.pdf

Watson, D., Wiese, D., Vaidya, J., & Tellegen, A. E. (1999). The two general activation systems of affect: Structural findings, evolutionary considerations, and psychobiological evidence. *Journal of Personality and Social Psychology, 76*, 820–838.

6

How to Think About and Measure Psychological Well-Being

Peter Warr

Like the broader construct of health, well-being takes many forms and has no single index. Effective measurement clearly requires instruments that are technically sound, but particularly important is clarity about the form of well-being to be measured. Only when we are clear what we are trying to assess can we decide whether we have been successful. This chapter will focus on conceptual themes underlying effective measurement.

In all cases we are in effect applying an *operational* definition; the variable is being defined in terms of the measure applied—its operationalization. However, that operational definition may well or poorly represent a corresponding *conceptual* definition—the meaning of the construct independent of any measure. In this way, observed correlations reflect empirical associations between specific operationalizations but not necessarily between underlying concepts. Since many different operationalizations exist for constructs such as well-being, different researchers' operational definitions may be addressing different versions of supposedly the same construct.

This situation is unavoidable and can be acceptable. However, scientific understanding is likely to increase more rapidly if investigators make sustained efforts to be clear about each variable's conceptual definition and recognize the ways in which their operational measurements do and do not match that definition.

Eight Issues in the Measurement of Well-being

The first task of measurement is thus to develop a clear and appropriate conceptual definition. The two halves of the term *well-being* point directly to its meaning. It involves "being" and "well"—living in a state that is in some sense good. Measures of well-being thus record evaluations of some kind about a person's life, usually through judgments made by the individual himself or herself. Beyond those generalities, however, the construct has many possible forms (different ways to be "well"), and measures necessarily vary between one form and another.

This chapter addresses eight issues about the effective conceptualization and measurement of well-being. Choices in one respect often depend on decisions in regard to others, and in many cases a measurement decision is likely to be between possibilities which are similarly desirable. Decisions should depend on the aims of a current study, and not derive solely from an investigator's previous routines or local institutional norms.

Issue 1: Psychological, Physiological, or Social Emphasis

Psychologists have understandably focused on well-being that is "psychological" or "subjective" (e.g., Diener, Suh, Lucas, & Smith, 1999). However, as a first step in a particular study we need to decide how much emphasis should also be given to physiological or social aspects. For example, certain forms of low or high well-being (e.g., during prolonged strain or in strong emotions) are closely linked to bodily processes, and in examining those from a psychophysiological perspective may be appropriate. In other cases, research objectives might suggest that a measure of *social* well-being would be helpful to examine interpersonal relatedness and integration (e.g., Keyes, 1998; Peterson, Park, & Sweeney, 2008). Yet again, an entirely psychological focus might best permit detailed attention to an investigator's primary topic. This chapter will emphasize well-being that is psychological.

Issue 2: State or Trait Well-Being

Both health and well-being are moderately stable over time and are sometimes examined in dispositional terms; someone might be described as "a healthy person" or "a cheerful individual." Investigators have differed in their concern for either trait or state, or in the temporal focus of their state measures. For example, Watson (1988) contrasted feelings of six durations, through a trait measure (asking about "in general") and five durations of state (during the past year, during the past few weeks, during the past few days, today, and at the present moment).

Whatever the specific form of measurement to be used, it is essential to review in advance alternative target durations to ensure that a chosen duration matches that of the construct and question being investigated. For example, a form of well-being expected to be linked only briefly to other variables should be assessed through a similarly brief target period. In addition, possible limitations of mental processing should be considered. Events occurring some time ago may be poorly recalled, and contrasting experiences within an extended period can be difficult to aggregate into an overall judgment. Conversely, a brief period ("in the past hour" for instance) may encourage accurate recall and help respondents to average their experiences, but that brief period may not be typical of usual conditions. One compromise is to ask about a moderate duration such as "the past week." The important point is that one should carefully reflect in advance about a study's need for trait versus state measurement and (in the latter case) about the appropriateness of different target durations.

Issue 3: The Scope of Measurement

In setting out to measure well-being, one must also specify the construct's desired scope. The broadest scope is in terms of life in general without restriction to a particular setting; that is "context-free" well-being, recorded in studies of life satisfaction, global happiness, and similar constructs. A medium-range focus is directed at one segment of a life-space, such as one's job, family, health, leisure, or oneself; that is "domain-specific" well-being. For example, *job-related* psychological well-being is a domain-specific form which reflects positive or negative evaluations of one's work. Third, we might examine "facet-specific" well-being, targeted at one particular aspect of one domain such as satisfaction with the pay received from one's job.

As expected from their conceptual overlap, experiences at the three levels of abstractness are empirically interrelated. Nevertheless, subjective well-being at different levels of scope is influenced in part by different factors (e.g., Warr, 1987, 2007). For example, job-related well-being is more responsive to conditions and activities in the domain of employment than other domains, and context-free well-being is additionally influenced by factors in health, family, community, and other domains.

It is clearly essential to make an explicit decision about which level of scope is theoretically and practically most relevant to a planned study. Well-being that is job-related is often of primary interest to occupational health and industrial-organizational psychologists, but some studies instead examine context-free well-being. That can of course be important in occupational samples, but its conceptual relevance to a current objective should be determined and the two levels of scope should not be conflated in analyses and interpretations.

Issue 4: Positive or Negative Emphasis

Like some other members of their profession, occupational health psychologists have tended to emphasize the negative—examining mental and physical strain in job settings. However, research attention has recently turned more to positive job experiences, in terms of constructs such as perceived meaningfulness, thriving, engagement in a job, resilience in the face of adversity, and sense of accomplishment (e.g., Cameron, Dutton, & Quinn, 2003; Linley, Harrington, & Garcea, 2010). An adequate conceptual definition of psychological well-being must extend to cover a wide range of elements, and it is often desirable to include both positive and negative themes when constructing a suite of instruments.

In that case, decisions are required about whether to examine positive and negative forms separately or instead to combine them into a single overall index. In part, this depends on which particular aspects of well-being are under investigation. When positive and negative variables are clearly distinct in conceptual terms, possibly with different associated predictions (e.g., perceived meaningfulness versus anxiety), they deserve separate examination and interpretation. However, conceptually similar forms of well-being, such as experienced core feelings, might be studied either overall or through separate positive and negative indicators.

Opinions remain divided about this issue. Arguments against the combined scoring of differently valenced affects and in favor of separate analyses of positive versus negative feelings include the potentially greater predictive specificity of two components rather than one, and an observed positive–negative separation in statistical analyses (e.g., Lucas, Diener, & Suh, 1996). However, positive and negative factors can in practice be artifacts arising from response acquiescence and other biases rather than from genuine conceptual bifurcation (e.g., Warr, 2007, chapter 2), and it may be preferable instead to examine affect scores with an alternative combined content. This topic is revisited under Issue 6.

Issue 5: Affective Well-Being and Cognitive-Affective Syndromes

Two principal perspectives and associated measuring instruments in this area may be identified, which are very different in their content and theoretical background. First are models and measures of well-being entirely in terms of people's feelings, more formally described as "affects" (above). Affects are experiences that are "primitive, universal, and simple, irreducible on the mental plane" (Russell, 2003, p. 148) and range along a bad-to-good continuum. They occur throughout waking life as components of emotions, moods, values, attitudes, orientations, prejudices, and ideologies, and are central to well-being in any setting.

However, well-being has also been examined through composites or syndromes which comprise thoughts as well as feelings. Organized around a particular theme, these take the form of satisfaction, strain, engagement, burnout, and so on. Well-being syndromes embody interlinked ideas, recollections, perspectives, and mental networks as well as merely affect. For instance, a measure of job-related burnout (a composite of experiences) might ask about feeling drained at the end of a workday or a person's reluctance to set off for work in the morning. Cognitive-affective syndromes thus differ from basic affects in having a variety of elements and involving thoughts and memories in addition to feelings. Self-descriptions in those terms, for example when responding to a job satisfaction scale, call for more reflection and mental processing than do affects: attending to and remembering particular elements and episodes, interpreting, evaluating, and integrating what is recalled, and perhaps making comparisons with other people or other jobs (e.g., Staw & Cohen-Charash, 2005).

It is often appropriate to assess both kinds of well-being, depending on current practical needs and conceptual relevance. "Affective well-being" concerns merely a person's feelings, whereas "syndrome well-being" is also a question of perceptions, recollections, comparisons, and anticipations as well as affects. As in other cases, the key requirement is for investigators in advance to review possibilities and choose those which may best meet their objectives.

Issue 6: Measuring Affective Well-Being

In studying the first of these well-being forms, one established perspective is in terms of the circumplex shown in Figure 6.1. This specifies experiences not only in terms of displeasure-to-pleasure (valence) but also through low-to-high mental arousal or activation (e.g., Remington, Fabrigar, & Visser, 2000; Russell, 1980, 2003).

Feelings in terms of those two axes are illustrated around the outside of Figure 6.1, and summary labels for each quadrant's content are indicated as Anxiety (activated negative affect), Enthusiasm (activated

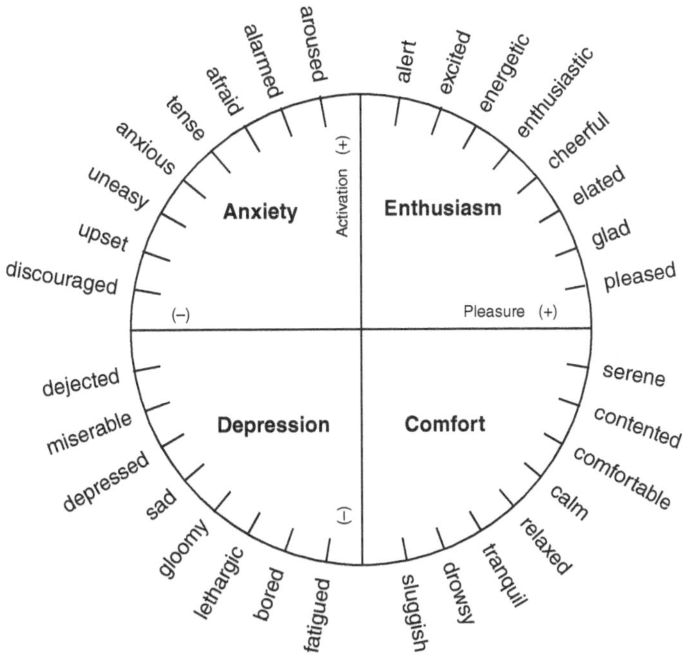

Figure 6.1 Some feelings and their locations within the affect circumplex

positive affect), Depression (low-activation negative affect), and Comfort (low-activation positive affect). As used here, those labels are shorthand descriptors for affect-sets within more complex mental and behavioral constructs; they do not denote the entirety of "Anxiety" and the other three constructs.

Studies in job settings have often been imprecise about the location of studied feelings in terms of Figure 6.1. For example, research into "positive affect" might be expected to cover all feelings on the right-hand side of the figure, involving both low and high activation. However, that has rarely been the case, in part because many measures have been based on the Positive and Negative Affect Schedule (PANAS; Watson, Clark, & Tellegen, 1988). The PANAS contains 20 items in two scales, with 10 positive terms, including, for example, *enthusiastic* and *excited* and 10 negative terms such as *distressed* and *nervous.*

Although the two scales were originally referred to as "positive affect" (PA) and "negative affect" (NA), they in fact tap only certain kinds of those—high-arousal feelings in the two upper segments of Figure 6.1 (e.g., Remington et al., 2000; Tellegen, Watson, & Clark, 1999). The construct of positive affect has thus often been incompletely operationalized merely as positive affect that is activated.

Instead, measures of affective well-being must reflect the conceptual nature of their domain, extending across those quadrants of Figure 6.1 which are considered appropriate in a particular investigation. Four-quadrant measures have been presented by Burke, Brief, George, Roberson, and Webster (1989), Warr (1990), and Van Katwyk, Fox, Spector, and Kelloway (2000), containing 20, 12, and 30 items respectively. Burke et al. (1989) emphasized that a model based on the four quadrants together was superior to one with positive-alone and negative-alone constructs, and Van Katwyk et al. (2000) advocated the use of an overall score across all quadrants as "the most comprehensive assessment" (p. 204).

Consistent with the PANAS conceptualization, Warr's analyses focused on opposite-quadrant diagonal axes from anxiety to comfort and from depression to enthusiasm (e.g., Warr, 2007). These are the bipolar dimensions labeled by Watson et al. (1988) as "negative affect" and "positive affect" respectively, although (as noted above) those two constructs have more often been assessed only in their activated forms. Sixteen affect items presented by Warr, Bindl, Parker, and Inceoglu (2012) can be scored in alternative combinations: for each quadrant singly, as pairs of quadrants, or as overall affect. The choice to be made in any setting will depend on circumstances and objectives;

for instance a concern for simplicity in reporting versus sensitivity of measurement.

A second issue to be resolved when measuring affective well-being concerns the response options to be offered. Instruments differ in their use of either "intensity" or "frequency" responses, asking either about how strongly (e.g., from *not at all* to *extremely*) or how often (e.g., from *never* to *nearly all the time*) something was felt. Which of those should be used?

Both involve mental averaging, as a person makes a summary judgment about his or her feelings. Intensity-averaging may be more appropriate in respect of a brief period of work such as a single meeting, but it can be difficult over longer periods (see also Issue 2 above). Frequency ratings may then be more suitable, and those also have the advantage of providing information that may be potentially useful in studies of ambivalence (see Issue 8).

Intensity responses were requested by Burke et al. (1989), whereas Van Katwyk et al. (2000), Warr (1990), and Warr et al. (2012) recorded the frequency of feelings. Warr's (1990) questionnaire is instead presented in terms of response-intensity by Warr and Clapperton (2010). Research findings appear very similar between the methods, but empirical comparisons between them are still required, as is a comprehensive analysis of their theoretical implications. Preferences will depend on local circumstances, but (as with the other issues here) the key requirement is for response alternatives to be explicitly reviewed in advance of an investigation.

Issue 7: Measuring Syndrome Well-being

The second approach to psychological well-being (above) is very different, focusing on compounds of thoughts as well as feelings. Such syndromes are illustrated throughout the book, in presentations about the measurement of stress disorders, job engagement, and other constructs. Two other possibilities will be considered here, in respect of job satisfaction and themes that are sometimes described as "eudaimonic."

For more than half a century, researchers and practitioners have been interested in workers' job satisfaction, conceptualizing this in different ways. One distinction is between overall job satisfaction (which is "domain-specific" in the terms introduced under Issue 3) and separate satisfactions with individual aspects of a job ("facet-specific" job satisfactions). For example, facet-specific scores are provided by five separate scales in the Job Descriptive Index (Smith, Kendall, & Hulin, 1969; Stanton et al., 2001).

Overall job satisfaction has been examined in two different ways, either asking only about satisfaction itself or in wider terms by covering a range of positive experiences. In the first case, a single question has sometimes been used (e.g., "all things considered, how satisfied are you with your job in general"), but more often instruments ask separately about satisfaction with each of a variety of job features so that several facet-specific satisfactions can be aggregated into an overall index (e.g., Warr, Cook, & Wall, 1979; Weiss, Dawis, England, & Lofquist, 1967). Such scales, focusing entirely on satisfaction itself, have the advantage of directly and exclusively operationalizing the essence of the concept.

Alternatively, instead of asking about that construct alone, different types of reaction to a job as a whole have been obtained and combined into an overall score. For example, widely used scales labeled as measures of overall job "satisfaction" include statements like: "Most days I am enthusiastic about my work" and "I find real enjoyment in my work" (Brayfield & Rothe, 1951) or request job-related ratings of, for instance, "worthwhile," "ideal," and "waste of time" (Ironson, Smith, Brannick, Gibson, & Paul, 1989).

Item-sets of this second kind may of course be valuable, but they extend beyond the definition of satisfaction. This term derives from the Latin *satis*, meaning "enough," and is restricted to a relatively passive acceptance that something is adequate—"satisfactory" rather than "outstanding." Being "satisfied with" something is more limited than actively "enjoying" it. Mixed-reaction scales, potentially asking about enjoyment, enthusiasm, involvement, feelings of worth, and so on, thus have a wider and more activated coverage. A broad perspective of that kind can be useful, but the "satisfaction" label should be reserved for indicators that remain within the conceptual definition, examining only satisfaction itself.

At the context-free level, syndrome well-being has sometimes been examined broadly as life satisfaction. For example, the Satisfaction with Life Scale (Diener, Emmons, Larsen, & Griffin, 1985) asks for assessments through items like "If I could live my life over, I would change almost nothing." Other context-free syndrome measures in this area include inventories like the General Health Questionnaire (GHQ; Goldberg & Williams, 1988), which covers a range of thoughts and feelings in terms of unhappiness, nervousness, and so on.

The General Health Questionnaire was developed to estimate psychiatric morbidity, and several other context-free and job-related instruments have a similarly negative emphasis, characterizing poor rather than good mental health. In recent years, some organizational researchers have also taken more positive approaches within a long-established philosophical framework that distinguishes between "hedonic" and "eudaimonic" perspectives on happiness.

Hedonic (pleasure-related) happiness concerns feelings and thoughts in ways illustrated above. However, the concept of happiness involves additional themes beyond hedonic experiences. It extends into a sense that one is fulfilling oneself, is exploiting one's potential, is fully functioning, is making good use of one's attributes, or that one is doing something worthwhile or meaningful. Other notions of this kind include a sense of wholeness, self-realization, being authentic (as one "should be"), being true to oneself, or acting in some way that is harmonious or morally desirable.

Such themes date back to early Greek philosophers such as Aristotle (384–322 BC), who was concerned with eudaimonia, a form of happiness that reflected a "good life." From a psychologist's standpoint, Seligman (2002) emphasized that a good life was not the same as a pleasant life, and he developed the notion of "authentic happiness" deriving from the use of one's personal strengths and virtues to achieve goals that had intrinsic value beyond mere pleasure. Paradoxically, eudaimonic happiness can be empirically unrelated to hedonic happiness, as it may be experienced in different ways: enjoyable, unpleasant, or devoid of feeling. For example, undertaking one's duty may involve pleasant activities or instead require stressful intervention in a difficult situation.

Despite their intrinsic significance, eudaimonic forms of happiness are difficult to specify and measure. Nevertheless, psychologists have focused on particular themes that may be applied in organizations. For example, Spreitzer, Sutcliffe, Dutton, Sonenshein, and Grant (2005) examined "thriving" as a sense of vitality and forward movement, with new learning and personal development. Consistent with earlier ideas of psychological growth and self-actualization, they emphasized that people in their jobs as well as elsewhere want to make progress toward achieving their potential. Linked to that, difficult expenditure of effort can itself be rewarding if it is directed at personally valued outcomes (e.g., Ferguson, 1767/1966; Warr & Inceoglu, 2012).

A related eudaimonic syndrome concerns the perception of meaningfulness in one's job. This involves awareness of some match between a job and one's personal values and self-identity, linked to the perception that the job has personal significance beyond providing mere transitory feelings. Meaningfulness may be perceived in everyday activities such as the use of valued skills or effective coping with personal challenges, or it may be a question of applying personally important moral principles. Perceived meaningfulness of a job thus reflects the degree to which in some (often ill-defined) way it matters to you. As reviewed by Ben-Shahar (2007), both pleasure and perceived meaning are essential for happiness in its full sense.

It is thus often desirable in investigations of psychological well-being to include measures of constructs like thriving and perceived meaningfulness. Those cognitive-affective syndromes overlap in part with established notions such as job involvement and organizational commitment (e.g., Cooper-Hakim & Viswesvaran, 2005) and job engagement (see chapter 10). Operationalizations of job-related thriving in terms of vitality and learning have been described by Porath, Spreitzer, Gibson, and Garnett (2012), and perceived meaningfulness scales have been applied by King, Hicks, Krull, and Del Gairo (2006; context-free) and May, Gibson, and Harter (2004; job-related).

Issue 8: Examining Ambivalence

The chapter's final issue concerns mixed patterns of well-being rather than single-aspect levels. Most publications associated with the previous issues are based on workers' assessments of their average feelings and thoughts across a period, and information about those average levels is of course essential for research and practice. However, differences between well-being of different forms and at different times can also be important, and for them measurement procedures require expansion.

The similarity or difference between well-being elements may be studied in two ways—across time or between those elements on a single occasion. In the first case, we may set out to learn how feelings differ in their level from period to period. In the second case, research can examine how those are similar or different between different forms of well-being; a person can feel good in some respects and feel bad in others within the same period of time.

Those patterns reflect the construct of ambivalence, sometimes referred to more positively as "balance." Ambivalence, literally being "valued in two ways" (but there can be more than two), is widely experienced in jobs and elsewhere (e.g., Smelser, 1998), but its few operationalizations have been partial and inconsistent. For example, Schwarz, Reynolds, Thase, Frank, and Fasiczka (2002) computed the proportion of all affects that were positive, Fredrickson and Losada (2005) examined the ratio of positive and negative emotions, and Sheldon and Niemiec (2006) recorded the sum of absolute differences between different need satisfactions.

Those investigations concerned experiences outside the workplace, and there is a great need for studies of ambivalence in job settings. Fluctuations across time might be studied through between-level contrasts in repeated surveys or diary completions, and investigations of ambivalence within a single period need to focus on ways in which different kinds of feeling coexist. The relative merits of different procedures to measure ambivalence remain unclear, and Issue 8 is included here to emphasize the need to develop and apply those procedures.

Content Validation and Related Concepts

This chapter has argued for explicit conceptualization as the basis for effectively measuring well-being. Conceptual definitions need to be more clearly specified than is often the case, and selected operational definitions should explicitly strive to match those in ways illustrated here. Several current themes might be viewed in terms of "content validation" as construed in respect of psychometric instruments for personnel selection.

Content validation is "the process of ensuring that test content is related to and representative of the attribute(s) of interest. This involves developing a clear theory of the attribute to be measured and creating items based on this theory" (Highhouse, 2009, p. 495). In doing that, we need also to examine the placement of particular concepts in relation to others—sometimes referred to as "discriminant validity." For instance, psychological well-being is a key aspect of happiness (e.g., Diener & Biswas-Diener, 2008), and is also central to the construct of mental health which also extends into affective-behavioral themes such as competence, aspiration, autonomy and integrated functioning (e.g., Warr, 2007). In creating precise conceptual definitions and assessing the content validity of their operationalizations, we need to place psychological well-being within, and partially distinct from, those broader notions.

References

Ben-Shahar, T. (2007). *Happier.* New York: McGraw-Hill.

Brayfield, A. H., & Rothe, H. F. (1951). An index of job satisfaction. *Journal of Applied Psychology, 35*, 307–311.

Burke, M. J., Brief, A. P., George, J. M., Roberson, L., & Webster, J. (1989). Measuring affect at work: Confirmatory analyses of competing mood structures with conceptual linkage to cortical regulatory systems. *Journal of Personality and Social Psychology, 57*, 1091–1102.

Cameron, K. S., Dutton, J. E., & Quinn, R. E. (2003). *Positive organizational scholarship.* San Francisco, CA: Barrett-Koehler.

Cooper-Hakim, A., & Viswesvaran, C. (2005). The construct of work commitment: Testing an integrative framework. *Psychological Bulletin, 131*, 241–259.

Diener, E., & Biswas-Diener, R. (2008). *Happiness: Unlocking the mysteries of psychological wealth.* Malden, MA: Blackwell.

Diener, E., Emmons, R. A., Larsen, R. J., & Griffin, S. (1985). The Satisfaction with Life Scale. *Journal of Personality Assessment, 49*, 71–75.

Diener, E., Suh, E. M., Lucas, R. E., & Smith, H. L. (1999). Subjective well-being: Three decades of progress. *Psychological Bulletin, 125*, 276–302.

Ferguson, A. (1966). *An essay on the history of civil society.* Edinburgh, Scotland: Edinburgh University Press. (Original work published 1767)

Fredrickson, B. L., & Losada, M. F. (2005). Positive affect and the complex dynamics of human flourishing. *American Psychologist, 60*, 678–686.

Goldberg, D. P., & Williams, P. (1988). *A user's guide to the General Health Questionnaire.* Windsor, England: NFER-Nelson.

Highhouse, S. (2009). Tests don't measure jobs: The meaning of content validation. *Industrial and Organizational Psychology, 2*, 493–495.

Ironson, G. H., Smith, P. C., Brannick, M. T., Gibson, W. M., & Paul, K. B. (1989). Construction of a Job in General scale: A comparison of global, composite, and specific measures. *Journal of Applied Psychology, 74*, 193–200.

Keyes, C. L. M. (1998). Social well-being. *Social Psychology Quarterly, 61*, 121–140.

King, L. A., Hicks, J. A., Krull, J. L., & Del Gairo, A. K. (2006). Positive affect and the experience of meaning in life. *Journal of Personality and Social Psychology, 90*, 179–196.

Linley, A., Harrington, S., & Garcea, N. (Eds.). (2010). *Oxford handbook of positive psychology and work.* New York: Oxford University Press.

Lucas, R. E., Diener, E., & Suh, E. (1996). Discriminant validity of well-being measures. *Journal of Personality and Social Psychology, 71*, 616–628.

May, D. R., Gilson, R. L., & Harter, L. (2004). The psychological conditions of meaningfulness, safety and availability, and the engagement of the human spirit at work. *Journal of Occupational and Organizational Psychology, 77*, 11–27.

Peterson, C., Park, N., & Sweeney, P. J. (2008). Group well-being: Morale from a positive psychology perspective. *Applied Psychology: An International Review, 57*, 19–36.

Porath, C., Spreitzer, G., Gibson, C., & Garnett, F. (2012). Thriving at work: Toward its measurement, construct validation, and theoretical refinement. *Journal of Organizational Behavior, 33,* 250–275.

Remington, N. A., Fabrigar, L. R., & Visser, P. S. (2000). Re-examining the circumplex model of affect. *Journal of Personality and Social Psychology, 79,* 286–300.

Russell, J. A. (1980). A circumplex model of affect. *Journal of Personality and Social Psychology, 39,* 1161–1178.

Russell, J. A. (2003). Core affect and the psychological construction of emotion. *Psychological Review, 110,* 145–172.

Schwarz, R. M., Reynolds, C. F., Thase, M. E., Frank, E., & Fasiczka, A. L. (2002). Optimal and normal affect balance in psychotherapy of major depression: Evaluation of the balanced states of mind model. *Behavioural and Cognitive Psychotherapy, 30,* 439–450.

Seligman, M. E. P. (2002). *Authentic happiness.* New York: Free Press.

Sheldon, K. M., & Niemiec, C. P. (2006). It's not just the amount that counts: Balanced need satisfaction also affects well-being. *Journal of Personality and Social Psychology, 91,* 331–341.

Smelser, N. J. (1998). The rational and the ambivalent in the social sciences. *American Sociological Review, 63,* 1–16.

Smith, P. C., Kendall, L. M., & Hulin, C. L. (1969). *The measurement of satisfaction in work and retirement.* Chicago, IL: Rand-McNally.

Spreitzer, G. M., Sutcliffe, K., Dutton, J., Sonenshein, S., & Grant, A. M. (2005). A socially embedded model of thriving at work. *Organization Science, 16,* 537–549.

Stanton, J. M., Sinar, E. F., Balzer, W. K., Julian, A. L., Thoresen, P., Aziz, S., & Smith, P. C. (2001). Development of a compact measure of job satisfaction: The abridged Job Descriptive Index. *Educational and Psychological Measurement, 61,* 1104–1122.

Staw, B. M., & Cohen-Charash, Y. (2005). The dispositional approach to job satisfaction: More than a mirage, but not yet an oasis. *Journal of Organizational Behavior, 26,* 59–78.

Tellegen, A., Watson, D., & Clark, L. A. (1999). On the dimensional and hierarchical structure of affect. *Psychological Science, 10,* 297–303.

Van Katwyk, P. T., Fox, S., Spector, P. E., & Kelloway, E. K. (2000). Using the Job-Related Affective Well-Being Scale (JAWS) to investigate affective responses to work stressors. *Journal of Occupational Health Psychology, 5,* 219–230.

Warr, P. B. (1987). *Work, unemployment, and mental health.* Oxford, England: Oxford University Press.

Warr, P. B. (1990). The measurement of well–being and other aspects of mental health. *Journal of Occupational Psychology, 63,* 193–210.

Warr, P. B. (2007). *Work, happiness, and unhappiness.* New York: Routledge.

Warr, P. B., Bindl, U., Parker, S., & Inceoglu, I. (2012). Four-quadrant investigation of job-related affects and behaviours. Submitted for publication.

Warr, P. B., & Clapperton, G. (2010). *The joy of work? Jobs, happiness, and you.* New York: Routledge.

Warr, P. B., & Inceoglu, I. (2012). Job engagement, job satisfaction, and contracting associations with person-job fit. *Journal of Occupational Health Psychology, 17,* 129–138.

Warr, P. B., Cook, J. D., & Wall, T. D. (1979). Scales for the measurement of some work attitudes and aspects of psychological well-being. *Journal of Occupational Psychology, 52,* 129–148.

Watson, D. (1988). The vicissitudes of mood measurement: Effects of varying descriptors, time frames, and response formats on measures of positive and negative affect. *Journal of Personality and Social Psychology, 55,* 128–141.

Watson, D., Clark, L. A., & Tellegen, A. (1988). Development and validation of brief measures of positive and negative affect: The PANAS scales. *Journal of Personality and Social Psychology, 54,* 1063–1070.

Weiss, D. J., Dawis, R. V., England, G. W., & Lofquist, L. H. (1967). *Manual for the Minnesota Satisfaction Questionnaire.* Minneapolis: University of Minnesota.

7

Measurement of Interpersonal Mistreatment in Organizations

Lilia M. Cortina and Lisa A. Marchiondo

Ostracism, incivility, deviance, undermining, aggression, bullying, and insidious workplace behavior: interpersonal mistreatment has many behavioral faces in organizations. Within the last two decades, occupational health psychology (OHP) has witnessed an explosion of interest in mistreatment phenomena. Along with that interest has come an array of survey tools for assessing mistreatment, both the "doing" and "receiving" of it. These tools are the focus of this chapter.

Before beginning, we outline the boundary conditions of this article. Our principal focus is quantitative methods of assessing interpersonal mistreatment *in the workplace*, not in schools, neighborhoods, or other contexts. We specifically concentrate on measuring mistreatment targeted *at persons*, for the most part excluding behaviors directed toward property (e.g., theft, vandalism) or organizations (e.g., organizational deviance). Finally, we pay particular attention to the assessment of *generic* mistreatment that makes no overt reference to gender, race, or any other social category. Instruments are available for assessing gender- and race-based mistreatment in detail (e.g., the Sexual Experiences Questionnaire: Fitzgerald, Gelfand, & Drasgow, 1995; or the Ethnic Harassment Experiences Scale: Schneider, Hitlan, & Radhakrishnan, 2000) but are beyond the scope of our chapter.

We organize this chapter into three primary sections. First, we outline 15 commonly used instruments for assessing both the enactment and experience of workplace mistreatment. Next comes an outline of the types of research questions that can be addressed with these scales, with examples from OHP. Finally, we alert researchers to issues that should guide their selection of the appropriate measurement strategy.

Specific Instruments for Assessing Workplace Mistreatment

Interpersonal mistreatment in organizations is largely assessed using multi-item behavioral self-report instruments. Below we review 15 such instruments that, to our knowledge, are still in active use at the time of this writing. Each has established psychometric properties. Table 7.1 details each scale's stem, response options, and sample items.

Aggression

Aggressive Experiences Scale (AES). The Aggressive Experiences Scale (AES; Glomb, 2010) measures interpersonal forms of workplace aggression, defined as "efforts by individuals to harm others with whom they work, or have worked ..." (Neuman & Baron, 1998, p. 395). The instrument was designed to capture both subtle and severe forms of aggression that carry a clear intent to harm. It contains two 20-item scales that tap the frequency (1 = *never* to 6 = *once a week or more*) with which respondents have been both actors ("Engaged In" scale) and recipients ("Target" scale) of the behavior within the last 6 months. Testing supports good convergent validity between the AES and other measures of aggression.

Workplace Aggression Research Questionnaire (WAR-Q). The Workplace Aggression Research Questionnaire (WAR-Q; Neuman & Keashly, 2004; see also Neuman & Keashly, 2010) is a 60-item instrument that captures the frequency (from *never* to *daily*) and source (e.g., from customers, supervisors) of employees' experiences of aggression within the last year. Respondents can describe up to three "other" behaviors they have encountered and rate how strongly, overall, the aggressive experiences bothered them. The WAR-Q has not been published but can be requested from its authors, Joel Neuman (at SUNY-New Palz) and Loraleigh Keashly (at Wayne State University).

Bullying

Leymann Inventory of Psychological Terror (LIPT). The Leymann Inventory of Psychological Terror (LIPT; 1990) measures workplace experiences of psychological terror or "mobbing" (termed *bullying* by some researchers), which "involves hostile and unethical

Table 7.1 Common Measures of Workplace Mistreatment

Behavior	Measure	Pers-pective	Stem/Opening Language	Response Scale	Sample Item(s)
Aggression	*Aggressive Experiences Scale (AES)—Target* (Glomb, 2010)	Target	Have you ever been involved in a situation in which you were the target of any of the following behaviors?	1 = Never 6 = Once a week or more	• Using an angry tone of voice • Avoiding you • Withholding information from you • Physically assaulting you • Interrupting or "cutting you off" while speaking
	Aggressive Experiences Scale—Engaged In (Glomb, 2010)	Actor	Have you ever been involved in a situation while on the job where you engaged in the following behaviors?	1 = Never 6 = Once a week or more	• Identical to Aggressive Experiences Scale—Target items (above); only change is in stem
	Workplace Aggression Research Questionnaire (Neuman & Keashly, 2004)	Target	How often have you been subjected to this behavior in your workplace over the past 12 months?	1 = Never 7 = Daily	• Been excluded from work-related social gatherings • Had others consistently arrive late for meetings that you called • Been sworn at in a hostile manner • Not been given the praise for which you felt entitled • Had your personal property defaced, damaged, or stolen • Been subjected to unwanted attempts to touch, fondle, kiss, or grab you
Bullying	*Leymann Inventory of Psychological Terrorization* (Leymann, 1990)	Target	Have you been subjected to one or more of the following actions during the last 12 months?	1 = Daily 5 = Seldom/ never	• You are silenced • Colleagues do not talk to you any longer • Others ridicule you • You are given meaningless work tasks • You are given dangerous work tasks • Others threaten you physically

(*continued*)

Table 7.1 Continued

Behavior	Measure	Perspective	Stem/Opening Language	Response Scale	Sample Item(s)
	Negative Acts Questionnaire (Einarsen & Raknes, 1997)	Target	During the last 6 months, how often have you been subjected to the following negative acts in the workplace?	1 = Never 5 = About daily	• Someone withholding necessary information so that your work gets complicated • Neglect of your opinion or views • Gossip or rumor about you • Devaluing of your work and efforts • Unwanted sexual attention
Incivility	*Workplace Incivility Scale* (Cortina et al., 2001)	Target	During the past [insert time-frame] years, while employed at X, have you been in a situation where any of your superiors or coworkers…?	1 = Never 5 = Many times	• Put you down or was condescending to you • Paid little attention to your statements or showed little interest in your opinion • Made demeaning or derogatory remarks about you • Doubted your judgment on a matter over which you have responsibility • Addressed you in unprofessional terms, either publicly or privately
	Instigated Workplace Incivility Scale (Blau & Andersson, 2005)	Actor	How often have you exhibited the following behaviors in the past year to someone at work (e.g., coworker, other employee, supervisor)...?	1 = Hardly ever (once every few months or less) 4 = Frequently (at least once a day)	• Put down others or were condescending to them in some way • Paid little attention to a statement made by someone or showed little interest in their opinion • Made demeaning, rude, or derogatory remarks about someone • Doubted someone's judgment on a matter over which they have responsibility • Addressed someone in unprofessional terms either privately or publicly

	Uncivil Workplace Behavior Questionnaire (Martin & Hine, 2005)	Target	During the past year, while employed at X, have you been in a situation where any of your superiors or coworkers…?	1 = Never 5 = Very often	• Used an inappropriate tone when speaking to you • Took items from your desk without prior permission • Did not consult you in reference to a decision you should have been involved in • Talked about you behind your back
Misc.	*Abusive Supervision Scale* (Tepper, 2000)	Target	My boss…	1 = I cannot remember him or her ever using this behavior with me 5 = He or she uses this behavior very often with me	• Gives me the silent treatment • Invades my privacy • Doesn't give me credit for jobs requiring a lot of effort • Blames me to save himself or herself embarrassment • Breaks promises he or she makes
	Counter-productive Work Behavior – Person (Spector et al., 2006)	Actor	How often have you done each of the following things on your present job?	1 = Never 5 = Every day	• Started or continued a damaging or harmful rumor at work • Been nasty or rude to a client or customer • Ignored someone at work • Threatened someone at work with violence • Looked at someone at work's private mail/property without permission
	Generalized Workplace Abuse (Richman et al., 1999)	Target	During the past year at X or a different workplace, how often have you been in a situation where someone in your work setting…?	1 = Never 3 = More than once	• Gossiped about you or spread rumors about you behind your back • Humiliated or belittled you in front of others • Took credit for your work or ideas • Ignored you or your work contributions • Pushed you or grabbed you

(*continued*)

Table 7.1 Continued

Behavior	Measure	Pers-pective.	Stem/Opening Language	Response Scale	Sample Item(s)
	Interpersonal Conflict at Work Scale (Spector & Jex, 1998)	Target (+ one actor item)	N/A	1 = Never 5 = Very often	• How often are people rude to you at work • How often do other people do nasty things to you at work • How often do you get into arguments with others at work
	Interpersonal Deviance Scale (Bennett & Robinson, 2000)	Actor	How often have you engaged in each of these behaviors in the last year?	1 = Never 7 = Daily	• Made fun of someone at work • Made an ethnic, religious, or racial remark at work • Acted rudely toward someone at work • Publicly embarrassed someone at work
	Social Undermining Scale (Duffy et al., 2002)	Target	In the past month, how often has your supervisor intentionally…? In the past month, how often has the coworker closest to you intentionally…?	1 = Never 6 = Every day	• Hurt your feelings • Let you know that they did not like something about you • Gave you the silent treatment • Talked down to you (Supervisor) • Competed with you for status and recognition (Coworker)
	Workplace Ostracism Scale (Ferris et al., 2008)	Target	Respondents are asked whether they have experienced the items within the previous year.	1 = Never 7 = Always	• Others ignored you at work • Others avoided you at work • Others would shut you out of the conversation • Others left the area when you entered

communication, which is directed in a systematic way by one or a few individuals mainly towards one individual who … is pushed into a helpless and defenceless position …" (Leymann, 1996, p. 168). Employees are categorized as having been "mobbed" when they experience at least one of the 45 LIPT behaviors at least once per week (i.e., frequently), for a period of at least 6 months (i.e., over an extended period of time). Leymann identified a typology of five domains in which these mobbing behaviors can affect a victim: *communication, maintenance of social contacts, maintenance of reputation, success at work,* and *physical health.*

Negative Acts Questionnaire (NAQ). The Negative Acts Questionnaire (NAQ; Einarsen & Raknes, 1997) is a 22-item measure of general negative work interactions and harassment from a target's perspective. It encompasses a wide range of direct and indirect behaviors, including verbal, physical, and sexual mistreatment, as well as behaviors that are either personal or work-specific. NAQ items fall into five factors: *personal derogation, work-related harassment, social exclusion, social control,* and *physical abuse.* Respondents report the frequency of their negative experiences at work within the last 6 months, from *never to about daily.*

Incivility

Workplace Incivility Scale (WIS). The Workplace Incivility Scale (WIS; Cortina, Magley, Williams, & Langhout, 2001) is a 7-item measure of experienced incivility, as defined by Andersson and Pearson (1999): "low-intensity deviant behavior with ambiguous intent to harm the target, in violation of workplace norms for mutual respect" (p. 457). Respondents indicate the frequency with which they have experienced specific uncivil behaviors on the job, within a particular time-frame, using a scale from 1 (never) to 5 (many times). Many studies support the reliability and content and discriminant validity of the WIS, making it a popular measure for experienced workplace incivility.

Instigated Workplace Incivility Scale. Blau and Andersson (2005) modified the WIS to measure instigated, rather than experienced, workplace incivility. Beginning with the stem "How often have you exhibited the following behaviors in the past year to someone at work …," each WIS behavior is rated on a scale from 1 = hardly ever to 4 =

frequently (at least once a day). Although this scale raises unique social-desirability challenges, it is distinctive and valuable in its examination of workplace incivility from the actor's perspective.

Uncivil Workplace Behavior Questionnaire (UWBQ). The Uncivil Workplace Behavior Questionnaire (UWBQ; Martin & Hine, 2005) offers a more detailed assessment of experienced workplace incivility. This measure contains 17 items that map onto a four-factor structure, differentiating hostility, privacy invasion, exclusionary behavior, and gossiping. The response scale mirrors that of the WIS. Tests of the UWBQ have shown good convergent, divergent, and concurrent validity. It is not (yet) as widely used as the WIS, perhaps owing to its greater length.

Miscellaneous Constructs

Abusive Supervision Scale. Tepper (2000) defines abusive supervision as "subordinates' perceptions of the extent to which supervisors engage in the sustained display of hostile verbal and nonverbal behaviors, excluding physical contact" (p. 178). Though conceptually similar to other forms of workplace mistreatment, abusive supervision is a subjective assessment and includes manifestations of both indifference and willful hostility. Following the stem "My boss …," respondents rate the frequency with which they have experienced 15 abusive behaviors (from 1 = I cannot remember him/her ever using this behavior with me to 5 = He/she uses this behavior very often with me).

Counterproductive Work Behavior Checklist (CWB-C). One of the broadest forms of workplace misconduct, counterproductive work behavior (CWB) refers to actions committed with the intent of harming an organization or its constituents (Spector, Fox, Penney, Bruursema, Goh, & Kessler, 2006). Two versions of the CWB Checklist have been tested: a 45-item version can be either scored as a composite measure, or divided into two subscales based on whether the target is an organization (CWB-O) or person (CWB-P). Alternatively, a 33-item version with five factors (abuse, production deviance, sabotage, theft, and withdrawal) can be used to differentiate antecedents and outcomes. The CWB-P is most relevant to this chapter; with this scale, respondents rate the frequency (from never to every day) with which they have committed 22 person-targeted behaviors, most of which are categorized as abuse.

Generalized Workplace Abuse (GWA) Scale. Through their Generalized Workplace Abuse (GWA) Scale, Richman and colleagues (Richman, Rospenda, Flaherty, & Freels, 2001; Richman, Rospenda, Nawyn et al., 1999) tap abusive interpersonal interactions. The five dimensions of their 29-item instrument include verbal aggression, disrespectful behavior, isolation/exclusion, threats/bribes, and physical aggression. Employees indicate the frequency with which they have experienced abusive interactions during the past year by indicating either 1 (never), 2 (once), or 3 (more than once). The GWA contains high internal reliability, and its factors are intercorrelated but assess discrete phenomena.

Interpersonal Conflict at Work Scale (ICAWS). The Interpersonal Conflict at Work Scale (ICAWS; Spector & Jex, 1998) measures stressful interpersonal conflicts between coworkers. Respondents indicate the frequency of these stressful interactions, including arguments, yelling, and rudeness, from 1 (never) to 5 (very often). Tested across multiple samples, the 4-item ICAWS shows good internal reliability and construct validity.

Interpersonal Deviance Scale. Workplace (or employee) deviance refers to "voluntary behavior that violates significant organizational norms and in so doing threatens the well-being of an organization, its members, or both" (Robinson & Bennett, 1995, p. 556). It captures both minor and egregious violations of dominant organizational social expectations (not breaches of workgroup or subculture norms). Bennett and Robinson's (2000) measure contains seven interpersonal deviance items and 12 organizational deviance items. Employees indicate the frequency with which they have committed each behavior within the last year (from 1 = never to 7 = daily). Tests of the measure's convergent and discriminant validity support its overall construct validity and alleviate social desirability concerns.

Social Undermining Scale. Duffy, Ganster, and Pagon (2002) developed the Social Undermining Scale to assess experiences of "behavior intended to hinder, over time, the ability to establish and maintain positive interpersonal relationships, work-related success, and favorable reputation" (p. 332). Encompassing verbal, physical, direct, and indirect behaviors, the conduct must be perceived as intentionally hindering, and its effects on targets must be insidious. Respondents indicate how often during the past month they have been targeted with 13 behaviors from their supervisor and 13 behaviors from the coworker closest to them, from 1 (never) to 6 (every day). The measure shows good discriminant validity and internal reliability.

Workplace Ostracism Scale (WOS). As defined by Williams (2001), ostracism is a person's perception that others are ignoring or excluding him or her. The exclusion may or may not be intentional, but this ambiguity from the target's perspective often underlies ostracism's aversive impact. Ferris, Brown, Berry, and Lian's (2008) Workplace Ostracism Scale (WOS) measures employees' experiences of ostracism within the past year, using a unidimensional 10-item scale (frequency rated from 1 = never to 7 = always). The scale is internally reliable and possesses good convergent, discriminant, and criterion-related validity.

Types of Research Questions Addressed with These Instruments

The instruments described above have appeared across different domains of OHP. The choice of instrument depends in part on the research questions at hand. For some researchers (or their organizational sponsors), the main questions revolve around prevalence rates (what proportion of the workforce has experienced the mistreatment?) and characteristics of targets (who is most at risk for being targeted?). Though these questions might be criticized as overly exploratory or atheoretical, answering them helps raise awareness about a problem. Powerful examples of this come from research on violence against women: neither acquaintance rape nor intimate partner violence met with public outcry until the 1980s, when prevalence surveys revealed that large proportions of the female population had endured these abuses. For examples of OHP research posing these types of questions, see Cortina et al. (2001), Einarsen and Raknes (1997), and Richman et al. (1999).

In addition to establishing prevalence estimates, research on mistreatment experiences often examines correlates (sometimes framed as "outcomes") of those experiences. In other words, how does the experience of that behavior relate to (or "affect") the target's workplace attitudes, withdrawal cognitions and behaviors, or psychological and physical strains? These questions are posed at times to demonstrate the harmful impact of mistreatment and at times to support the validity of the mistreatment instrument. Illustrations of this research approach appear in Cortina et al. (2001), Duffy et al. (2002), Ferris et al. (2008), Tepper (2000), Richman et al. (1999), and Spector and Jex (1998). Some OHP researchers take these questions a step further by investigating what mediates and moderates relationships between experienced mistreatment and "outcomes" (e.g., Duffy et al., 2002; Lim, Cortina, & Magley, 2008; Richman et al., 2001; Tepper, 2000).

The above research questions relate to experiences of mistreatment. Enactment of mistreatment is a different issue, raising different research questions. Most commonly, psychologists have investigated factors that predict this misbehavior. Put differently, what drives employees to mistreat others at work? What types of persons, under what kinds of conditions, engage in interpersonal mistreatment? Or, what organizational policies, programs, and characteristics reduce the enactment of mistreatment? Answers to such questions can inform prevention and intervention efforts in organizations. One can see examples of this approach in research on aggression (Baron & Neuman, 1996; Douglas & Martinko, 2001; Glomb, 2010), incivility (Blau & Andersson, 2005), and counterproductive behavior in organizations (Spector et al., 2006).

Issues to Consider When Using These Instruments

When embarking on a study of interpersonal mistreatment in organizations, the OHP researcher has a number of decisions to make. The results of those decisions should then determine the measurement strategy. First and foremost, will the focus of the research be the experience of mistreatment (from the target's perspective) or the enactment of mistreatment (from the actor's perspective)? From a psychological standpoint, these are very different phenomena and call for different instruments. Table 7.1 details the perspective (target vs. actor) taken by each assessment tool.

A second key issue is the features of the mistreatment that are central to the research question; those features should be captured somehow by the measurement. As seen in Table 7.1, occupational health psychologists have developed an array of tools for assessing myriad mistreatment constructs. But, as Hershcovis (2010, p. 505) notes, "while each of these constructs differentiates itself theoretically, these differences are *assumptions of the definition and conceptualization* [emphasis in original]. Researchers have not tended to measure the factors (i.e., persistence, power, intent, intensity) that make these constructs different." In other words, existing instruments (including those reviewed in this article), often fail to measure these factors, even when they are central to the construct definition. If these factors are vital to one's questions or hypotheses, one should adapt existing scales or supplement them with additional measures.

A third question to ask is whose perspective is most valuable and valid, given the research question. In other words, should details of the mistreatment be reported by oneself or by one's peers, supervisors, subordinates, or customers/clients? Common advice is to avoid purely self-report measurement, due to beliefs that self-report data are inherently flawed and inevitably produce inflated correlations, and therefore are inferior to other-report measures. Chan (2009), however, refers to these ideas as "urban legends" and goes on to debunk the myths behind them. Moreover, psychologists have argued that for some research questions (e.g., questions about mistreatment that is perceptual by definition or that rarely takes place in view of others; questions about the impact of mistreatment on employees' emotional states), self-report is a valid and necessary assessment strategy (e.g., Spector, 1994; Spector & Jex, 1998). That said, many OHP researchers have strengthened their conclusions by supplementing self-report data with data collected from other sources (for examples, see Glomb & Liao, 2003; Raver & Gelfand, 2005).

Fourth, before adopting any of the instruments described in this chapter, one must ask whether the research question at hand is best answered using a quantitative survey paradigm. Other methodological approaches are less common in this topic domain, but when undertaken have yielded interesting results. See, for instance, studies of incivility using laboratory experiments (e.g., Porath & Erez, 2009; Porath, Overbeck, & Pearson, 2008) and qualitative interviews (Pearson, Andersson, & Wegner, 2001). James et al. (2005) have advocated the conditional reasoning paradigm in the study of aggression, while Miner, Glomb, and Hulin (2005) have investigated negative events at work using experience sampling methods. Beyond OHP, innovative experimental protocols have emerged from social-psychological research on aggression (e.g., Cohen, Nisbett, Bowdle, & Schwarz, 1996) and ostracism (e.g., Williams, Cheung, & Choi, 2000; Zadro, Williams, & Richardson, 2004). In other words, although the focus of this chapter is quantitative survey instruments, scientists interested in workplace mistreatment need not limit their research to the quantitative survey paradigm.

A final challenge applies to any attempt (quantitative or otherwise) to assess interpersonal mistreatment: much of this conduct entails "insidious workplace behavior," taking stealthy, seemingly benign, and entirely legal forms (Edwards & Greenberg, 2010). The "smallness" of the behaviors can defy precise measurement. For instance, the ambiguity inherent in some mistreatment (e.g., incivility) can make it difficult for targets to recognize and report, and subtle isolated acts may not seem "severe" enough to warrant mention (Cortina, 2008). Even some instigators fail to see their behavior as harmful or abusive (e.g., Brief, Dietz, Cohen, Pugh, & Vaslow, 2000; Cortina, 2008). In seeking to assess mistreatment experience or enactment, the OHP researcher is sometimes chasing an elusive phenomenon.

Conclusion

In sum, OHP researchers have a variety of tools at their disposal for measuring interpersonal mistreatment in organizations, both the "doing" and "receiving" of it. Some scholars have cautioned against the proliferation of constructs (and associated measures) in this domain, fearing that this will foster fragmentation and impede science. In our opinion, however, the variety of instruments available is helping rather than hindering the field. There are countless behaviors one can use to subordinate, derogate, and relegate people to the margins of organizational life. It seems only appropriate, then, that we use numerous tools to assess those behaviors.

References

Andersson, L. M., & Pearson, C. M. (1999). Tit for tat? The spiraling effect of incivility in the workplace. *Academy of Management Review, 24*, 452–471.

Baron, R. A., & Neuman, J. H. (1996). Workplace violence and workplace aggression: Evidence on their relative frequency and potential causes. *Aggressive Behavior, 22*, 161–173.

Bennett, R. J., & Robinson, S. L. (2000). Development of a measure of workplace deviance. *Journal of Applied Psychology, 85*, 349–360.

Blau, G., & Andersson, L. (2005). Testing a measure of instigated workplace incivility. *Journal of Occupational and Organizational Psychology, 78*, 595–614.

Brief, A. P., Dietz, J., Cohen, R. R., Pugh, S. D., & Vaslow, J. B. (2000). Just doing business: Modern racism and obedience to authority as explanations for employment discrimination. *Organizational Behavior and Human Decision Processes, 81,* 72–97.

Chan, D. (2009). So why ask me? Are self-report data really that bad? In C. E. Lance & R. J. Vandenberg (Eds.), *Statistical and methodological myths and urban legends: Doctrine, verity and fable in the organizational and social sciences* (pp. 309–336). New York: Routledge.

Cohen, D., Nisbett, R. E., Bowdle, B. F., & Schwarz, N. (1996). Insult, aggression, and the Southern culture of honor: An "experimental ethnography." *Journal of Personality and Social Psychology, 70,* 945–960.

Cortina, L. M. (2008). Unseen injustice: Incivility as modern discrimination in organizations. *Academy of Management Review, 33,* 55–75.

Cortina, L. M., Magley, V. J., Williams, J. H., & Langhout, R. D. (2001). Incivility in the workplace: Incidence and impact. *Journal of Occupational Health Psychology, 6,* 64–80.

Douglas, S. C., & Martinko, M. J. (2001). Exploring the role of individual differences in the prediction of workplace aggression. *Journal of Applied Psychology, 86,* 547–559.

Duffy, M. K., Ganster, D. C., & Pagon, M. (2002). Social undermining in the workplace. *The Academy of Management Journal, 45,* 331–351.

Einarsen, S., & Raknes, B. I. (1997). Harassment in the workplace and the victimization of men. *Violence and Victims, 12,* 247–263.

Ferris, D. L., Brown, D. J., Berry, J. W., & Lian, H. (2008). The development and validation of the Workplace Ostracism Scale. *Journal of Applied Psychology, 93,* 1348–1366.

Fitzgerald, L. F., Gelfand, M. J., & Drasgow, F. (1995). Measuring sexual harassment: Theoretical and psychometric advances. *Basic and Applied Social Psychology, 17,* 425–445.

Glomb, T. M. (2010). Predicting workplace aggression: Reciprocal aggression, organizational, and individual antecedents. *International Journal of Organization Theory and Behavior, 132,* 249–291.

Glomb, T. M., & Liao, H. (2003). Interpersonal aggression in workgroups: Social influence, reciprocal, and individual effects. *Academy of Management Journal, 46,* 486–496.

Hershcovis, M. S. (2010). Incivility, social undermining, bullying … Oh my! A call to reconcile constructs within workplace aggression research. *Journal of Organizational Behavior, 32, 499–519.*

James, L. R., McIntyre, M. D., Glisson, C. A., Green, P. D., Patton, T. W., LeBreton, J. M., … Williams, L. J. (2005). A conditional reasoning measure for aggression. *Organizational Research Methods, 8,* 69–99.

Leymann, H. (1990). Mobbing and psychological terror at workplaces. *Violence and Victims, 5,* 119–126.

Leymann, H. (1996). The content and development of mobbing at work. *The European Journal of Work and Organizational Psychology, 5*, 165–184.

Lim, S., Cortina, L. M., & Magley, V. J. (2008). Personal and workgroup incivility: Impact on work and health outcomes. *Journal of Applied Psychology, 93*, 95–107.

Martin, R. J., & Hine, D. W. (2005). Development and validation of the Uncivil Workplace Behavior Questionnaire. *Journal of Occupational Health Psychology, 10*, 477–490.

Miner, A. G., Glomb, T., & Hulin, C. (2005). Experience sampling mood and its correlates at work. *Journal of Occupational and Organizational Psychology, 78, 171–193.*

Neuman, J. H., & Baron, R. A. (1998). Workplace violence and workplace aggression: Evidence concerning specific forms, potential causes, and preferred targets. Journal of Management, 24, 391–419.

Neuman, J. H., & Keashly, L. (2004, April 4). Development of the Workplace Aggression Research Questionnaire (WAR–Q): Preliminary data from the Workplace Stress and Aggression Project. In R. J. Bennett & C. D. Crossley (Chairs), *Theoretical advancements in the study of anti-social behavior at work.* Symposium conducted at the meeting of the Society for Industrial and Organizational Psychology, Chicago, IL.

Neuman, J. H., & Keashly, L. (2010). Means, motive, opportunity, and aggressive workplace behavior. In J. Greenberg (Ed.), *Insidious workplace behavior* (pp. 31–76). New York: Routledge.

Pearson, C. M., Andersson, L. M., & Wegner, J. W. (2001). When workers flout convention: A study of workplace incivility. *Human Relations, 54*, 1387–1419.

Porath. C. L., & Erez, A. (2009). Overlooked but not untouched: How rudeness reduces onlookers' performance on routine and creative tasks. *Organizational Behavior and Human Decision Processes, 109*, 29–44.

Porath, C. L., Overbeck, J. R., & Pearson, C. M. (2008). Picking up the gauntlet: How individuals respond to status challenges. *Journal of Applied Social Psychology, 38*, 7, 1945–1980.

Raver, J. L., & Gelfand, M. J. (2005). Beyond the individual victim: Linking sexual harassment, team processes, and team performance. *Academy of Management Journal, 48*, 387–400.

Richman, J. A., Rospenda, K. M., Flaherty, J. A., & Freels, S. (2001). Workplace harassment, active coping, and alcohol-related outcomes. *Journal of Substance Abuse, 13*, 357–366.

Richman, J. A., Rospenda, K. M., Nawyn, S. J., Flaherty, J. A., Fendrich, M., Drum, M. L., & Johnson, T. P. (1999). Sexual harassment and generalized workplace abuse among university employees: Prevalence and mental health correlates. *American Journal of Public Health, 89*, 358–363.

Robinson, S. L., & Bennett, R. J. (1995). A typology of deviant workplace behaviors: A multidimensional scaling study. *The Academy of Management Journal, 38*, 555–572.

Schneider, K. T., Hitlan, R. T., & Radhakrishnan, P. (2000). An examination of the nature and correlates of ethnic harassment experiences in multiple contexts. *Journal of Applied Psychology, 85,* 3–12.

Spector, P. E. (1994). Using self-report questionnaires in OB research: A comment on the use of a controversial method. *Journal of Organizational Behavior, 15,* 385–392.

Spector, P. E., Fox, S., Penney, L. M., Bruursema, K., Goh, A., & Kessler, S. (2006). The dimensionality of counterproductivity: Are all counterproductive behaviors created equal? *Journal of Vocational Behavior, 68,* 446–460.

Spector, P. E., & Jex, S. M. (1998). Development of four self-report measures of job stressors and strain: Interpersonal Conflict at Work Scale, Organizational Constraints Scale, Quantitative Workload Inventory, and Physical Symptoms Inventory. *Journal of Occupational Health Psychology, 3,* 356–367.

Tepper, B. J. (2000). Consequences of abusive supervision. *The Academy of Management Journal, 43,* 178–190.

Williams, K. D. (2001). *Ostracism: The power of silence.* New York: Guilford.

Williams, K. D., Cheung, C. K. T., & Choi, W. (2000). Cyberostracism: Effects of being ignored over the Internet. *Journal of Personality and Social Psychology, 79,* 748–762.

Zadro, L., Williams, K. D., & Richardson, R. (2004). How low can you go? Ostracism by a computer is sufficient to lower self-reported levels of belonging, control, self-esteem, and meaningful existence. *Journal of Experimental Social Psychology, 40,* 560–567.

8

The Measurement of Depression and Anxiety in Occupational Health Psychology

Jay C. Thomas, Björn Bergström, and Johan Rosqvist

Depression and anxiety are pervasive and potentially debilitating conditions that endanger the individual's quality of life and life itself. They also have an enormous economic impact; in fact, Greenberg and colleagues (Greenberg, Sisitsky, et al., 1999) estimated the economic burden of anxiety disorders in the United States alone as $42.3 billion annually, and earlier they estimated the annual costs of depressive illnesses in the United States to be $43.7 billion (Greenberg, Stiglin, et al., 1993). Beyond the clinical and economic implications, depression and anxiety are themselves of interest as psychological phenomena. As Herbert Simon (1969) pointed out, we begin to understand systems best when they break down. Anxiety at its extreme levels and depression represent forms of breakdown of the human system. As such they tell us about the limits of the human as an individual, social person, member of a larger organization, and greater environment. As constructs, anxiety and depression receive most attention as outcomes, yet they can also serve as causes for other effects. However, as important as these constructs are, they present measurement challenges for researchers and practitioners alike. This chapter examines some of the most common issues and possible solutions to the measurement problems encountered by researchers in both clinical and occupational health psychology (OHP).

Depression and Anxiety across Cultures

Depression and clinically elevated levels of anxiety are common and commonly co-occur; about 50% of people with one are also diagnosed with the other (Kessler, Nelson et al., 1996). This has led to speculation about shared etiologies and physiological foundations (Middeldorp, Cath, Van Dyck, & Boomsma, 2005). It is clear, however, that even if there are physiological underpinnings, there is much more underlying both conditions than simple biology. This is most importantly illustrated by the fact that the manifestation and experience of depression varies across time and culture. Research on depression in East Asian populations has identified some distinct differences from Western patterns in symptom expression. Chinese populations have been found to associate depression with higher levels of physical ailments and somatic symptoms, and fewer reported symptoms of internal psychological distress (Ryder et al., 2008). Comparatively, this contrasts with many Western populations who report higher levels of sadness and guilt, and other internal psychological mood-state descriptors, in addition to physiological symptoms (Ryder et al., 2008). Similar results have been found in Japanese populations, which tend to emphasize external and somatic terms when describing depression (Waza, Graham, Zyzanski, & Inoue, 1999). Researchers have found analogous results in some African (e.g., Zimbabwe) populations and developing third world nations, as well (Patel, Abas, Broadhead, Todd, & Reeler, 2001). Research across cultures suggests that somatization is not so much unique to Eastern populations, but more globally pervasive. It appears that cognitive emphasis on depressive symptoms is, in fact, more unique to Western societies (Ryder et al., 2008). This highlights the importance of culture in shaping individual experiences in that "the onset of depression triggers a biological response that takes place within a specific social context, resulting in a cascade of somatic and psychological experiences that are interpreted through a particular cultural lens" (Ryder et al., 2008, p. 310).

Researchers have consistently found that different cultures have diverse ways of expressing and emphasizing their psychological experiences. Multiple cross-cultural studies indicate that while there are many psychological fundamentals which appear more or less universal, the clusters of specific symptoms which define a disorder can vary from culture to culture (Kleinman, 1998). In addition to symptom expression, research has found that the prevalence of depression varies across ethnicities, nations, and cultures (Ahmed & Bhugra, 2006; Castro-Costa et al., 2007; Dunlop, Song, Lyons, Manheim, & Chang, 2003; Josephson, 2009). Other researchers have suggested a similar pattern for anxiety disorders; that globally, there is variation in the prevalence of anxiety, and how it is experienced and defined across cultures, suggesting that pathological behavior and assessment should be understood within the context of the culture itself (Friedman, 2002; Guarnaccia, 1997).

Although it is popularly believed that many psychological disorders are more or less universally present across populations, the manifestation, beliefs, and experiences associated with those disorders can be markedly different. Historically, the constellation of symptoms commonly used to define psychological disorders has been predominately developed in the West, based on Western populations (Ardila, 1995). Thus, the current definitions of psychopathology are heavily influenced by Western culture (Ardila, 2005). Further research is needed to determine the generalizability of these constructs across cultures, and the generalizability of the assessment measures which are founded on these constructs.

Measuring Stress, Burnout, Depression, and Anxiety: What's the Difference?

It is particularly important that the test be normed to the population in question when evaluating appropriate assessment measures intended to identify depression or anxiety within the workplace or clinical setting. If a test is not normed to the population in question, the responses may be over- or underpathologized by the test itself, risking false positives and negatives. As a result, it is critical to know whether or not an assessment has been culturally normed. Also, the purpose and design of the assessment measure determines the appropriateness of its use. For instance, there are multiple assessment and test measures designed for clinical and organizational use. However, these assessments vary greatly in purpose, and have varied to limited construct overlap (Shirom, 2011). For instance, many organizational assessments test for "job stress" and general "anxiety." However, these constructs originate from different fields, designed for different purposes that derive from clinical and organizational research and practice. Clinical assessments are meant to identify clinical levels of dysfunction (based on a specific set of diagnostic criteria). In contrast, many organizational assessments are often intended to identify problem areas within the workplace, or to screen for individual employee functioning, in order to improve overall employee performance. These are two very different ends of the continuum of functioning. For example, trying to evaluate someone who experiences suicidal thoughts and does not get out of bed in the morning or evaluating someone who is having difficulty performing at optimal levels in a complex work environment are two very different situations. Attempting to utilize a universal construct, like "stress," across clinical and organizational contexts potentially dilutes these very tangible differences.

Assessment of Functioning in Clinical and Organizational Populations

The functional roots and origins of a construct determine how the construct is measured and applied. It is important to know at the outset whether or not the goal of the assessment is to identify clinical levels of depression or anxiety, or identify how "stress" or stressors are impacting otherwise healthy employees. The question at hand determines the appropriateness of the assessment measure, as well as any organizational or individual interventions that might follow. Different assessment measures will yield differing types of data (e.g., personal vs. group/environmental factors, clinical vs. organizational, etc.). For instance, Table 8.1 provides a list of commonly used clinical and organizational assessments of depression, anxiety, and stress. Each one of these assessments has unique advantages and qualities: length, multiple forms, international norms, specificity vs. breadth, and administration and interpretation ease or complexity. Job stress surveys typically identify forms of physical and psychological stressors within the occupational setting which can contribute to job stress and strain (Spielberger & Reheiser, 2005). However, this is uniquely different from burnout, depression, or anxiety, which can be conceptualized as reactions to stressors (Shirom, 2003). Although there are many overlapping variables within emotional distress (e.g., burnout, job stress, anxiety, or depression), these are also very different constructs both in type and intensity (Schaufeli & Van Dierendonck, 1993), which have historically advanced within related but different fields of study. As such, using clinical measures for organizational purposes needs to be guided by professional standards of care and ethics, given that further research is needed to discriminate the construct validity between these instruments or surveys and the concepts (e.g., depression, anxiety vs. stress, or burnout) which they are founded on, rather than treat them as overlapping or synonymous constructs (Shirom, 2011). For instance, an organizational measure of stress may not identify clinical levels of distress. Likewise, clinical measures may miss organizational problem areas, since they are typically normed for individuals experiencing clinically significant levels of distress. Depending on the specificity of the assessment measure, and the appropriateness of the underlying constructs, assessments utilized in the workplace can potentially overpathologize or miss key data. However, tailoring and norming clinical assessments to organizational environments could greatly enhance the specificity of current organizational constructs and data gathering as a whole.

Thus, clinical psychology can offer a significant contribution toward enhancing the assessment and practice of organizational health. The success of this endeavor will likely rely on the translation and mutual application of organizational and clinical research, resulting in increased specificity in operational definitions and their respective forms of measurement. The assessments found within Table 8.1 could be a valuable asset for future assessment and research in this area.

Defining Anxiety and Depression

Since the manifestations of depression and anxiety are dependent on external circumstances there is a continual evolution in both the formal diagnostic systems (ICD and DSM) as well as our conceptualization of the conditions (American Psychiatric Association [DSM-IV-TR], 2000; World Health Organization [ICD-10], 1994). Depression may or may not have functional roots. As a behavioral phenomenon we see a shutdown in behavior with an inability to "get started," to set and meet goals, and to more broadly examine the self, others, or the world and the future with a set expectation of ineffectiveness, failure, and a lack of hope, as opposed to examining life through a realistic lens or at least one anchored in neutrality. Loss of interests and an inability to take pleasure in events big and small characterize depression as a disorder. Anxiety is clearly functional when within nonpathological limits and when environmentally appropriate. Some theorists see inborn tendencies toward certain types of anxiety; for example, phobias directed at potentially dangerous creatures such as spiders (de Silva, Rachman, & Seligman, 1977). A functional level of anxiety can lead to escape or avoidance behavior that benefits the organism. Anxiety loses its functionality when it leads to generalized avoidance or when it becomes so overwhelming that the person is no longer able to engage in coherent

Table 8.1 Commonly Used Clinical and Organizational Assessments of Depression, Anxiety, and Stress

Assessment Measures of Employee Functioning	Depres-sion	Anxiety	Organi-zational Assess ment	Clinical Assess ment	Language Trans lations	Typical Content	Number of Items	Sample Study
(X = Yes, * = Related Content)								
The Work Limitations Questionnaire (WLQ)			X		X	Limitations on-the-job due to health problems, and health-related productivity loss.	25 or 8	Lerner et al. (2001)
Health and Productivity Questionnaire (HPQ)			X		X	Workplace costs of health problems in terms of reduced job performance, sickness absence, and work-related accidents/injuries.	24	Mills, Kessler, Cooper, & Sullivan (2007)
Work Productivity Short Inventory (WPSI)	*	*	X			Productivity losses associated with 15 common disease conditions.		Ozminkowski, Goetzel, Chang, & Long (2004)
Work Productivity and Activity Impairment Questionnaire (WPAI)			X		X	Time and productivity losses.		Giovannetti, Wolff, Frick, & Boult (2009)
Health Risk Assessments (HRA)	*	*	X		X	Impact of health conditions on productivity.	21	Mills (2005)

Hospital Anxiety and Depression Scale (HADS)	X	X		X	X	Psychological health of cancer patients.	14	Sanne, Mykletun, Dahl, Moen, & Tell (2003)
Center for epidemiological Studies Depression Scale (CESD)	X			X	X	Individual symptoms and levels of depression.	20	Giovannetti et al. (2009)
Semi-structured Interview (SCID)	X	X		X	X	DSM Clinical Diagnosistics.	124	Kessler et al. (2006)
Composite International Diagnostic Interview (CIDI)	X	X		X	X	DSM & ICD-10 Clinical Diagnosistics.	59	Kessler et al. (2006)
Single Item Depression or Anxiety Questions	X	X		X	X	e.g., “Are you anxious?”	1	Skoogh et al. (2010)
Job Stress Survey (JSS)			X		X	Sources of workplace stress and productivity loss.	60 or 30	Spielberger & Reheiser (1994)
Endicott Work Productivity Scale (EWPS)			X		X	Evaluate the degree to which medical and psychological problems impact work performance.	25	Ericson (2009)
Job Content Questionnaire (JCQ)	*	*	X		X	Physical & mental health and social supports .	18	Hurrell & Simmons (1998)
General Health Questionnaire (GHQ 12)	X	X		X	X	General distress and nonpsychotic psychiatric disorders.	12 to 60	Henkel et al. (2004)

(continued)

Table 8.1 Continued

Assessment Measures of Employee Functioning	Depression	Anxiety	Organizational Assessment	Clinical Assessment	Language Translations	Typical Content	Number of Items	Sample Study
Brief Patient Health Questionnaire (PHQ-9)	X	X		X	X	Clinical symptoms of depression.	10	Sanderson, Nicholson, Graves, Tilse, & Oldenburg (2008)
Beck Depression Inventory-II (BDI-II)	X			X	X	Clinical symptoms of depression.	21	Forman, Herbert, Moitra, Yeomans, & Geller (2007)
Beck Anxiety Inventory (BAI)		X			X	Clinical symptoms of depression.	21	Bender & Kennedy (2010)
WHO Major Depression Inventory (MDI)	X			X	X	Clinical symptoms of depression.	10	Bech et al. (2005)
Montgomery-Åsberg Depression Rating Scale	X			X	X	Clinical symptoms of depression.	10	Wahlberg et al. (2009)
Hamilton Anxiety Rating Scale (HAM-A)		X		X		Clinical symptoms of anxiety.	14	Franulic, Carbonell, Pinto, & Sepulveda (2004)
Zung's Self Rating Score	X			X	X	Clinical symptoms of depression.	10	Kawakami, Takatsuka, Shimizu, & Ishibashi (1999)

and rational behavior. In general, pathological anxiety (Rothbaum, 2005) can be suggested to be anxiety that occurs either at the wrong time (e.g., sitting in one's office, during a "regular" work day when nothing much, if anything, is truly "amiss") or in the wrong amount (e.g., responding to a job interview with such an overwhelming amount of anxiety that one jumps up and runs from the interview room).

Anxiety comes in many forms, but in broad terms it can be said to exhibit emotional, cognitive, and behavioral markers in both common and abnormal variations. Emotionally, anxiety is characterized by feelings of fear that, at normal levels, may motivate a person to action, but at too intense levels may in fact serve to freeze, confuse, and make a person literally unable to make basic decisions. Such persons may in fact become so stuck that all of life's day-to-day activities halt. Cognitively, normal anxiety facilitates problem-solving and actively considering circumstances rationally, while in extreme forms it instead makes it impossible to come to reasoned decisions, as basic thinking processes are slowed and skewed and atypical and nonrepresentative thinking patterns instead dominate (promoting maladaptive thinking patterns). Behaviorally, normal anxiety leads to prosocial and adaptive coping behaviors (e.g., completing the job interview), but at pathological levels it most often leads to various escape, avoidance, and reassurance (maladaptive) patterns of behaving; for example, abruptly leaving the interview, not showing up for the interview, or asking the interviewer for inappropriate feedback during the interview (Rosqvist, 2005).

Methods of Measurement

Therapists and theorists have developed a variety of means for conceptualizing both depression and anxiety and these heavily influence the measurement of both constructs. A behavioral approach results in items that focus on particular aspects of behavior. The BDI-II (Beck, Steer, & Brown, 1996), for example, has items that parallel the DSM-IV-TR criteria for depression, asking about such behaviors as difficulty sleeping, changes in eating habits, and change in sexual habits. Cognitively oriented conceptualizations concentrate on the nature of thoughts based on the expectation that how one thinks about one's situation reveals both internal interpretations and mechanisms of maintaining the dysfunction. An example is when an individual who suffers from the pathological form of anxiety or depression often looks at challenging/difficult/negative events with a flawed explanatory perspective.

Specifically, people who are anxious or depressed often blame themselves (i.e., *internal* causes) when things go badly (e.g., "It's my fault, I'm to blame, I'm responsible") versus assigning cause to the outside world or onto others (e.g., "It's someone else's fault, it's just bad luck"). They also tend to take a very *global* perspective on problems (e.g., "My entire life is over, everything is ruined") versus thinking about the specific time or circumstance (e.g., "Okay, this hurts, but it's temporary, this only hurts right now and will end"). Lastly, they persist in *stable* perspectives (e.g., "This will never end, it will go on forever") versus focusing on the temporary nature of life (e.g., "Things will change with time"). A third approach concentrates on affective responses, identifying specific mood states (e.g., low mood, sadness, decreased motivation, guilt or worthlessness, stress, agitation, irritability, anger, mood instability, etc.) indicative of depression or anxiety. The Montgomery-Åsberg Depression Rating Scale (MADRS) taps into many of these constructs. A fourth conceptualization focuses on physiological symptoms (fatigue, sleep, decreased motor activity, muscle tightness, cardiovascular, gastrointestinal, and autonomic symptoms, etc.). An example of this form of measurement is the Hamilton Rating Scale for Depression (HRSD). Virtually all assessment measures utilize overlapping constructs related to physiology, cognition, mood, and overt behaviors for anxiety and depression. Different assessment measures possess unique qualities and advantages (e.g., language availability, validity and reliability, length, content, research-, clinical-, and organizational/applied-foci) (see Table 8.1 for examples).

Issues in Measurement Practice

Streiner (2003a, b) made a distinction between measurement instruments that are designed as *scales* and those that are designed as *indices*. In a scale it is believed that the items are sampling a population of potential items and that some underlying condition (latent trait) caused the responses reported on the items. For example, an item such as "I feel sad all the time" might be derived from a population of possible affective statements. Since the latent trait, depression, is causing the responses to all the items, the instrument is a scale and it is reasonable and sensible to calculate internal consistency reliability. For other instruments the items may be used to define the condition. An inability to sleep, change in eating habits, and feelings of worthlessness may combine to form an index used to establish that the individual is depressed. As Streiner points out, there is no necessary underlying connection between the items and it makes no sense to calculate internal consistency reliability or related indices. Steiner illustrates some of the consequences of ignoring this distinction; researchers should be clear about their theory of item generation and the theorized causal relationship between either traits and items or items and labels, as the case may be.

Although we see a great advantage in understanding the behavioral approach to depression, anxiety, and other human conditions, there are other inherent problems for the researcher in behaviorally oriented measurement. Depression may imply disrupted sleep patterns, but the opposite does not necessarily logically hold. This example is complicated by the fact that continually disrupted sleep may lead to depressionlike symptoms if not depression itself. Thus, people who are not depressed may indicate agreement with this item.

Sensitivity and Specificity

We normally expect measures of a condition to have excellent specificity and sensitivity. The term *sensitivity* refers to the measure's ability to detect a condition when it is present. *Specificity* is the measure's ability to only identify those who have the condition as having it. People without the condition should not obtain scores leading to a diagnosis of the condition (cf., Thomas & Christiansen, 2011). This can prove difficult when working with nonclinical populations because there is, by definition, no standard with which to assign subjects to groups with or without a disorder. OHP researchers are often interested in job situations that increase symptoms or feelings of anxiety or depression, but not necessarily to clinical levels. Researchers in other fields have a similar dilemma and the measures they choose may not agree with other measures when it comes to identifying those at risk or with a condition. There is no universal or fool-proof solution. Researchers need to be able to explain the rationale for choosing the measure and associated cutoffs that go beyond traditional criteria related or convergent validity.

Problems with Measurement Over Time

Depression and anxiety are not constant traits; both fluctuate over time with depression being particularly noted for occurring in cycles. Thus, measures of depression should be expected to demonstrate relatively low stability and those of anxiety moderately low stability in nonclinical populations. To complicate matters further, repeated measurement may result in changes in depression scores, but not the actual affective state (Longwell & Truax, 2005). Another concern that is sometimes voiced is the problem of regression to the mean, particularly of extreme scores. Fortunately, this does not appear to be as much of an issue as its proponents fear (Thomas & Truax, 2008). If a person has an extreme score on a depression or anxiety measure, regression effects would generally not be enough to change the diagnosis.

The issues described in this chapter illustrate how the researchers of depression, anxiety, and other mental disorders must be knowledgeable about the workplace, mental disorders versus stress or other emotional states, different populations, and research and statistical methodology. Simply following a research checklist without regard to what is being studied and its context will lead to results that do not inform either science or practice. There are continual trade-offs and these require that the researchers be knowledgeable about all of these areas. Such mastery can lead to exciting and useful research findings that inform science and practice.

References

Ahmed, K., & Bhugra, D. (2006). Diagnosis and management of depression across cultures. *Psychiatry, 5*(11), 417–419.

American Psychiatric Association, (2000). *Diagnostic and statistical manual of mental disorders* (4th ed., rev.). Washington, DC: Author.

Ardila, A. (2005). Cultural values underlying psychometric cognitive testing. *Neuropsychology Review, 15*(4), 185–195.

Ardila, A. (1995). Directions of research in cross-cultural neuropsychology. *Journal of Clinical and Experimental Neuropsychology, 17*(1), 143–150.

Bech, P., Andersen, M., Bech–Andersen, G., Tønnesen, S., Agnarsdottir, E., & Borg, V. (2005). Work-related stressors, depression and quality of life in Danish managers. *European Psychiatry, 20*(3), S318–S325.

Beck, A. T., Steer, R. A., & Brown, G. K. (1996). *Beck depression inventory-II (BDI-II).* Upper Saddle River, NJ: Pearson.

Bender, A., & Kennedy, S. (2010). Mental health and mental illness in the workplace: Diagnostic and treatment issues. *Healthcare Papers, 5*(2), 54–67.

Castro-Costa, E., Dewey, M., Huppert, F., Mendonca-Lima, C., Bula, C., Reisches, F., … Prince, M. (2007). Prevalence of depression symptoms and syndromes in later life in ten European countries. *The British Journal of Psychiatry, 191,* 393–401.

de Silva, P., Rachman, S., & Seligman, M. E. (1977). Prepared phobias and obsessions: Therapeutic outcome. *Behaviour Research and Therapy, 15*(1), 65–77.

Dunlop, D., Song, J., Lyons, J., Manheim, L., & Chang, R. (2003). Racial/ethnic differences in rates of depression among preretirement adults. *American Journal of Public Health, 93*(11), 1945-1952.

Ericson, S., Guthrie, S., Van Etten-Lee, M., Himle, J., Janeck, A., Zivin, K., & Abelson, J. (2009). Severity of anxiety and work-related outcomes of patients with anxiety disorders. *Depression and Anxiety, 26*(12), 1165–1171.

Forman, E., Herbert, J., Moitra, E., Yeomans, P., & Geller, P. (2007). A randomized controlled effectiveness trial of acceptance and commitment therapy and cognitive therapy for anxiety and depression. *Behavior Modification, 31*(6), 772–799.

Franulic, A., Carbonell, C., Pinto, P., & Sepulveda, I. (2004). Psychosocial adjustment and employment outcome 2, 5 and 10 years after TBI. *Brain Injury, 18*(2), 119–129.

Friedman, S. (2002). Cultural issues in the assessment of anxiety disorders. In M. Anthony (Ed.), *Practitioner's guide to empirically based measures of anxiety* (pp. 37-41). New York: Springer-Verlag.

Giovannetti, E., Wolff, J., Frick, K., & Boult, C. (2009). Construct validity of the Work Productivity and Activity Impairment Questionnaire across informed caregivers of chronically ill older patients. *Value in Health, 12*(6), 1011–1017.

Greenberg, P. E., Sisitsky, T., Kessler, R. C., Finkelstein, S. M., Berndt, E. R., Davidson, J., … Fyer, A. (1999). The economic burden of anxiety disorders in the 1990s. *Journal of Clinical Psychiatry, 60*(7), 427–435.

Greenberg, P. E., Stiglin, L. E., Finkelstein, S. N., Berndt, E. R., & Ernst, R. (1993). The economic burden of depression in 1990. *Journal of Clinical Psychiatry, 54*(11), 405–418.

Guarnaccia, P. (1997). A cross-cultural perspective on anxiety disorders. In S. Friedman (Ed.), *Cultural issues in the treatment of anxiety* (pp. 3-20). New York: Guilford.

Henkel, V., Mergl, R., Kohnen, R., Allgaier, A.-K., Moller, H.-J., & Hegerl, U. (2004). Use of brief depression screening tools in primary care: Consideration of heterogeneity in performance in different patient groups. *General Hospital Psychiatry, 26*, 190–198.

Hurrell, J., & Simmons, N. B. (1998). Measuring job stressors and strains: Where we have been, where we are, and where we need to go. *Journal of Occupational Health Psychology, 3*(4), 368–389.

Josephson, S. (2009). Primary care physicians' diagnosis of major depression has modest accuracy. *AccessMedicine from McGraw-Hill.* Retrieved from http://www.medscape.com/viewarticle/708600.

Kawakami, N., Takatsuka, N., Shimizu, H., & Ishibashi, H. (1999). Depressive symptoms and occurrence of Type 2 diabetes among Japanese men. *Diabetes Care, 22*(7), 1071–1076.

Kessler, R., Akiskal, H., Ames, M., Birnbaum, H., Greenberg, P., Hirschfeld, R., … Wang, P. (2006). Prevalence and effects of mood disorders on work performance in a nationally representative sample of U.S. workers. *American Journal of Psychiatry, 163*, 1561–1568.

Kessler, R., Nelson, C., McGonagle, K., Liu, J., Swartz, M., & Blazer, D.G. (1996). Comorbidity of DSM-III=R major depressive disorder in the general population: Results from the US National Comorbidity Survey. *British Journal of Psychiatry, 168*(Suppl. 30), 17–30.

Kleinman, A. (1998). *Rethinking psychiatry: From cultural category to personal experience.* New York: Free Press.

Lerner, D., Amick, B. C., III, Rogers, W. H., Malspeis, S., Bungay, K., & Cynn, D. (2001). The Work Limitations Questionnaire. *Medical Care, 39*, 72–85.

Longwell, B. T., & Truax, P. (2005). The differential effects of weekly, monthly, and bimonthly administrations of the Beck Depression Inventory-II: Psychometric properties and clinical implications. *Behavior Therapy, 36*(3), 265–275.

Middeldorp, C., Cath, D., Van Dyck, R., & Boomsma, D. (2005). The co-morbidity of anxiety and depression in the perspective of genetic epidemiology: A review of twin and family studies. *Psychological Medicine, 35*, 611–624.

Mills, P. R. (2005). The development of a new corporate specific health risk measurement instrument, and its use in investigating the relationship between health and well-being and employee productivity. *Environmental Health, 4*(1), 1-9.

Mills, P., Kessler, R., Cooper, J., & Sullivan, S. (2007). Impact of a health promotion program on employee health risks and work productivity. *American Journal of Health Promotion, 22*, 1.

Ozminkowski, R., Goetzel, R., Chang, S., & Long, S. (2004). The application of two health and productivity instruments at a large employer. *Journal of Occupational and Environmental Medicine, 46*, 635–648.

Patel, V., Abas, M., Broadhead, J., Todd, C., & Reeler, A. (2001). Depression in developing countries: Lessons from Zimbabwe. *British Medical Journal, 322*, 482. doi: 10.1136/bmj.322.7284.482

Rosqvist, J. (2005). *Exposure treatments for anxiety disorders: A practitioner's guide to concepts, methods, and evidence-based practice.* New York: Routledge.

Rothbaum, B. O. (Ed.). (2005). *Pathological anxiety: Emotional processing in etiology and treatment.* New York: Guilford.

Ryder, A., Yang, J., Zhu, X., Yao, S., Yi, J., Heine, S., & Bagby, R. (2008). The cultural shaping of depression: Somatic symptoms in China, psychological symptoms in North America? *Journal of Abnormal Psychology, 117*(2), 300–313.

Sanderson, K., Nicholson, J., Graves, N., Tilse, E., & Oldenburg, B. (2008). Mental health in the workplace: Using the ICF to model the prospective associations between symptoms, activities, participation, and environmental factors. *Disability and Rehabilitation, 30*(17), 1289–1297.

Sanne, B., Mykletun, A., Dahl, A., Moen, B., & Tell, G. (2003). Occupational differences in levels of anxiety and depression: The Hordaland health study. *Journal of Occupational and Environmental Medicine, 45*, 628–638.

Schaufeli, W. B., & Dierendonck, D. Van (1993). The construct validity of two burnout measures. *Journal of Organizational Behavior, 14,* 631–647.

Shirom, A. (2011). Job-related burnout: A review of major research foci and challenges. In J. C. Quick & L. E. Tetrick (Eds.), *Handbook of occupational health psychology* (2nd ed., pp. 223–241). Washington, DC: American Psychological Association.

Simon, H. (1969). *The sciences of the artificial.* Cambridge, MA: MIT Press.

Skoogh, J., Ylitalo, N., Larsson Omerov, P., Hauksdottir, A., Nyberg, U., Wildrang, U., … Steineck, G. (2010). "A no means no"— Measuring depression using a single-item question versus Hospital Anxiety and Depression Scale (HADS–D). *Annals of Oncology, Oxford Journals,* 21(9), 1905–1909.

Spielberger, C. D., & Reheiser, E. C. (1994). Job stress in university, corporate, and military personnel. *International Journal of Stress Management, 1,* 19–31.

Spielberger, C. D., & Reheiser, E. C. (2005). Occupational stress and health. In A. Antoniou & C. Cooper (Eds.), *Research companion to organizational health psychology* (pp. 441–454). Northampton, MA: Edward Elgar.

Streiner, D. L. (2003a). Being inconsistent about consistency: When coefficient alpha does and doesn't matter. *Journal of Personality Assessment, 80*(3), 217–222.

Streiner, D. L. (2003b). Diagnosing tests: Using and misusing diagnostic and screening tests. *Journal of Personality Assessment, 81,* 209–219.

Thomas, J. C., & Christiansen, L. (2011). Measurement theory in research. In J. C. Thomas & M. Hersen (Eds.), *Understanding research in clinical and counseling psychology* (2nd ed., pp. 29–55). New York: Routledge.

Thomas, J. C., & Truax, P. (2008). Assessment and analysis of clinically significant change. In D. McKay (Ed.), *Handbook of research methods in clinical and counseling psychology* (pp. 317–336). Thousand Oaks, CA: Sage.

Wahlberg, K., Ghatan, P., Modell, S., Nygren, Å., Ingvar, M., Åsberg, M., & Heilig, M. (2009). Suppressed neuroendocrine stress response in depressed on job-stress related long-term sick-leave: A stable maker potentially suggestive of pre-existing vulnerability. *Biological Psychiatry, 65*(9), 742–747.

Waza, K., Graham, A., Zyzanski, S., & Inoue, K. (1999). Comparison of symptoms in Japanese and American depressed primary care patients. *Family Practice, 16*(5), 528–533.

Williams, R., Pruitt, S., Doctor, J., Epping-Jordan, J., Wahlgren, D., Grant, I., … Atkinson, J. (1998). The contribution of job satisfaction to the transition from acute to chronic low back pain. *Archives of Physical Medicine and Rehabilitation, 79*(4), 366–374.

World Health Organization (WHO). (1994). *The international statistical classification of diseases and related health problems* (ICD-10, 10th Rev.). Geneva, Switzerland: Author.

9

Measurement of Posttraumatic Stress Disorder in an Occupational Health Context[1]

Amy B. Adler, Terence M. Keane, and Paul D. Bliese

Posttraumatic stress disorder (PTSD) is a critical construct to measure for occupations that routinely place individuals in high-risk situations. Prevalence rates in the general population are 4% (Kessler, Berglund, Demler, Jin, & Walters, 2005) whereas studies with individuals in high-risk occupations have, for example, documented rates for at-risk occupations ranging from about 13% in police officers (Robinson, Sigman, & Wilson, 1997) and 17% in firefighters (Perrin et al, 2007), to 35% in U.S. reservists returning from a combat deployment (Thomas et al., 2010). These rates signify the personal toll that high-risk occupations can have on the well-being of individuals and their families. Studies have also documented the impact of PTSD on work productivity (Zatzick et al., 2008) and health care utilization (Dobie et al., 2006).

According to the fourth edition of the *Diagnostic and Statistical Manual of Mental Disorders* (DSM-IV; American Psychiatric Association, 1994), there are several criteria for the diagnosis of PTSD. First, the individual has to be exposed to some traumatic event that involves "... actual or threatened death or serious injury, or a threat to the physical

1. The views expressed in this chapter are those of the authors and do not reflect the official position of the Walter Reed Army Institute of Research, the U.S. Army or Department of Defense. This chapter was authored or coauthored by an employee of the U.S. government as part of official duty and is considered to be in the public domain. Any views expressed herein do not necessarily represent the views of the U.S. government, and the author's participation is not meant to serve as an official endorsement.

integrity of self or others" (p. 427). The second criterion requires the individual to respond to that event with fear, helplessness, or horror. Then there are three symptom clusters: reexperiencing, avoidance, and hyperarousal. There are five reexperiencing symptoms and an individual must have at least one of these in order to meet diagnostic criteria. These symptoms include (a) intrusive and distressing memories of the traumatic event, (b) distressing dreams of the traumatic event, (c) the sense that the traumatic event is recurring, (d) intense psychological distress in response to cues that remind the individual of the traumatic event, and (e) physiological reactivity to cues that remind the individual of the traumatic event. There are seven avoidance symptoms and at least three need to be present to meet criteria in this cluster. The symptoms are (a) attempts to avoid thoughts, feelings, or conversations about the trauma, (b) attempts to avoid activities, places, or people that result in memories of the trauma, (c) gaps in recall regarding aspects of the traumatic event, (d) reduced interest in personal activities, (e) feeling detached from others, (f) limited range of emotions, and (g) a sense of a foreshortened future. There are five hyperarousal symptoms, and at least two must be present for diagnosis and must not have been present prior to the traumatic event. These symptoms include (a) difficulty sleeping, (b) irritability or anger, (c) difficulty concentrating, (d) hypervigilance, and (e) an exaggerated startle response. In addition, symptoms need to be present for at least one month and need to cause significant distress or problems in daily functioning.

Common Research Questions

In occupational health psychology, PTSD rates are typically assessed in order to determine the psychological impact of a particular traumatic event. These rates can then be used to guide appropriate resourcing and to target intervention efforts. Such intervention efforts traditionally focus on treatment after the development of symptoms although there has recently been a movement to assess preexposure resilience training (e.g., Arnetz, Nevedal, Lumley, Backman, & Lublin, 2008) and psychoeducation efforts (Mulligan, Fear, Jones, Wessely, & Greenberg, 2010).

Another focus of PTSD research with at-risk occupations has been to establish risk factors predictive of developing PTSD. These risk factors emphasize degree of exposure to the traumatic event (e.g., Adler, Bliese, McGurk, Hoge, & Castro, 2009; Jordan et al., 2004; Perrin et al., 2007), relevant training and work experience (e.g., McCarroll,

Ursano, Fullerton, & Lundy, 1995; Perrin et al., 2007), previous history of trauma (e.g., Fullerton, Ursano, & Wang, 2004), and other individual difference variables (see Ozer, Best, Lipsey, & Weiss, 2003 for a meta-analytic review).

Measurement: Assessing Symptoms and Occupational Health Interventions

In measuring PTSD, occupational health researchers need to consider whether the key variable of interest is the actual diagnosis or the presence of symptoms. This is a key issue for measuring PTSD because while cut-offs can be established to determine whether an individual meets criteria for PTSD ("caseness"), these values will not represent the degree of subclinical distress individuals may be experiencing. For research purposes, caseness can be useful for comparisons across occupational groups and for determining treatment outcomes; however, symptom levels may be more useful for assessing the impact of prevention or early intervention initiatives (e.g., Adler et al., 2009).

The benefit of examining symptom levels is that continuous measures are generally more sensitive to changes than dichotomous measures such as caseness. That is, power is enhanced by examining changes in symptom levels, and this is important because universally applied public health-style interventions administered to large groups that include both those with symptoms and those without are not likely to yield large effect sizes for at least two reasons. First, those with few or no symptoms (often the majority of the sample) are unlikely to be impacted by the intervention and this lack of change among what generally constitutes more than 80% of the population lowers effect sizes substantially. Second, the short duration of universally applied interventions cannot be expected to produce large magnitude changes even among those with symptoms. The value associated with conducting universally applied prevention programs hinges on the observation that such programs can teach positive skills that accumulate over situations and time. Under these circumstances, even small effect sizes associated with changes in overall symptom levels at a single postintervention time may be expected to produce meaningful effects over time (Bliese, Adler, & Castro, 2011).

Another key advantage to conducting universally applied prevention programs is that these programs need not be particularly lengthy or time-demanding if they are able to capitalize on small-group cohesion and positive leadership. Given that members of small cohesive work groups usually know each other well and are motivated to look out for one another, public-health style interventions can build on this cohesion by encouraging team members to be aware of common reactions in themselves and each other and by training team members in how to best support one another. Such interventions can also encourage members of small cohesive groups to pay attention to the training not only for their own sake but also so that they can provide effective support to their colleagues who may be struggling. In the same way, good leaders can encourage the effectiveness of these interventions by personally attending and participating in the training, and otherwise leading by example.

In contrast, when programs are implemented in nonintact groups without a preexisting foundation of cohesion or integration of leaders, it is not realistic to expect that short duration interventions will be effective. However, when such interventions lead to the development of group cohesion, the effectiveness may increase. For instance, programs such as the Penn Resiliency Program, have been implemented by assembling students into newly created intervention groups that meet multiple times (e.g., 12 sessions). Researchers have suggested that the effectiveness of these programs may, in part, be attributed to the sense of group cohesion that develops over time (Brunwasser, Gillham, & Kim, 2009).

While the efficacy of these interventions is likely to be most easily detected using continuous symptom measures, there can be value in determining the rates of individuals scoring above established criteria for PTSD. In a practical sense, using a cut-off value (even if it reduces power) produces findings that are relatively simple to communicate. For instance, finding that an intervention condition had a 10% rate of individuals meeting the criteria for PTSD while a control condition had a 15% rate can be easy to convey to a lay audience. In contrast, stating that an intervention condition had a PCL mean score of 30 while a control condition had a PCL mean score of 35 is less transparent. Clear descriptions of study outcomes can be particularly important when communicating results to organizational members who make decisions about whether to support such programs.

Cut-off rates are also important when the goal of the intervention is to use screening-based instruments to identify high-risk individuals for potential subsequent interventions or treatment. Cut-off rates are part of the decision algorithm used to determine whether an individual is regarded as high risk. In these situations, however, the challenge is to identify optimal cut-off scores for that occupational population. Establishing appropriate cut-offs can be difficult because the context of the evaluation may influence the optimal cut-off score. For example, lower cut-off scores on the PTSD Checklist have been reported for primary care samples than treatment-seeking samples (Bliese et al., 2008). Such findings demonstrate that optimal psychometric cut-offs in terms of sensitivity and specificity for PTSD measures are likely to be lower for relatively healthy individuals (e.g., primary care samples and occupational settings). Thus, occupational health psychology researchers assessing a group of employees may need to either determine the correct cut-off values for their specific sample and/or consider using cut-offs reported in the literature that are based on non-treatment seeking samples.

Given the confusion around various cut-off scores as well as how best to use the DSM symptom algorithm in determining cut-off scores, researchers may decide to report different methods of calculating cut-offs in the same paper (e.g., Perrin et al., 2007; Terhakopian, Sinaii, Engel, Schnurr, & Hoge, 2008; Thomas et al., 2010). Note also that since the criteria for PTSD require symptom duration of at least 4 weeks, it would be inappropriate to measure PTSD if fewer than 4 weeks have elapsed since the traumatic event. Instead, it would be appropriate to measure acute stress disorder (e.g., Fullerton et al., 2004).

Assessment Methods

In determining the best assessment of PTSD, occupational health psychologists need to consider the goals of the assessment (for review of available techniques for the assessment of PTSD within various contexts, see Wilson & Keane, 2004). If diagnostic accuracy is the most critical goal, then a multimethod approach for assessing PTSD is recommended (Keane, Fairbank, Caddell, Zimering, & Bender, 1985).

These different methods may include a structured diagnostic interview to assess PTSD and other comorbid clinical domains (Keane, Kolb et al., 1998), self-report psychological questionnaires, and psychophysiological measures. Although differences in biologically based responses have been identified in those with and without PTSD, there is no reliable physiological assessment currently available. The measures that have been used in research to assess, for example, psychophysiological reactivity to trauma-related cues, are also expensive and require specialized equipment and training (Orr, Metzger, Miller, & Kaloupek, 2004). If the goal is to determine the overall presence of PTSD or PTSD symptoms, then self-report questionnaires alone can be an efficient assessment technique.

Structured Diagnostic Interviews

The benefit of conducting a structured interview, the "gold standard" in psychological research, is that results are considered an accurate representation of an individual's mental health status. The potential limitation is that structured interviewing is time consuming and requires training and expertise in clinical interviewing. Ideally, within an occupational health context, structured interviews would be used for randomized treatment trials and for confirming optimal cut-off scores for self-report questionnaires.

Clinician Administered PTSD Scale (CAPS). Developed by the National Center for PTSD (Blake et al., 1990), the CAPS is the most widely used structured interview for diagnosing and measuring the severity of PTSD (Weathers, Keane, & Davidson, 2001). The CAPS assesses all DSM-IV (American Psychiatric Association, 1994) diagnostic criteria for PTSD, as well as the associated symptoms of guilt and dissociation. Importantly, the CAPS contains separate ratings for the frequency and intensity of each symptom; this permits flexibility in scoring and analyses. It also promotes uniform administration and scoring through carefully phrased prompt questions and explicit rating scale anchors with clear behavioral referents. There is also flexibility built into the administration of the CAPS. Interviewers can administer only the 17 core symptoms, all DSM-IV criteria, and/or the associated symptoms. Administration time is approximately 30 minutes to an hour, depending on those sections the interviewer chooses to utilize. Once trained, interviewers are able to ask their own follow-up questions and use clinical judgment in arriving at a diagnosis.

Weathers, Ruscio, and Keane (1999) examined the reliability and validity data of the CAPS across five samples of male Vietnam veterans collected at the National Center for PTSD in Boston. Robust estimates were found for interrater reliability over a 2- to 3-day interval for each of the three symptom clusters and all 17 symptoms. Test–retest reliability for a CAPS-based PTSD diagnosis was also excellent. Thus, the data indicate that trained raters can achieve a high degree of consistency in using the CAPS to rate symptom severity and diagnose PTSD. Weathers et al. (1999) also found excellent internal consistency across all 17 items in research and clinical samples, which supports the use of the CAPS in both settings.

These investigators also reported strong evidence for validity of the CAPS. They reported that the CAPS total severity score correlated highly with other measures of PTSD (Weathers, Litz, Herman, Huska, & Keane, 1993). They also found strong evidence for the diagnostic utility of the CAPS using three different CAPS scoring rules for predicting a PTSD diagnosis.

The CAPS has been used successfully in a wide variety of trauma populations (e.g., combat veterans, Cambodian and Bosnian refugees, and victims of rape, crime, motor vehicle accidents, incest, the Holocaust, torture, and cancer), has served as the primary diagnostic or outcome measure in more than 250 empirical studies on PTSD, and has been translated into at least 12 languages (Weathers et al., 2001). Thus, the existing data strongly support its continued use across clinical and research settings.

Structured Clinical Interview for DSM-IV (SCID-IV). The SCID-IV (First, Spitzer, Gibbon, & Williams, 1996; First, Spitzer, Gibbon, Williams, & Benjamin, 1997) assesses a broad range of Axis I and II psychiatric conditions. It is divided into separate modules corresponding to DSM-IV diagnostic criteria, with each module providing the interviewer with prompts and follow-up inquiries intended to be read verbatim to respondents. The SCID can be administered by clinicians and highly trained interviewers. Although the administration of the full SCID-IV can be time consuming, the modular structure allows clinicians to tailor their assessment appropriately.

The SCID-PTSD module is considered psychometrically sound. Keane et al. (1998) reported that the SCID-PTSD had adequate reliability, and McFall, Smith, Roszell, Tarver, and Malas (1990) reported evidence of convergent validity, finding significant correlations between the SCID-PTSD and other measures of PTSD. The SCID-PTSD module also had good diagnostic utility (Kulka et al., 1990).

Although the SCID is a good diagnostic tool, several limitations exist. First, most clinicians agree that psychological symptoms occur in a dimensional rather than dichotomous fashion, but the SCID permits only a dichotomous rating of PTSD (e.g., presence or absence of symptoms (Keane, Weathers, & Foa, 2000). Second, the SCID does not assess for the frequency or severity of symptoms. Third, only those symptoms associated with the "worst event" are assessed; the effects of other traumas are not evaluated.

Other Interview Methods. There are several other interview measures of PTSD that are beyond the scope of this chapter but that are identified briefly here. In terms of diagnostic interviews, Foa, Riggs, Dancu, and Rothbaum (1993) developed the PTSD Symptom Scale Interview (PSS-I), in which interviewers rate the severity of 17 symptoms. However, the symptoms correspond to the DSM-III-R and measures symptoms over the past 2 weeks, rather than one month (Cusack, Falsetti, & de Arellano, 2002). The Mini International Neuropsychiatric Interview (MINI) is a brief but comprehensive interview schedule that has been validated against the SCID (Sheehan et al., 1998).

Self-Report Measures

Self-report measures are typically used for large scale survey-based studies. They are more time and cost efficient than structured interviews; they also allow for anonymity. Furthermore, self-report measures trauma symptom severity which most interview schedules do not. The key issue in using a self-report measure is to ensure it has been normed on the target population for optimal accuracy and efficiency (Keane & Barlow, 2002).

Impact of Event Scale-Revised (IES-R). Developed by Horowitz, Wilner, and Alvarez (1979), the IES is one of the most widely used self-report measures to assess psychological responses to a traumatic event. The revised 22-item version was developed to accommodate the DSM-IV criteria, containing items on hyperarousal symptoms and flashback experiences (IES-R; Weiss & Marmar, 1997). Respondents complete the measure by rating on a Likert scale "how distressed or bothered" they were by each symptom during the past week. The IES has been translated into several languages, has been used with many different trauma populations, and takes approximately 10 minutes to complete.

Data on the psychometric properties of the revised IES-R are preliminary in nature. In two studies that incorporated four samples of emergency workers and earthquake survivors, Weiss and Marmar (1997) reported satisfactory internal consistency for each of the subscales, while test–retest reliability data from two samples yielded a range of reliability coefficients. Weiss and Marmar (1997) suggest that the shorter interval between assessments and the greater recency of the traumatic event for one sample contributed to higher coefficients of stability for that sample. Still, it remains difficult to make determinations regarding reliability.

Convergent and discriminant validity data are not yet available for the IES-R. There were many questions raised about the validity of the original scale, in part because it did not assess all DSM criteria for PTSD (see Joseph, 2002). Although it now more closely parallels DSM-IV, items measuring numbing are considered limited by some investigators (Foa, Cashman, Jaycox, & Perry, 1997). In a review of psychometric studies on the IES, Sundin and Horowitz (2002) reported a wide range of correlations between the IES subscales and other self-report measures and diagnostic interviews. Additional studies with the revised instrument are clearly needed to establish its reliability and validity and insure its continued use in clinics.

PTSD Checklist (PCL). Developed by researchers at the National Center for PTSD in Boston (Weathers et al., 1993), the PCL is a 17-item self-report measure of PTSD symptoms. Different scoring procedures may be used to yield either a continuous measure of symptom severity or a dichotomous indicator of diagnostic status. Dichotomous scoring methods include either an overall cutoff score or a symptom cluster scoring approach. The original scale was based on the DSM-III-R criteria for PTSD (American Psychiatric Association, 1987) and has been updated to reflect the 17 diagnostic criteria outlined in DSM-IV (American Psychiatric Association, 1994). Respondents are asked to rate, on a Likert scale, “how much each problem has bothered them” during the past month. The time frame can be adjusted as needed to suit the goals of the assessment. There is a civilian (PCL-C) and a military version (PCL-M) of the measure. On the PCL-C, reexperiencing and avoidance symptoms apply to any lifetime stressful event, while for the PCL-M, reexperiencing and avoidance symptoms apply to stressful events that are military-related only. The PCL has been used extensively in both research and clinical settings and takes 5 to 10 minutes to administer.

The PCL was validated in a sample of Vietnam and Persian Gulf War veterans and found to have strong psychometric properties (Weathers et al., 1993). Keen, Kutter, Niles, and Krinsley (2008) examined the psychometric properties of the updated PCL in veterans with both combat and noncombat traumas and found evidence for high internal consistency. Other investigators have also documented adequate test–retest reliability of this measure over a 2-week time frame (Ruggiero, Del Ben, Scotti, & Rabalais, 2003).

With respect to validity, Keen et al. (2008) found that the scale was highly correlated with other measures of PTSD including the Mississippi Scale and the CAPS, and had good diagnostic power. Additionally, using the CAPS as the gold standard, Dobie et al. (2006) reported that the PCL had good diagnostic utility.

Several studies provide evidence for the reliability and validity of the PCL in nonveteran samples (e.g. primary care patients, severely mentally ill adults), although the optimal cut-off score varies across samples (Cook, Elhai, & Arean, 2005; Grubaugh, Elhai, Cusack, Wells, & Frueh, 2007; Walker, Newman, Dobie, Ciechanowski, & Katon, 2002). In addition, there is evidence that different scoring options for the PCL (e.g., an absolute cutoff score vs. symptom cluster scoring) yield differences in sensitivity, specificity, and diagnostic efficiency.

Other Self-Report Measures. There are several other self-report measures of PTSD that are beyond the scope of this chapter but that are identified briefly here. The 35-item Mississippi Scale for Combat Related PTSD, developed by Keane, Caddell, and Taylor (1988), has been widely used to assess combat-related PTSD symptoms; the 46-item Keane PTSD Scale is empirically drawn from the MMPI-2 (Lyons & Keane, 1992) and the scale is typically administered as part of the full MMPI-2; the 49-item Posttraumatic Diagnostic Scale (PDS) developed by Foa et al. (1997) is designed to measure DSM-IV PTSD criteria and symptom severity, including subjective response and functional impairment. In contrast to these longer measures, the Primary Care-PTSD screen (PC-PTSD; Prins et al., 2004) consists of only four dichotomously scored items. The PC-PTSD was originally validated in primary care (Prins et al., 2004) and cut-off scores have been established with soldiers returning from a combat deployment (Bliese et al., 2008). While the PC-PTSD has the benefit of being short, its utility as a continuous measure is limited.

In terms of short scales, Bliese et al. (2008) also report the results of an Item Response Theory (IRT) analysis of the 17-items of the PCL that produced a 4-item version of the PCL. This short version was psychometrically equivalent to the 17-item scale in terms of diagnostic efficiency, and was also similar to a short version identified by Lang and Stein (2005). A 4-item version of the PCL may be useful in situations where survey length precludes the inclusion of the 17-item scale. Alternative versions of a scale may also be useful if self-report measures are used repeatedly with the same group of workers in order to reduce the likelihood that the psychometric properties of the scale will be negatively affected by rote responses.

Occupationally Related PTSD

What differentiates a good study from an outstanding study when it comes to measuring PTSD? The gold standard in PTSD assessment is the CAPS structured interview, given that it is a sound measure, is widely used, and possesses excellent psychometric properties. However, for most occupational health studies, self-report measures are completely acceptable. Many of the self-report measures can be used interchangeably; we recommend that researchers consider the available psychometric data for their population used.

Finally, we consider future directions in the field of occupationally related PTSD. Although PTSD may occur because of an unexpected workplace event or accident, much of the occupational health research on PTSD has focused on at-risk occupations in which individuals are trained to encounter potentially traumatic events. This emphasis suggests the need to shift the current diagnostic understanding of PTSD in an occupational context (e.g., Creamer, McFarlane, & Burgess, 2005; Weathers & Keane, 2007). Currently, the diagnosis is based on a victim perspective in which the constellation and development of symptoms is seen as occurring after a sudden, unexpected, and unwanted event. In contrast, an occupational health model of PTSD, as proposed by Castro and Adler (2011), suggests that individuals who are trained for a traumatic event as part of their occupation may not find the event unexpected and may instead at least partially interpret the event as an opportunity to use their training.

Adopting the occupational health model potentially changes the way in which PTSD diagnostic criteria are understood. For example, instead of subjectively responding with helplessness, horror, and fear, individuals may report responding to a traumatic event with an occupationally relevant reaction (e.g., "My training kicked in"; Adler, Wright, Bliese, Eckford, & Hoge, 2008). In this context, measures like the CAPS potentially need to be calculated both with and without subjective responses to the trauma to accurately gauge PTSD reactions to events.

Examining PTSD from an occupational health perspective has several other implications. For instance, certain symptoms typically seen as being part of PTSD may be better understood as normal reactions to occupational demands. Symptoms such as hypervigilance may occur prior to the traumatic event and be a consequence of training, rather than a consequence of the traumatic event. For example, service members who have spent 12 months on a forward operating base in a combat environment in which they have to constantly monitor their environment for improvised explosive devices, snipers, and other signs of danger may develop hypervigilance as an occupationally relevant and adaptive skill. Upon returning home, the service members may find that although hypervigilance is not as necessary because the threat level is different, the combat skill remains part of their way of relating to their environment. In the context of home, this reaction can be labeled as a psychological symptom; however, it may be more useful for the individuals to understand their reaction as an expected outcome of an occupational demand rather than a sign of their psychological ill-health. Besides the fact that many PTSD symptoms can be conceptualized as combat-related skills, other emotions such as anger, guilt, and grief, are currently neglected in the criteria for PTSD. These other emotions may be paramount in describing the kinds of adjustment difficulties facing individuals trained for serving in high-risk occupations such as soldiers, police officers, and other rescue personnel. In these occupations, group cohesion and good leadership may provide an important buffer that is overlooked in the current conceptualization of PTSD (Castro & Adler, 2011).

In short, the reconceptualization of PTSD within an occupational framework has direct implications for the diagnostic criteria for occupationally-related PTSD, its concomitant assessment, and addressing the group-level properties of the data. This reconceptualization also has implications for early interventions and public-health style initiatives that target individuals in high-risk occupations. Such programs can be used to clarify misperceptions about PTSD symptoms and place them in an occupational framework while leveraging group and leader support (e.g., Adler et al., 2009). Future occupational health research should be sure to consider the larger occupational context and the meaning of symptoms from an occupational perspective when assessing and providing early interventions to address PTSD symptoms.

References

Adler, A. B., Bliese, P. B., McGurk, D., Hoge, C. W., & Castro, C. A. (2009). Battlemind Debriefing and Battlemind Training as early interventions with soldiers returning from Iraq: Randomization by platoon. *Journal of Consulting and Clinical Psychology, 77*, 928–940.

Adler, A. B., Wright, K. M., Bliese, P. D., Eckford, R., & Hoge, C. W. (2008). A2 diagnostic criterion for combat-related posttraumatic stress disorder. *Journal of Traumatic Stress, 21*, 301–308.

American Psychiatric Association. (1987). *Diagnostic and statistical manual of mental disorders* (3rd ed., rev.). Washington, DC: Author.

American Psychiatric Association. (1994). *Diagnostic and statistical manual of mental disorders* (4th ed.). Washington, DC: Author.

Arnetz, B. B., Nevedal, D. C., Lumley, M. A., Backman, L., & Lublin, A. (2009). Trauma resilience training for police: Psychophysiological and performance effects. *Journal of Police and Criminal Psychology, 24*, 1–9.

Blake, D. D., Weathers, F. W., Nagy, L. M., Kaloupek, D. G., Klauminzer, G., Charney, D. S., & Keane, T. M. (1990). A clinician rating scale for assessing current and lifetime PTSD: The CAPS-1. *Behavior Therapist, 13*, 187–188.

Bliese, P. B., Adler, A. B., & Castro, C. A. (2011). The deployment context: Psychology and implementing mental health interventions. In A. B. Adler, P. B. Bliese, & C. A. Castro (Eds.), *Deployment psychology: Evidence-based strategies to promote mental health in the military* (pp. 103–124). Washington, DC: American Psychological Association.

Bliese, P. D., Wright, K. M., Adler, A. B., Cabrera, O., Castro, C. A., & Hoge, C. W. (2008). Validating the PC–PTSD and the PTSD checklist with soldiers returning from combat. *Journal of Consulting and Clinical Psychology, 76*, 272–281.

Brunwasser, S. M., Gillham, J. E., & Kim, E. S. (2009). A meta-analytic review of the Penn Resiliency Program's effect on depressive symptoms. *Journal of Consulting and Clinical Psychology, 77*, 1042–1054.

Castro, C. A., & Adler, A. B. (2011). Re-conceptualizing combat-related posttraumatic stress disorder as an occupational hazard. In A. B. Adler, P. B. Bliese, & C. A. Castro (Eds.), *Deployment psychology: Evidence-based strategies to promote mental health in the military* (pp. 217–242). Washington, DC: American Psychological Association:

Cook, J. M., Elhai, J. D., & Arean, P. A. (2005). Psychometric properties of the PTSD Checklist with older primary care patients. *Journal of Traumatic Stress, 18*, 371–376.

Creamer, M., McFarlane, A. C., & Burgess, P. (2005). Psychopathology following trauma: The role of subjective experience. *Journal of Affective Disorders, 86*, 175–182.

Cusack, K., Falsetti, S., & de Arellano, M. (2002). Gender considerations in the psychometric assessment of PTSD. In R. Kimerling, P. Ouimette, & J. Wolfe (Eds.), *Gender and PTSD* (pp. 150–176). New York: Guilford.

Dobie, D. J., Maynard, C., Kivlahan, D. R., Johnson, K. M., Simpson, T. L., David, A. C., & Bradley, K. A. (2006). Posttraumatic stress disorder screening status is associated with increased VA medical and surgical utilization in women. *Journal of General Internal Medicine, 21,* 58–64.

First, M. B., Gibbon, M., Spitzer, R. L., Williams, J. B. W, & Benjamin, L. S. (1997). *Structured clinical interview for DSM–IV Axis II personality disorders (SCID–II).* Washington, DC: American Psychiatric Press.

First, M. B., Spitzer, R. K., Gibbon, M., & Williams, J. B. W. (1996). *Structured clinical interview for DSM-IV Axis I disorders, clinician version (SCID–CV).* Washington, DC: American Psychiatric Press.

Foa, E. B., Cashman, L., Jaycox, L., & Perry, K. (1997). The validation of a self-report measure of post-traumatic stress disorder: The posttraumatic diagnostic scale. *Psychological Assessment, 9,* 445–451.

Foa, E. B., Riggs, D. S., Dancu, C. V., & Rothbaum, B. O. (1993). Reliability and validity of a brief instrument for assessing post-traumatic stress disorder. *Journal of Traumatic Stress, 6,* 459-473.

Fullerton, C. S., Ursano, R. J., & Wang, L. (2004). Acute Stress Disorder, Posttraumatic Stress Disorder, and depression in disaster or rescue workers. *American Journal of Psychiatry, 161,* 1370-1376.

Grubaugh, A. L., Elhai, J. D., Cusack, K. J., Wells, C., & Frueh, B. C. (2007). Screening for PTSD in public-sector mental health settings: The diagnostic utility of the PTSD Checklist. *Depression and Anxiety, 24,* 124–129.

Horowitz, M. J., Wilner, N. R., & Alvarez, W. A. (1979). Impact of Event Scale: A measure of subjective stress. *Psychosomatic Medicine, 41,* 209–218.

Jordan, N. N., Hoge, C. W., Tobler, S. K., Wells, J., Dyrdek, G. J., & Egerton, W. E. (2004). Mental health impact of 9/11 Pentagon attack: Validation of a rapid assessment tool. *American Journal of Preventive Medicine, 26,* 284–93.

Joseph, S. A. (2002). Psychometric evaluation of Horowitz's Impact of Event Scale: A review. *Journal of Traumatic Stress, 13,* 101–113.

Keane, T. M., & Barlow, D. H. (2002). Posttraumatic stress disorder. In D. H. Barlow (Ed.), *Anxiety and its disorders* (pp. 418–453). New York: Guilford.

Keane, T. M., Caddell, J. M., & Taylor, K. L. (1988). Mississippi Scale for combat-related posttraumatic stress disorder: Three studies in reliability and validity. *Journal of Consulting and Clinical Psychology, 56,* 85–90.

Keane, T. M., Fairbank, J. A., Caddell, J. M., Zimering, R. T., & Bender, M. E. (1985). A behavioral approach to assessing and treating post-traumatic stress disorder in Vietnam veterans. In C. R. Figley (Ed.), *Trauma and its wake: Vol. 1. The study and treatment of post-traumatic stress disorder* (pp. 257–294). New York: Brunner/Mazel.

Keane, T. M., Kolb, L. C, Kaloupek, D. G., Orr, S. P., Blanchard, E. B., Thomas, R. G., … Lavori, P. W. (1998). Utility of psychophysiology measurement in the diagnosis of posttraumatic stress disorder: Results from a Department of Veterans Affairs cooperative study. *Journal of Consulting and Clinical Psychology, 66,* 914–923.

Keane, T. M., Weathers, F. W., & Foa, E. B. (2000). Diagnosis and assessment. In E. B. Foa, T. M. Keane, & M. J. Friedman (Eds.), *Effective treatments for PTSD* (pp. 18–36). New York: Guilford.

Keen, S. M., Kutter, C. J., Niles, B. L., & Krinsely, K. E. (2008). Psychometric properties of PTSD Checklist in sample of male veterans. *Journal of Rehabilitation Research & Development, 45,* 465–474.

Kessler, R. C., Berglund, P., Demler, O., Jin, R., & Walters, E. E. (2005). Lifetime prevalence and age-of-onset distributions of *DSM-IV* disorders in the National Comorbidity Survey Replication. *Archives of General Psychiatry, 62,* 593–602.

Kulka, R. A., Schlenger, W. E., Fairbank, J. A., Hough, R. L., Jordan, B. K., Marmar, C. R., … Grady, D. A. (1990). *Trauma and the Vietnam War generation: Report of findings from the national Vietnam veterans' readjustment study.* New York: Brunner/Mazel.

Lang, A. J., & Stein, M. B. (2005). An abbreviated PTSD checklist for use as a screening instrument in primary care. *Behaviour Research and Therapy, 43,* 585–594.

Lyons, J. A., & Keane, T. M. (1992). Keane PTSD Scale: MMPI and MMPI-2 update. *Journal of Traumatic Stress, 5,* 111–117.

McCarroll, J. E., Ursano, R. J., Fullerton, C. S., & Lundy, A. (1995). Anticipatory stress of handling human remains from the Persian Gulf War: Predictors of intrusion and avoidance. *Journal of Nervous and Mental Disease, 183,* 698–703.

McFall, M. E., Smith, D. E., Roszell, D. K., Tarver, D. J., & Malas, K. L. (1990). Convergent validity of measures of PTSD in Vietnam combat veterans. *American Journal of Psychiatry, 147*(5), 645–648.

Mulligan, K., Fear, N. T., Jones, N., Wessely, S., & Greenberg, N. (2010). Psycho-educational interventions designed to prevent deployment-related psychological ill-health in Armed Forces personnel: A review. *Psychological Medicine,* doi:10.1017/S003329171000125X

Orr, S. P., Metzger, L. J., Miller, M. W., & Kaloupek, D. G. (2004). Psychophysiological assessment of PTSD. In J. P. Wilson & T. M. Keane (Eds.), *Assessing psychological trauma and PTSD* (pp. 289–343). New York: Guildford.

Ozer, E. J., Best, S. R., Lipsey, T. L., & Weiss, D. S. (2003). Predictors of posttraumatic stress disorder and symptoms in adults: A meta-analysis. *Psychological Bulletin, 129*(1), 52–73.

Perrin, M. A., DiGrande, L., Wheeler, K., Thorpe, L., Farfel, M., & Brackbill, R. (2007). Differences in PTSD prevalence and associated risk factors among World Trade Center disaster rescue and recovery workers. *American Journal of Psychiatry, 164,* 1385–1394.

Prins, A., Ouimette, P., Kimerling, R., Cameron, R. P., Hugelshofer, D. S., Shaw-Hegwer, J., … Sheikh, J. I. (2004). The primary care PTSD screen (PCPTSD): Development and operating characteristics. *Primary Care Psychiatry, 9,* 9–14.

Robinson, H. M., Sigman, M. R., & Wilson, J. P. (1997). Duty-related stressors and PTSD symptoms in suburban police officers. *Psychological Reports, 81,* 835–845.

Ruggiero, K. J., Del Ben, K. S., Scotti, J. R., & Rabalais, A. E. (2003). Psychometric properties of the PTSD Checklist Civilian Version. *Journal of Traumatic Stress, 16,* 495–502.

Sheehan, D. V., Lecrubier, Y., Sheehan, K. H., Amorim, P., Janavs, J., Weiller, E., … Dunbar, G. C. (1998). The MINI-International Neuropsychiatric Interview (M. I. N. I): The development and validation of a structured interview for *DSM–IV* and *ICD–10. Journal of Clinical Psychiatry, 59*(Suppl. 20), 22–33.

Sundin, E. C., & Horowitz, M. J. (2002). Impact of Event Scale: Psychometric properties. *British Journal of Psychiatry, 180,* 205–209.

Terhakopian, A., Sinaii, N., Engel, C. C., Schnurr, P. P., & Hoge, C. W. (2008). Estimating population prevalence of posttraumatic stress disorder: An example using the PTSD Checklist. *Journal of Traumatic Stress, 21,* 290–300.

Thomas, J. L., Wilk, J. E., Riviere, L. A., McGurk, D., Castro, C. A., & Hoge, C. W. (2010). Prevalence of mental health problems and functional impairment among active component and National Guard Soldiers 3 and 12 months following combat in Iraq *Archives of General Psychiatry, 67,* 614–623.

Walker, E. A., Newman, E., Dobie, D. J., Ciechanowski, P., & Katon, W. (2002). Validation of the PTSD Checklist in an HMO sample of women. *General Hospital Psychiatry, 24,* 375–380.

Weathers, F. W., & Keane, T. M. (2007). The criterion A problem revisited: Controversies and challenges in defining and measuring psychological trauma. *Journal of Traumatic Stress, 20,* 107–121.

Weathers, F. W., Keane, T. M., & Davidson, J. R. T. (2001). Clinician-Administered PTSD Scale: A review of the first ten years of research. *Depression and Anxiety, 13,* 132–156.

Weathers, F. W., Litz, B. T., Herman, D. S., Huska, J. A., & Keane, T. M. (1993, October). *The PTSD checklist (PCL): Reliability, validity and diagnostic utility.* Paper presented at the 9th annual meeting of the International Society for Traumatic Stress Studies, San Antonio, TX.

Weathers, F. W., Ruscio, A. M., & Keane, T. M. (1999). Psychometric properties of nine scoring rules for the Clinician-Administered Posttraumatic Stress Disorder Scale. *Psychological Assessment, 11,* 124–133.

Weiss, D. S., & Marmar, C. R. (1997). The Impact of Event Scale—Revised. In J. P. Wilson & T. M. Keane (Eds.), *Assessing psychological trauma and PTSD* (pp. 131–148). New York: Guilford.

Wilson, J. P., & Keane, T. M. (2004). *Assessing psychological trauma and PTSD* (2nd ed.). New York: Guilford.

Zatzick, D., Jurkovich, G. J., Rivara, F. P., Wang, J., Fan, M., Joesch, J., & Mackenzie, E. (2008). A national US study of posttraumatic stress disorder, depression, and work and functional outcomes after hospitalization for traumatic injury. *Annals of Surgery, 248,* 429–437.

10

The Measurement of Work Engagement

Wilmar B. Schaufeli

Introduction

Although the origin of the term is not entirely clear, most likely in its present usage *employee engagement* was coined by the Gallup Organization in the 1990s as a result of 25 years of interviewing and surveying employees (Buckingham & Coffman, 1999). Please note that the terms *employee engagement* and *work engagement* are typically used interchangeably, albeit that the former seems to be preferred in business and consultancy, and the latter in academia. In this chapter the term *work engagement* is used because it is more specific and refers to the relationship of the employee with his or her *work*. Employee engagement, in contrast, also includes the relationship with the *organization*, which might explain its popularity in business and consultancy.

This chapter starts with the questions What is work engagement? and How can it be measured? Next, the relationship of work engagement with associated concepts such as workaholism and job burnout is discussed. Like engaged employees, workaholics also work very hard, yet both constructs refer to different psychological states. In addition, work engagement has been characterized as the opposite of job burnout, but they seem to have different antecedents and consequences. The chapter concludes with a section on the practical use of engagement questionnaires.

What is Work Engagement?

The Merriam-Webster dictionary describes engagement as "emotional involvement or commitment" and as "the state of being in gear." This reflects our everyday, colloquial connotation of engagement which refers to involvement, commitment, passion, enthusiasm, absorption, zeal, vitality, focused effort, vigor, immersion, and energy. However, neither scholars nor practitioners agree on a conceptualization of engagement in relation to work.

Engagement in Business

Based on an overview of the business literature Attridge (2009) concludes that employee engagement is a major topic for managers and human resources (HR) professionals. So unsurprisingly, virtually all major human resources consultancy firms offer tools to assess and to improve levels of employee engagement. Almost without exception these firms claim conclusive and compelling evidence that employee engagement increases profitability through higher productivity, sales, customer satisfaction, effectiveness, employee retention, and so on. However, with the exception of the Gallup Organization (Harter, Schmidt, & Hayes, 2002) this claim is not *evidenced* by publications in peer-reviewed journals.

Although the descriptions of work engagement that are used in business may differ at first glance, a closer look reveals that in essence engagement is defined in terms of: (a) extrarole behavior (i.e., discretionary behavior that promotes the organization's effective functioning) and (b) commitment to the organization and its goals (i.e., affective commitment), and the desire to stay with the organization (i.e., continuance commitment). Hence, the way business conceptualizes engagement comes close to putting old (extrarole and commitment) wine in new (engagement) bottles.

Engagement in Academia

The first scholar who conceptualized engagement at work was Kahn (1990), an ethnographic researcher who described it as the "... harnessing of organization members' selves to their work-roles: in engagement, people employ and express themselves physically, cognitively, emotionally and mentally during role performances" (p. 694). In other words, engaged employees put a lot of effort into their work role because they identify with it.

A different approach is followed in occupational health psychology where work engagement is considered as the positive counterpart of burnout (Maslach, Schaufeli, & Leiter, 2001). Work engagement is considered to be indicative for employee health, because it is related to perceived health (Schaufeli, Taris, & Van Rhenen, 2008), proactive behavior (Salanova & Schaufeli, 2008), low sickness absence (Schaufeli, Bakker & Van Rhenen, 2009), and reactivity of the hypothalamic-pituitary-adrenal-axis (Langelaan, Bakker, Schaufeli, Van Rhenen, & Van Doornen, 2006).

According to Maslach and Leiter (1997) engagement is characterized by energy, involvement, and efficacy—the direct opposites of the three burnout dimensions exhaustion, cynicism, and reduced professional efficacy, respectively. By implication, the reversed pattern of scores on the three dimensions of the Maslach Burnout Inventory (MBI; Maslach, Jackson, & Leiter, 1996) is indicative for engagement: *low* scores on exhaustion and cynicism, and *high* scores on professional efficacy. Hence, a one-dimensional approach is adopted that considers engagement and burnout as two ends of a continuum that is assessed by the MBI. However, this approach has a serious drawback; both constructs cannot be studied independently from each other.

In contrast, work engagement is also considered as an independent, distinct concept that is defined in its own right, namely as "… a positive, fulfilling, work-related state of mind that is characterized by vigor, dedication, and absorption" (Schaufeli, Salanova, González-Romá, & Bakker, 2002, p. 74). Vigor refers to high levels of energy and mental resilience while working, the willingness to invest effort in one's work, and persistence even in the face of difficulties. Dedication refers to being strongly involved in one's work, and experiencing a sense of significance, enthusiasm, inspiration, pride, and challenge. Finally, absorption is characterized by being fully concentrated and happily engrossed in one's work, whereby time passes quickly and one has difficulties with detaching oneself from work. Vigor and dedication are considered the core aspects of work engagement that constitute the opposites of exhaustion and cynicism, respectively, the two core symptoms of burnout. Moreover, vigor and exhaustion span a continuum that is dubbed "energy," whereas the dedication–cynicism continuum is dubbed "identification" (González-Romá, Schaufeli, Bakker, & Lloret, 2006). It follows that work engagement is characterized by a high level of energy and strong identification with one's work, whereas burnout is characterized by a low level of energy and poor identification with one's work.

In sum, the key reference of engagement for Kahn (1990) is the work *role*, whereas for occupational health psychologists it is the employee's work *activity*, or the work itself. In contrast, in business contexts engagement refers to the *organization*, including its mission, goals, and values.

A Model of Work Engagement

Macey and Schneider (2008) presented an all-inclusive conceptual framework that subsumes a broad variety of meanings of engagement. They distinguish between: (a) engagement as a *disposition* (e.g., proactive personality, autotelic personality, positive affectivity); (b) engagement as a *psychological state* (e.g., satisfaction, involvement, energy, absorption); and (c) engagement as a *performance construct* (e.g., personal initiative, organizational citizenship behavior, role expansion). However, by including such a wide range of concepts that all refer to "engagement," the term serves as an umbrella for whatever one wants it to be. Or put differently, when the meaning of engagement is stretched to the limit the concept ends up meaning nothing at all.

In contrast, Schaufeli and Bakker (2010) proposed a model of employee motivation that integrates work engagement into a nomological net. A slightly adapted version of this model is displayed in Figure 10.1, in which work engagement is defined as a psychological state that

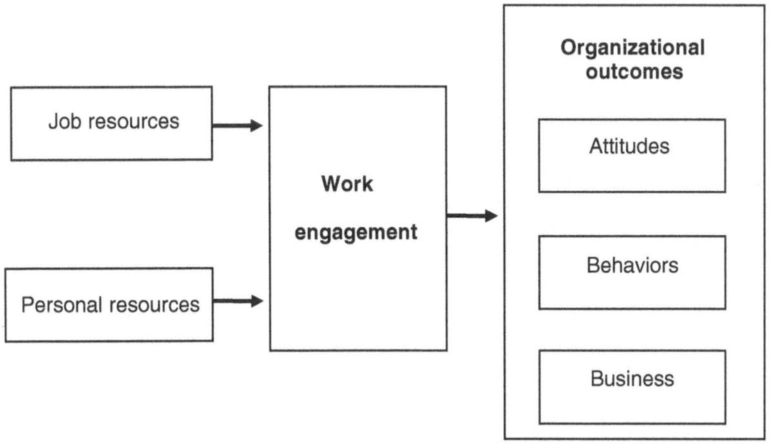

Figure 10.1 A model of work engagement.

mediates the impact of job resources and personal resources on organizational outcomes.

In fact, Figure 10.1 represents the motivational process of the job demands-resources (JD-R) mode (Bakker & Demerouti, 2007) that assumes that job resources have motivational potential and lead to positive organizational outcomes, via work engagement. According to the JD-R model job resources may play either an intrinsic motivational role because they foster employees' growth, learning, and development, or they may play an extrinsic role because they are instrumental in achieving work goals. Based on Fredrickson's broaden-and-built theory of positive emotions, Bakker and Demerouti (2008) argued that work engagement boosts performance because it has the capacity to broaden the employee's momentary thought and action repertoires and to mobilize job and personal resources.

Hence, work engagement is distinguished from various organizational outcomes, which is at odds with the view of most consultancy firms who define engagement *in terms of* these very outcomes. In contrast, this chapter considers work engagement as a psychological state that drives employee's behavioral investment of personal energy and that neither coincides with the performance behavior itself nor with the concomitant attitudes or business outcomes.

How is Work Engagement Measured?

Since no psychometric data are available from engagement questionnaires that have been used by consultancy firms in business contexts, these instruments cannot be reviewed. There is one exception though: Gallup's Workplace Audit (GWA) or Q^{12}.

The Gallop Q^{12}

After an iterative process of item formulation and testing that took several decades, the final wording of the Gallup questionnaire was established in 1998. It was dubbed Q^{12} because it includes 12 items. In the development of the Q^{12}, practical considerations regarding its usefulness for managers in creating change in the workplace have been the leading principles.

A closer look at the item content reveals that the Q^{12} basically assesses the employee's perceived job resources. For instance, items like "In the last six months, has someone at work talked to you about your progress?", and "In the last seven days, have you received recognition or praise for doing good work?" tap performance feedback and social support, respectively. Hence, rather than the employee's actual *experience* of engagement, the Q^{12} assesses the *antecedents* of engagement in terms of perceived job resources (see Figure 10.1).

Nevertheless, a very high correlation of .91 (after controlling for measurement error) was observed between the Q^{12} and a single item measure of job satisfaction (Harter et al., 2002). Moreover, correlations with a composite measure of business unit performance were exactly the same for satisfaction and engagement ($r = .22$; Harter et al., 2002). Taken together this means that Gallup's employee engagement concept is virtually identical with overall job satisfaction. The Q^{12} has excellent internal consistency ($\alpha = .88$), whereas several test–retest reliability studies reveal stability coefficients around .80 (Harter et al., 2002). Finally, a strong association of the Q^{12} with the Utrecht Work Engagement Scale (see below) indicates its convergent validity (Harter & Schmidt, 2008).

The Utrecht Work Engagement Scale (UWES)

Currently, the Utrecht Work Engagement Scale (UWES) is the most widely used engagement questionnaire and it is available in 22 languages (see www.schaufeli.com). It is a three-dimensional questionnaire that includes subscales for vigor, dedication, and absorption (Schaufeli, Salanova et al., 2002). In addition to the original UWES that contains 17 items, a shortened version of nine items is available (Schaufeli, Bakker, & Salanova, 2006) as well as a student version (Schaufeli, Martinez, Marques Pinto, Salanova, & Bakker, 2002).

Factorial Validity. Confirmatory factor analyses convincingly show that the hypothesized three-factor structure of the UWES is superior to the one-factor model that assumes an undifferentiated engagement factor. This has been demonstrated in samples from China, Finland, Italy; Norway, Portugal, Spain, South Africa, Sweden, and The Netherlands. However, it appears that the three dimensions of engagement are very closely related with intercorrelations exceeding .65.

Factorial Invariance. Confirmatory factor analyses in which samples of two or more countries are simultaneously included, showed that the three-factor structure of the UWES is invariant across nations such as Australia, Belgium, Canada, Finland, France, Germany, The Netherlands, Norway, South Africa, and Spain (Schaufeli et al., 2006). More specifically, the three-factor structure of the UWES is similar and does not differ between countries, but the values of the factor loadings and the correlations between the latent factors slightly differ across nations. In a similar vein, Storm and Rothmann (2003) concluded that the equivalence of the UWES is acceptable for White, Black, Coloured, and Indian members of the South African Police Service, no evidence was found for item-bias in these groups.

In addition to cross-national invariance, factorial invariance was also demonstrated between various occupational groups, such as Dutch (Schaufeli & Bakker, 2004) white collar employees and health care professionals; Spanish workers and students (Schaufeli et al., 2002); Finnish health care workers, educators, and white and blue collar workers (Seppälä et al., 2009); and 10 different Norwegian occupational groups, including air traffic controllers, physiotherapists, and journalists (Nerstad, Richardsen, & Martinussen, 2010). Finally, Seppälä et al. (2009) demonstrated that the correlated three-factor structure of the UWES was invariant across a time interval of 3 years.

Internal Consistency. A meta-analysis1of the original and the short versions of the UWES indicated very good internal consistencies for vigor, dedication, and absorption. More particularly, analyses across 33 samples (total N = 19,940) from eight different countries (i.e., Australia, Belgium, Finland, Greece, the Netherlands, Norway, Spain, South Africa, and Sweden) revealed that sample weighted values for Cronbach's α of all three scales of the original and short versions of the UWES exceeds .80. Moreover, Cronbach's α for the composite score exceeds .90. Hence, it can be concluded that the three scales of the UWES as well as the composite questionnaire are sufficiently internally consistent.

Stability. An analysis (see note 1) across five samples from three countries (i.e., Australia, the Netherlands, and Norway; total N = 1,057)

1. Details of the meta-analyses can be obtained from the author of this chapter.

revealed that the mean stability coefficient of the original and short versions of the UWES across a one-year time interval is .65 (ranging from .56 to .75). Seppälä et al. (2009) studied the rank-order stability of the UWES that reflects the degree to which the relative ordering of individuals within a group is maintained over time. They found high rank-order stability coefficients for the three scales of the short version of the UWES across a 3-year time interval, ranging from .82 to .86.

Construct Validity. Recently, Newman, Joseph, and Hulin (2010) presented the results of a meta-analysis of the relationships between work engagement—as measured with the UWES—and three job attitudes: job satisfaction, job involvement, and affective organizational commitment. They found work engagement shares between 21 and 29% of the variance with these three job attitudes, or between 28 and 37% of the variance after correction for reliability. Moreover, Newman et al. (2010) estimated a corrected correlation of .77 between work engagement and a composite measure of satisfaction, involvement, and commitment, dubbed as the "A-factor."

In a somewhat similar vein, Halbesleben (2010) carried out a meta-analysis that included 74 unique samples in order to assess the associations between work engagements and the components of the JD-R model. As expected, job demands are negatively associated with engagement ($-.07 < \rho < -.24$, depending on the dimension, whereby ρ is the correlation corrected for unreliability of measures). Job resources ($.30 < \rho < .35$) and organizational outcomes ($.12 < \rho < .20$) are positively related to work engagement. Although all associations are in the expected direction, their sizes are rather moderate. The results from the meta-analysis of Halbesleben (2010), who used the JD-R model as a heuristic model, agree with various studies that explicitly *tested* the JD-R model. These studies showed that work engagement is associated with (a) *attitude-based outcomes,* such as organizational commitment (Hakanen, Schaufeli, & Ahola, 2008) and turnover intention (Schaufeli & Bakker, 2004); (b) *behavior-based outcomes*, such as low levels of counterproductive work behavior (Balducci, Fraccaroli, & Schaufeli, 2010), in-role and extrarole behavior (Bakker, Demerouti, & Verbeke, 2004), few self-reported medical errors (Prins et al., 2010), supervisor-rated and coworker-rated performance (Halbesleben & Wheeler, 2008), personal initiative (Salanova & Schaufeli, 2008), and sickness absence frequency (Schaufeli et al.,2009); and (c) *business-based outcomes*, such as work-unit innovativeness (Hakanen, Perhomein, & Toppinen-Tanner,

2008), customer satisfaction (Salanova, Agut, & Peiró, 2005), and financial turnover (Xanthopoulou, Bakker, Demerouti, & Schaufeli, 2009).

In conclusion: (a) work engagement as assessed by the UWES seems to be a unitary construct that is constituted by three different yet closely related aspects (vigor, dedication, and absorption); (b) these three factors seems to be invariant across nations and occupational groups; (c) work engagement is relatively stable across time; (d) work engagement is rather strongly related, but is not identical to job satisfaction, job involvement, or affective organizational commitment; (e) work engagement is associated with job demands, resources, and outcomes in ways as predicted by the JD-R model.

Questionnaires with Limited Application

Based on Kahn's (1990) conceptualization of engagement May, Gilson, and Harter (2004) developed a 13-item scale that consists of three dimensions: cognitive, emotional, and physical engagement. The items of these three scales show a striking resemblance with those of the absorption, dedication, and vigor scales of the UWES, respectively. However, instead of three factors only one factor emerged, albeit that the total scale is sufficiently internally consistent ($\alpha = .77$).

Also basing herself on the work of Kahn (1990), Rothbard (2001) distinguished two separate but related components of role engagement: attention (4 items; $\alpha = .74$) and absorption (5 items; $\alpha = .65$). Attention refers to cognitive availability and the amount of time one spends thinking about one's work role, whereas absorption means being engrossed in one's work role. Although both aspects of engagement are moderately correlated ($r = .56$) they seem to play a different role in the dynamics of engagement in work and family roles.

Saks (2006) distinguished between job engagement (5 items; $\alpha = .82$) and organizational engagement (6 items; $\alpha = .90$) that are described as the employee's: "... psychological presence in their job and their organization" (p. 608), respectively. Both aspects of engagement are moderately related with each other ($r = .62$) and show different patterns of relationships with antecedents and outcomes, thus suggesting conceptual distinctness.

Recently, Hultell, and Gustavsson (2010) developed the 18-item Scale of Work Engagement and Burnout (SWEBO) that includes three engagement dimensions (vigor, dedication, and absorption) and three burnout dimensions (exhaustion, disengagement, and inattentiveness) of three items each. The SWEBO assesses work engagement and burnout by using mood adjectives (e.g. "active," "inspired," "immersed"). The engagement scales are moderately and less strongly interrelated ($.28 < r < .57$) as compared to the burnout scales ($.61 < r < .69$). As expected, the engagement and burnout scales are negatively related ($-.10 < r < -.58$). Except for absorption ($\alpha = .58$), all other scales have internal consistencies that exceed .75.

Obviously, all four operationalizations above agree that engagement is a multidimensional construct and that it includes absorption as its common denominator.

How Is Work Engagement Related to Workaholism and Job Burnout?

Evidently, both engaged workers as well as workaholics work hard. Despite this similarity, confirmatory factor analysis showed that engagement and workaholism (operationalized by working excessively and working compulsively) can indeed be assessed as two distinct constructs (Schaufeli et al., 2008), albeit that the absorption scale of the UWES has a weak double loading on the latent workaholism factor. This might indicate that absorption could also entail an obsessive tendency to work that is characteristic for workaholism. Moreover, Schaufeli et al. (2008) showed that work engagement and workaholism are related to different variables: both types of employees work hard and are loyal to the organization they work for, but in the case of workaholism this comes at the expense of poor mental health and few and unrewarding social contacts outside work, whereas engaged workers feel quite well, both mentally and socially. In a similar vein, Andreassen, Ursin, and Eriksen (2007) found that work engagement is predicted by enjoyment but not by drive, being the more typical workaholism component. Finally, Van Beek, Hu, Schaufeli, Taris, and Schreurs (2012) showed that the motivational regulation of work engagement and workaholism differs. The former is characterized by intrinsic motivation, whereas the latter is characterized by extrinsic (i.e., introjected and identified) regulation.

Because work engagement is supposed to be the positive antithesis of burnout, we also have to consider the way burnout is conceptualized and measured. Basically, two approaches exist that consider burnout to be a one-dimensional or a multidimensional concept (Maslach, Leiter, & Schaufeli, 2008). The one-dimensional conceptualization of burnout identifies exhaustion as the sole defining criterion, which is also described as wearing out, loss of energy, depletion, debilitation, or fatigue. Although distinctions between various aspects of exhaustion are made (e.g., physical fatigue, emotional exhaustion, and cognitive weariness; Shirom & Melamed, 2005), the corresponding measures like the Shirom-Melamed Burnout Measure (SMBM) inevitably produces a single overriding factor of exhaustion.

Based on a theoretical analysis, Schaufeli and Taris (2005) argue in favor of a two-dimensional model of burnout that is characterized by exhaustion and withdrawal. In their view, burnout is both the inability and the unwillingness to spend effort, reflecting its energetic and its motivational component, respectively. The unwillingness to perform manifests itself by increased resistance, reduced commitment, lack of interest, disengagement, mental distancing, cynicism, and so on—in short, psychological withdrawal. This withdrawal reaction serves as a protective mechanism against exhaustion, it prevents the employee from spending additional energy. The Oldenburg Burnout Inventory (OLBI) uses this conceptualization of burnout and includes two dimensions that are labeled "disengagement" and "exhaustion" (Demerouti, Bakker, Vardakou, & Kantas, 2002). Since these two dimensions also include positively framed items that refer to "dedication" and "vigor," respectively, the OLBI might be used to assess work engagement as well; namely by reversing the negatively worded items. A recent study found that the disengagement (burnout) and dedication (engagement) items of the OLBI constitute one "identification" dimension, whereas the exhaustion (burnout) and vigor (engagement) items represent two separate but highly related constructs (Demerouti, Mostert, & Bakker, 2010). The MBI, which is the most frequently used questionnaire to assess burnout, includes in addition to exhaustion and cynicism a third dimension: reduced professional efficacy (Maslach et al., 1996). Tellingly, a recent study on the convergent validity among various burnout instruments (e.g., SMBM, OLBI, and MBI) concludes that burnout is best conceived as a two-dimensional construct consisting of exhaustion and withdrawal, which are two related but conceptually distinct aspects (Hu & Schaufeli, 2011).

In accordance with the assumption that work engagement is the positive counterpart of burnout, the three dimensions of the UWES are negatively related to burnout. Studies using confirmatory factor-analyses revealed that the correlations between the latent MBI-burnout and engagement factors ranged from –.45 to –.66 (e.g., Schaufeli, Salanova et al., 2002; Schaufeli, Bakker, & Salanova, 2006; Schaufeli, Taris, & Van Rhenen, 2008). However, instead of loading on the burnout factor, reduced professional efficacy loads on the engagement factor. A possible explanation for this "wrong" loading is that lack of professional efficacy is measured with reversed positively formulated items. This explanation is supported by a study by Schaufeli and Salanova (2007) who showed that an *in*efficacy scale, consisting of negatively reworded MBI-efficacy items, loads on burnout, whereas the original positively worded MBI-efficacy scale loads on engagement.

In conclusion, although some overlap seems to exist with workaholism (notably absorption) this does not seriously call into question the conceptual distinctness of work engagement. The bottom line is that engaged workers are *pulled* to their work because for them work is fun, whereas workaholics are *pushed* to their work by an uncontrollable, compulsive inner drive that they cannot resist. Furthermore, as expected, engagement is negatively related with burnout, whereby the unexpected results regarding professional efficacy are likely to result from an artifact caused by the reversing positively phrased items. Therefore it is recommended for future research to use negatively worded inefficacy items to assess burnout (Schaufeli & Salanova, 2007).

What about the Practical Use of Engagement Questionnaires?

Based on the conceptual model that is displayed in Figure 10.1, as well as on its excellent psychometric features the UWES seems to be the most promising tool to assess work engagement, both in academia as well as in business, not the least because it is linked to meaningful organizational and business outcomes. The UWES may also be used in different national contexts. However, one should be cautious when comparing levels of work engagement *between* countries, particularly when Asian countries such as Japan are concerned. It is observed that Japanese employees score consistently lower on work engagement than employees from Western countries, most likely because of the prevailing tendency in Japan to suppress the expression of positive affect (Shimazu, Miyanaka, & Schaufeli, 2010). This underscores the necessity for establishing nation-specific norm-scores for the UWES that are based on representative samples. Such cut-off values are available in the Netherlands and are used as reference points for identifying employees who score "(very) low," "average," or "(very) high" on work engagement as compared with the national working population.

In Spain as well as in the Netherlands, the UWES (and the MBI) are integrated into a more comprehensive, online tool that is based on the job demands-resources model and that is also used commercially. This tool not only assesses the employee's levels of work engagement (and burnout), but also its drivers and consequences. Participants receive an individualized personal report with their level of engagement (and burnout) as well their scores on various job demands, job resources, personal resources, and organizational outcomes—as compared to a reference group. This report may be discussed with colleagues, supervisors, or professionals such as occupational physicians or psychologists. By aggregating the information of individual employees to the level of work teams, departments, or the organization as a whole, specific suggestions can be made as to how to enhance work engagement (and decrease burnout) and thus improve organizational outcomes (see Schaufeli & Salanova, 2010, for further details).

Conclusion

Work engagement is a popular concept, in business and academia alike. Although some alternatives of limited application exist, the UWES emerges as a valid and reliable indicator of work engagement, which can be considered the positive counterpart of job burnout. Not the least because engagement, as assessed with the UWES, is related to attitude-based, behavior-based, and business-based outcomes. Moreover, engagement is associated with, but nevertheless conceptually and empirically different from, job satisfaction, work involvement, organizational commitment, and workaholism. The UWES can be used as a screening instrument in organizations to identify those employees whose levels of engagement may be boosted.

References

Andreassen, C. S., Ursin, H., & Eriksen, H. R. (2007). The relationship between strong motivation to work, "workaholism," and health. *Psychology and Health, 22*, 615–629.

Attridge, M. (2009). Measuring and managing employee work engagement: A review of the research and business literature. *Journal of Workplace Behavioral Health, 24*, 383–398.

Bakker, A. B., & Demerouti, E. (2007). The job demands-resources model: State of the art. *Journal of Managerial Psychology, 22*, 309–328.

Bakker, A. B., & Demerouti, E. (2008). Toward a model of work engagement. *Career Development International, 13*, 209–223.

Bakker, A. B., Demerouti, E., & Verbeke, W. (2004). Using the job demands-resources model to predict burnout and performance. *Human Resource Management, 43*, 83–104.

Balducci, C., Fraccaroli, F., & Schaufeli, W. B. (2010). Psychometric properties of the Italian version of the Utrecht Work Engagement Scale (UWES-9): A cross-cultural analysis. *European Journal of Psychological Assessment, 26*, 143–149.

Buckingham, M., & Coffman, C. (1999). *First, break all the rules: What the world's greatest managers do differently.* New York: Simon & Schuster.

Demerouti, E., Bakker, A. B., Vardakou, I., & Kantas, A. (2002). The convergent validity of two burnout instruments: A multi-trait multi-method analysis. *European Journal of Personality Assessment, 18*, 296–307.

Demerouti, E., Mostert, K., & Bakker, A. B. (2010). Burnout and work engagement: A thorough investigation of the independence of both constructs. *Journal of Occupational Health Psychology, 15*, 209–222.

González-Roma, V., Schaufeli, W. B., Bakker, A. B., & Lloret, S. (2006). Burnout and engagement: Independent factors or opposite poles? *Journal of Vocational Behavior, 68*, 165–174.

Hakanen, J. J., Perhomeini, L., & Toppinen-Tanner, S. (2008). Positive gain spirals at work: From job resources to work engagement, personal initiative and work-unit innovativeness. *Journal of Vocational Behavior, 73*, 78–91.

Hakanen, J., Schaufeli, W. B., & Ahola, K. (2008). The job demands-resources model: A three-year cross-lagged study of burnout, depression, commitment, and work engagement. *Work & Stress, 22*, 224–241.

Halbesleben, J. R. B. (2010). A meta-analysis of work engagement: Relationships with burnout, demands, resources, and consequences. In A. B. Bakker & M. P. Leiter (Eds.), *Work engagement: A handbook of essential theory and research* (pp. 102–117). New York: Psychology Press.

Halbesleben, J. R. B., & Wheeler, A. R. (2008). The relative roles of engagement and embeddedness in predicting job performance and intention to leave. *Work & Stress, 22*, 242–256.

Hallberg, U., & Schaufeli, W. B. (2006). "Same same" but different: Can work engagement be discriminated from job involvement and organizational commitment? *European Journal of Psychology, 11*, 119–127.

Harter, J. K., & Schmidt, F. L. (2008). Conceptual versus empirical distinctions among constructs: Implications for discriminant validity. *Industrial and Organizational Psychology, 1*, 27–40.

Harter, J. K., Schmidt, F. L., & Hayes, T. L. (2002). Business-unit-level relationships between employee satisfaction, employee engagement, and business outcomes: A meta-analysis. *Journal of Applied Psychology, 87*, 268–279.

Hu, Q., & Schaufeli, W. B. (2011). The convergent validity of four burnout measures in a Chinese sample: A confirmative factor-analytic approach. *Applied Psychology: An International Review, 60*, 87–111.

Hultell, D., & Gustavsson, J. P. (2010). A psychometric evaluation of the Scale of Work Engagement and Burnout (SWEBO). *Work, 37*, 261–274.

Kahn, W. A. (1990). Psychological conditions of personal engagement and disengagement at work. *Academy of Management Journal, 33*, 692–724.

Langelaan, S., Bakker, A. B., Schaufeli, W. B., Van Rhenen, W., & Van Doornen, L. J. P. (2006). Do burned-out and engaged employees differ in HPA-axis functioning? *Scandinavian Journal of Work Environment and Health, 32*, 339–348.

Macey, W. H., & Schneider, B. (2008). The meaning of employee engagement. *Industrial and Organizational Psychology, 1*, 3–30.

Maslach, C., Jackson, S. E., & Leiter, M. P. (1996). *The Maslach burnout inventory* (3rd ed.). Palo Alto, CA: Consulting Psychologists Press.

Maslach, C., & Leiter, M. P. (1997). *The truth about burnout*. San Francisco, CA: Jossey-Bass.

Maslach, C., Leiter, M. P., & Schaufeli, W. B. (2008). Measuring burnout. In C. L. Cooper & S. Cartwright (Eds.), *The Oxford handbook of organizational well-being* (pp. 86–108). Oxford, England: Oxford University Press.

Maslach, C., Schaufeli, W. B., & Leiter, M. P. (2001). Job burnout. *Annual Review of Psychology, 52*, 397–422.

May, D. R., Gilson, R. L., & Harter, L. M. (2004). The psychological conditions of meaningfulness, safety and availability and the engagement of the human spirit at work. *Journal of Occupational and Organizational Psychology, 77*, 11–37.

Nerstad, C. G. L., Richardsen, A. M., & Martinussen, M. (2010). Factorial validity of the Utrecht Work Engagement Scale (UWES) cross occupational groups in Norway. *Scandinavian Journal of Psychology, 51*, 326–333.

Newman, D. A., Joseph, D. L., & Hulin, C. J. (2010). Job attitudes and employee engagement: Considering the attitude "A-factor". In S. Albrecht (Ed.), *The handbook of employee engagement: Perspectives, issues, research, and practices* (pp. 43–61). Northampton, MA: Edward Elgar.

Prins, J. T., Van der Heijden, F. M., Hoekstra-Weebers, J. E. Bakker, A. B., Jacobs, B., & Gazendam-Donofrio, S. M. (2010). Burnout, engagement and resident physicians' self-reported errors. *Psychology, Health & Medicine, 6*, 654–667.

Rothbard, N. P. (2001). "Enriching or depleting?" The dynamics of engagement in work and family roles. *Administrative Science Quarterly, 46*, 655–684.

Saks, A. M. (2006). Antecedents and consequences of employee engagement. *Journal of Managerial Psychology, 21*, 600–619.

Salanova, M., Agut, S., & Peiró, J. M. (2005). Linking organizational resources and work engagement to employee performance and customer loyalty: The mediation of service climate. *Journal of Applied Psychology, 90*, 1217–1227.

Salanova, M., & Schaufeli, W. B. (2008). A cross-national study of work engagement as a mediator between job resources and proactive behavior: A cross-national study. *International Journal of Human Resources Management, 19*, 226–231.

Schaufeli, W. B., & Bakker, A. B. (2004). Job demands, job resources and their relationship with burnout and engagement: A multi-sample study. *Journal of Organizational Behavior, 25*, 293–315.

Schaufeli, W. B., & Bakker, A. B. (2010). The conceptualization and measurement of work engagement. In A. B. Bakker & M. P. Leiter (Eds.), *Work engagement: A handbook of essential theory and research* (pp. 10–24). New York: Psychology Press.

Schaufeli, W. B., Bakker, A. B., & Salanova, M. (2006). The measurement of work engagement with a short questionnaire: A cross-national study. *Educational and Psychological Measurement, 66*, 701–716.

Schaufeli, W. B., Bakker, A. B., & Van Rhenen, W. (2009). How changes in job demands and resources predict burnout, work engagement, and sickness absenteeism. *Journal of Organizational Behavior, 30*, 893–917.

Schaufeli, W. B., Martínez, I., Marques Pinto, A., Salanova, M., & Bakker, A. B. (2002). Burnout and engagement in university students: A cross national study. *Journal of Cross- Cultural Psychology, 33*, 464–481.

Schaufeli, W. B., & Salanova, M. (2007). Efficacy or inefficacy, that's the question: Burnout and work engagement, and their relationship with efficacy beliefs. *Anxiety, Stress & Coping, 20*, 177–196.

Schaufeli, W. B., & Salanova, M. (2010). How to improve work engagement? In S. Albrecht (Ed.), *The handbook of employee engagement: Perspectives, issues, research, and practices* (pp. 399–415). Northampton, MA: Edward Elgar.

Schaufeli, W. B., Salanova, M., González-Romá, V., & Bakker, A. B. (2002). The measurement of engagement and burnout: A confirmative analytic approach. *Journal of Happiness Studies, 3*, 71–92.

Schaufeli, W. B., & Taris, T. W. (2005). The conceptualization and measurement of burnout: Common ground and worlds apart. *Work & Stress, 19*, 356–362.

Schaufeli, W. B., Taris, T. W., & Van Rhenen, W. (2008). Workaholism, burnout and engagement: Three of a kind or three different kinds of employee well-being? *Applied Psychology: An International Review, 57*, 173–203.

Seppälä, P., Mauno, S., Feldt, T., Hakanen, J., Kinnunen, U., Tolvanen, A., & Schaufeli, W. B. (2009). The construct validity of the Utrecht Work Engagement Scale: Multisample and longitudinal evidence. *Journal of Happiness Studies, 10*, 459–481.

Shimazu, A., Miyanaka, D., & Schaufeli, W.B. (2010). Work engagement from a cultural perspective. In S. Albrecht (Ed.), *The handbook of employee engagement: Perspectives, issues, research, and practices* (pp. 364–373). Northampton, MA: Edward Elgar.

Shirom, A., & Melamed, S. (2005). Does burnout affect physical health? A review of the evidence. In A. S. G. Antoniou & C. L. Cooper (Eds.), *Research companion to organizational health psychology* (pp. 599–622). Cheltenham, UK: Edward Elgar.

Storm, K., & Rothmann, I. (2003). A psychometric analysis of the Utrecht Work Engagement Scale in the South African police service. *South African Journal of Industrial Psychology, 29*, 62–70.

Van Beek, I., Hu, Q., Schaufeli, W. B., Taris. T., & Schreurs, B. H. (2010). For fun, love or money. What drives workaholic, engaged and burned-out employees at work? *Applied Psychology: An International Review, 61*, 30–55.

Xanthopoulou, D., Bakker, A. B., Demerouti, E., & Schaufeli, W. B. (2009). Work engagement and financial returns: A diary study on the role of job and personal resources. *Journal of Occupational and Organizational Psychology, 82*, 183–200.

Part II

Design and Analysis

11

Cognitive Assessment

Implications for Occupational Health Psychology

Gerald Matthews, April Rose Panganiban, and Kirby Gilliland

There are several reasons why occupational health psychologists should be concerned with the measurement of cognitive functioning. First, various factors including acute stress, fatigue, boredom, drug use, and illness may adversely influence information processing and performance. Such impairments may in turn lead to lost productivity, and an elevated risk of errors and accidents (Matthews, Davies, Westerman, & Stammers, 2000). Indeed, the operator's awareness of loss of competence may add to stress, with potentially severe consequences for safety-critical work. Performance assessments may indicate whether the person is fit for work, and what tasks he or she may adequately perform. Second, large numbers of individuals with chronic disabilities are available for work, and assessments of cognitive functioning are requisite for career guidance. The severity of both neurological conditions (e.g., Parkinson's disease) and disorders treated as psychological (e.g., mild depression) may vary from day to day, requiring further monitoring of acute status. The aging of the workforce also raises some concern for declines in cognitive functioning in some elderly employees. Third, tasks performed at work may be intrinsically stressful; recent research implicates both underload and overload as sources of job stress (Schultz, Wang, & Olsen, 2010). Countermeasures for job stress may require understanding of the cognitive demands of specific work tasks and activities.

Thus, several subdomains of basic psychology are relevant to cognitive assessments, including experimental studies of human performance, clinical neuropsychology, and the scientific traditions that underpin basic cognitive psychology and clinical/neuropsychological assessment, respectively. Even a cursory examination of basic texts in cognitive/neuroscience (e.g., Kandel, Schwartz, & Jessell, 2000) and clinical neurology/neuropsychology (Heilman & Valenstein, 2003) reveals that both traditions focus on such basic functions as attention, concentration, memory, visuomotor skill, spatial ability, decision making, language, computational ability, and higher-order executive functioning. This commonality suggests that the basic cognitive processes essential for optimal human performance and for optimal psychological health are either intimately related or one and the same. Not surprisingly, these are much the same processes that are targeted in many basic mental status-type examinations and cognitive performance assessment systems (Kabat, Kane, Jefferson, & DiPino, 2001; Kane & Kay, 1992; Schatz & Browndyke, 2002). It also illustrates the value in drawing ever closer the current discoveries in the fields of cognition and human performance to the evolution and practice of occupational health psychology.

Job performance may be supported by a wide range of cognitive functions. The general principle that task demands are critical for both performance and health is well-accepted (Hurrell & Murphy, 1992). Karasek's (1979) influential model of work stress identified mental workload, together with level of personal control, as a critical factor. However, as Schultz et al. (2010) point out, tests of the Karasek model have tended to investigate broad job demand factors rather than mental workload directly derived from task performance. In addition, measures of job performance such as supervisor evaluations or even self-ratings are often only indirectly related to cognitive functioning. Thus, it becomes important to investigate objective measurements of specific cognitive processes. Indeed, specialized neuropsychological testing for injury, illness, or disability, is often informed by cognitive models.

In this chapter, we will survey the range of assessments of cognitive function available to the occupational health psychologist with the appropriate expertise. We will cover three types of assessment: measurement of cognitive processes using performance tests, measurement of task-induced stress and fatigue, and use of psychophysiological measures of cognitive function. We conclude by summarizing some general issues in assessment and areas of application.

Assessment of Cognitive Performance

As previously mentioned, cognitive psychology assumes that task performance reflects a multitude of individual component processes supporting major functions such as perception, attention, working memory, and decision making. Experimental techniques for investigating such processes have become increasingly sophisticated, and understanding of the neural systems that support information processing is also advancing rapidly (Fan, McCandliss, Sommer, Raz, & Posner, 2002). Thus, there is no shortage of cognitive performance measures with good construct validity. As we will discuss, comprehensive performance-based assessments may be conducted using systematically developed batteries comprised of multiple performance tasks.

However, before proceeding further, some cautionary comments are necessary. First, debate continues over the roles of basic information-processing and learned skills in generating performance variation (Ericsson & Ward, 2007). On the one hand, key processing components such as working memory capacity may limit the performance of a wide range of complex skills (Ackerman, Beier, & Boyle, 2005). On the other hand, operators typically acquire task-specific expertise that may protect performance against variation in basic processing efficiency. Expert performance may attain autonomy of this kind especially when it relies on a fixed set of stimulus-response mappings that come to be processed automatically, without the need for conscious attention (Matthews, Davies et al., 2000). The technique of cognitive task analysis (CTA: Roth, 2008) may be useful for occupational health psychologists in determining the critical processing components for a given job skill.

Second, performance measures have both trait and state aspects. That is, variation across individuals depends in part on stable abilities that include not only general cognitive ability or intelligence, but a host of specific abilities also, such as verbal, spatial, and math abilities (Carroll, 1993). In addition, while trait factors may define a typical level of performance for the individual, the influence of stress, fatigue, and other temporary influences will lead to transient state variations around the long-term mean (Matthews & Campbell, 2010). Thus, a focus on chronic functioning (e.g., neurological conditions) suggests a need to assess trait-level functioning, whereas a focus on acute impairment due to stress or illness of short duration calls for state-level functioning. It will generally be difficult to separate trait and state influences on a single performance assessment, and a strategy of repeated performance testing is often desirable. For example, an assessment taken when the person is in good health may be used as a reference point for detecting subsequent impairments due to stress, illness, or injury.

Third, the landscape of cognitive assessment is sculpted by the definitions of cognitive domains, and there is not always a consensus for those definitions. For example, there are multiple models of "executive function" (Matthews, Gruszka, & Szymura, 2010), and drawing distinctions within the realm of broader complex decision making is challenging at best. Practitioners should strive to make explicit the specific cognitive models that underpin their assessment strategies.

Working Memory, Attention, and Executive Control

It is beyond our present scope to review all the tasks that might potentially be used in assessment of cognitive function. As examples, we will focus on tests of working memory, attention, and executive control. This family of cognitive constructs is important for several reasons (Matthews, Gruszka, & Szymura, 2010). Deficits in concentration and in controlling mental activities are likely to limit numerous operational tasks (Matthews, Davies, et al., 2000). Evidence from longitudinal studies allows us to separate trait and state factors in working memory psychometrically (Matthews & Campbell, 2010). These functions may also be localized through cognitive neuroscience studies in areas of the frontal lobe including prefrontal cortex and anterior cingulate. Frontal areas also seem to regulate interactions between emotion and cognition, including those contributing to decision making and emotion regulation (Ochsner & Gross, 2008).

The challenge for research has been to untangle three overlapping but potentially distinct constructs. The first is working memory, defined as a system for performing complex cognitive operations such as comprehension and reasoning, while maintaining relevant information in short-term storage. Typical tasks require the person to keep strings of words in memory whilst also reading text for comprehension or to perform math and word recall tasks simultaneously (Ilkowska & Engle, 2010). Modern techniques such as item-response theory have been used to enhance the psychometric qualities of working memory tests (e.g., Beckmann, Holling, & Kuhn 2007). Such tasks predict a range of criteria, including various aspects of skilled performance and general cognitive intelligence (Ackerman et al., 2005; Ilkowska & Engle, 2010).

A second construct is the person's overall attentional capacity or resources, defined as a general pool of processing capabilities that must be allocated in performance of a range of mentally demanding tasks

(Matthews, Davies, et al., 2000). Techniques for measuring capacity include various dual-task paradigms that will tend to overload capacity, although these are fraught with methodological difficulties (Matthews, Davies, et al., 2000; Pashler, 1998). Measures of sustained attention or vigilance may also index resource availability (Matthews, Warm, et al., 2010). The Continuous Performance Test (Egeland & Kovalik-Gran, 2010) is widely used in neuropsychological assessments of attention. One version requires detection of a zero (the target stimulus) each time this digit appears in a rapidly presented succession of single digits. More demanding tasks may be required for measurement of sustained attention in normal individuals.

A third construct, executive control, comes originally from neuropsychological observations of patients with frontal damage. Tasks such as the Wisconsin card-sorting task reveal relatively subtle impairments that suggest difficulties in keeping track of changing task requirements (Miyake, Friedman, Emerson, Witzki, & Howerter, 2000). More recent work has focused on executive control of attention in normal individuals without neurological deficit. For example, tasks that generate conflict between competing responses may provide indices of the executive control necessary for resolving such conflicts. A classic example is the Stroop task, which generates conflicting color-naming responses generated by the ink color of the word stimulus, and its lexical content. Another popular task is based on the Eriksen effect (Eriksen & Eriksen, 1974). Participants responding to a focal target (e.g., a left-pointing arrow) must suppress competing responses associated with irrelevant flanking stimuli (e.g., a right-pointing arrow). Both Stroop and Eriksen tasks provide quantitative indices of executive control based on reaction times. The Eriksen task provides the basis for a test that assesses three fundamental neural networks for attention, discriminating spatial orienting, alerting, and executive control of attention (Fan et al., 2002).

Psychometric models help to define more sharply potentially vague constructs such as "attentional resources." The performance tests allied to each model may also find applications in occupational settings. For example, the Miyake et al. (2000) model might be used to assess the worker's general distractibility (poor inhibition), capacity for flexible redirection of attention (shifting), and capacity of working memory for keeping track of a train of thought (updating). However, while such approaches are promising, more research on transitioning laboratory findings into workplace task environments is needed. Furthermore, executive functioning depends on both trait and state factors (Ilkowska

& Engle, 2010), suggesting the need for repeated assessments in practical settings.

Test Batteries for Cognitive Assessment

One of the limitations of much cognitive psychology is the reliance of researchers on bespoke tasks that are frequently modified to address immediate research goals. It is only fairly recently that standardized psychometric instruments (e.g., Fan et al., 2002) have started to emerge from the theory-driven research of the field. By contrast, the development of comprehensive test batteries for practical use developed in parallel with basic cognitive psychology and with personal computer technology (Schatz & Browndyke, 2002). This area of test development was, of course, often influenced by theoretical developments, but maintained a practical focus. The 1970s saw the rapid introduction of many affordable personal computers and with them came near immediate attempts to transfer electromechanical and pencil-and-paper tests to this new assessment medium (Bartram & Bayliss, 1984). Attempts to implement standard clinical tests of cognitive functioning were among the first technology translations, as well as many human factors or research-based tests of cognitive function (for reviews of test battery development, see Bartram & Bayliss, 1984; Kane & Kay, 1992; Schatz & Browndyke, 2002).

Automated cognitive testing has a number of advantages including: (a) uniform presentation of test stimuli, (b) enhanced capability for presenting complex stimuli, (c) rapid and accurate administration/scoring, (d) accuracy in data recording/storing, (e) objective assessment, (f) economical use, especially in high volume applications, (g) automated data management, and (h) in some cases, a reduction in staff resources. Along with these advantages are challenges as well: (a) computer experience level alone can influence test scores, (b) self-administration can lead to confusion and inaccurate assessment, (c) initial investment can be expensive, (d) mass testing can be limited by equipment costs, (e) effective use of hardware and software can require investment in time and practice, (f) continued updating of technology can be a hidden cost, and (g) automated versions of face-to-face or pencil-and-paper tests may not have comparable response characteristics or normative values.

Nonetheless, the era of computer-based testing is rapidly expanding. There is an increasing number of well-known, traditional, psychological tests of cognition, typically clinical in nature, that have been converted to computer-based form and distributed by private or commercial test providers. There are also specialized batteries of cognitive tests that serve to assess specific research or clinical needs; for example, batteries for concussion management in sports (Echemendia, 2006), posttraumatic stress (Vasterling et al., 2006), and underlying cognitive processes associated with flying (Kay & Spector, 1991).

There are also broadly constructed computer-based test systems that have been designed for multiple uses. The best among these are grounded in theoretical models and have preserved a comparative pathway to past bodies of clinical and experimental research literature. An example is the line of test batteries developed by the U.S. Department of Defense (DoD) over a period of three decades. This effort began in the early 1980s to improve drug testing, general human performance assessment, and clinical applications (Reeves, Winter, Bleiberg, & Kane, 2007). What began as several independent test batteries eventually formed the present day Automated Neuropsychological Assessment Metrics (ANAM; Englund et al., 1987; Reeves et al., 1992). ANAM is thus unique in that it rests on decades of test development work, it was built historically on a synthesis of theoretical models of information processing, it was designed to meet international standards for test battery development, and it incorporates a wide range of tests built on volumes of human performance research and clinical test construction/validation. ANAM and other batteries may also be constructed to

Table 11.1 Features of the Automated Neuropsychological Assessment Metrics (ANAM) Test Battery

- The ANAM test library provides a selection of approximately 30 tests diversified enough to permit a wide assessment range from basic tests of very fundamental processes (e.g., simple response time) to more complex tests of cognitive or neuropsychological function (e.g., executive function).
- Tests are easily selected from the ANAM library to form smaller batteries for specialized research and clinical applications.
- ANAM test stimuli can be varied session-to-session to provide an almost infinite number of repeated-measures testing sessions.
- Millisecond timing accuracy and large capacity data collection are provided on common contemporary hardware/software operating systems.
- ANAM tests are used by governmental agencies (e.g., DoD, FAA, NASA), as well as numerous businesses (e.g., major government contractors, drug companies, etc.) thereby adding to the versatility, comparability, and ability to generalize from ANAM test results.

measure both individual domains of cognitive functioning (e.g., attention, concentration, response speed, memory, etc.) and specific cognitive processes contributing to test performance.

As noted previously, test systems that provide libraries of tests can be configured for optimal use in assessing a wide variety of research and clinical applications. For example, ANAM is used by NASA for the medical assessment of cognition during spaceflight, by the FAA for studies of basic human performance during aviation, and in studies of risk factors such as hypoxia. Other test batteries were used for studies of "readiness to perform" (Gilliland & Schlegel, 1995), and by researchers studying Alzheimer's disease, multiple sclerosis, Parkinson's disease, and lupus. In what is now perhaps the largest automated cognitive assessment program ever attempted, the DoD has collected baseline cognitive assessments for over 750,000 service members, using the sports concussion management model of automated testing to enhance assessment of cognition often influenced by traumatic brain injuries in the recent war with Iraq. Finally, research projects using common ANAM cognitive metrics provide the unique opportunity for comparisons across research areas.

In closing this section on the behavioral assessment of cognition, it may be useful for occupational health psychologists to explore the growing use of automated cognitive tests for what is variously referred to as "effort testing" or "feigned cognitive impairment" (Boone, 2007; Larrabee, 2007). Individuals who attempt to feign or exaggerate dysfunction will often perform poorly on tests of cognitive function and many times much worse than individuals with actual injuries or disabilities. While clinical evaluations of this nature may be referred to clinical neuropsychologists, occupational health psychologists may be rewarded by having a functional understanding of the principles and methods of such assessments. Additionally, there may be emerging areas where occupational health psychologists may find similar testing useful, such as the validation of minimal performance compliance in the workplace and the verification of effortful performance in research investigations of occupational performance.

Transient Stress and Fatigue Factors

Performance measures may provide an indirect assessment of stress and fatigue, such as a decrement in sustained performance. However, direct measures may be preferable. Traditionally, the two types of measures used have been self-report of symptoms, and psychophysiological indices of autonomic and cortical arousal, which we will survey in the next two sections. Early studies of stress and fatigue states focused on their expression as moods. Thayer's (1989) Activation-Deactivation Adjective Checklist (AD-ACL), the Positive and Negative Affect Scale (PANAS: Watson & Clark, 1997), and the UWIST Mood Adjective Checklist (UMACL: Matthews, Jones, & Chamberlain, 1990) all provide ready assessment of 2 to 4 fundamental mood dimensions. The trait and state scales for anxiety, depression, and anger developed by Spielberger are also widely used (e.g., Spielberger & Reheiser, 2004).

As with performance assessments, it remains essential to distinguish trait and state factors. Measures designed specifically for occupational settings, such as scales for burnout (Worley et al., 2008), have tended to focus more on chronic stress than on the person's immediate experiences during performance. Theoretical analyses of stress and emotion (e.g., Lazarus, 1999) suggest chronic stress measures may aggregate numerous experiences, and are subject to the vagaries of recall of emotional state. Measures of chronic stress and burnout certainly have occupational relevance (e.g., Olsson, Roth, & Melin, 2010), but measures of acute state may be more directly informative about cognitive functioning on a specific performance test. We will illustrate the potential use of subjective state measures by briefly surveying recent work that aimed to develop a comprehensive stress state assessment for use in a variety of basic and applied settings (Matthews, Campbell, et al., 2002).

Assessment in the work performance context may need a scope broader than mood measurement. Stress may also be experienced as disturbances of cognition (e.g., worry) and of motivation (e.g., loss of task interest). Different forms of stress may differentially impact cognitive functioning. Extensive anxiety research suggests that worry may typically be more detrimental to performance than anxious emotion per se (Zeidner & Matthews, 2005).

A Multivariate Model For Stress and Fatigue States

Matthews, Campbell et al. (2002) developed the Dundee Stress State Questionnaire (DSSQ) to provide comprehensive measurement of the subjective states experienced in performance settings. The research literature on stress, fatigue, and performance was used as the basis for sampling state constructs in the domains of affect (e.g., mood), cognition, and motivation. Factor analysis of item responses identified 11 primary factors relating to one or other domain. A second-order factor analysis of the correlated primary scales found three broader factors, labeled task engagement, distress, and worry (see Table 11.2). Mental fatigue may correspond to the lower end of the task engagement dimension; that is, tiredness, loss of motivation, and distractibility. Distress and worry loosely correspond to affective and cognitive components of anxiety and other negative emotions.

The DSSQ was validated on the basis that the relationship between task performance and subjective state is bidirectional (Matthews, 2001). That is, task demands influence subjective state, a process in which the appraisal and coping processes described by the transactional model of stress play a central role (Lazarus, 1999). Several studies (e.g.,

Table 11.2 Correspondences between Second-Order Factors and First-Order Scales on the Dundee Stress State Questionnaire

Factor	*Scale*	*Items*	*Example item*	Scale α
Task engagement	Energetic arousal	8	I feel… Vigorous	80
	Task Interest	7	The content of the task is interesting	75
	Success Motivation	7	I want to perform better than most people do	87
	Concentration	7	My mind is wandering a great deal (-ve)	85
Distress	Tension	8	I feel … Nervous	82
	Hedonic Tone (low)	8	I feel … Contented	86
	Confidence-control (low)	6	I feel confident about my abilities	80
Worry	Self-focus	8	I am reflecting about myself	85
	Self-esteem	7	I am worrying about looking foolish (-ve)	87
	CI (task-relevant)	8	I have thoughts of … How much time I have left	78
	CI (task-irrelevant)	8	I have thoughts of … Personal worries	86

Note. CI = Cognitive Interference.

Matthews & Falconer, 2000; Shaw et al., 2010) have confirmed that the DSSQ dimensions relate to different patterns of appraisal and coping processes consistent with cognitive stress theory. In addition, in occupational samples, DSSQ scores relate modestly to standard criteria including job satisfaction and perceptions of job characteristics (Matthews, Campbell, et al., 2002).

Validation studies have confirmed that tasks making qualitatively different demands elicit different patterns of subjective state response (Matthews, Campbell, et al., 2002). For example, Matthews, Emo, et al. (2006) compared effects of three task stressors—vigilance, a high time-pressure working memory task, impossible anagrams—on state response, relative to a control condition of reading popular magazines. All three task stressors elevated distress, by more than 1 SD, but also evoked differing patterns of state change. The vigilance task was the only one to lower task engagement, and worry remained higher in the impossible anagrams condition than in the other task conditions. In the occupational context, we might expect to see differing state changes associated with monotonous monitoring tasks, with being overloaded by information, and with facing a seemingly impossible assignment. The DSSQ is also sensitive to standard stress manipulations including time pressure and loss of control and loud noise (see Matthews, Warm, Reinerman, Langheim, & Saxby, 2010). Furthermore, state factors that include engagement and distress, measured prior to performance, predict performance on a wide range of attentionally demanding tasks including vigilance, visual search, and working memory (Matthews, 2010; Matthews & Campbell, 2010).

Subjective States and Cognitive Functioning at Work

Some examples may show how the DSSQ may be used in an occupational health context. A growing trend in the workplace is increasing automation of tasks using computer technology, so that workers are required to monitor the automation rather than perform the task directly. This vigilance requirement may itself be a source of stress and fatigue (Matthews, Warm, Reinerman, Langheim, & Saxby, 2010; Matthews, Warm, Reinerman, Langheim, Washburn, & Tripp, 2010). Matthews, Warm, Shaw, and Finomore (2010) developed a two-phase strategy for assessing the operator's readiness to perform a simulation of a task that required sustained monitoring of a military tactical display. In the first phase, participants performed a short, 12-min vigilance task requiring character detection. Their stress response to this high-workload task was assessed using the DSSQ. Then, they performed the 60-min tactical monitoring task. A multivariate assessment strategy incorporating cognitive ability assessment, performance on the short task, and the DSSQ response afforded successful prediction of the criterion task; 30 to 40% of the variance was explained. Assessment of stress states may help to evaluate the employee's fitness to perform tasks requiring sustained attention (Matthews, Warm, Reinerman, Langheim, Washburn, & Tripp, 2010).

A second area of application is to vehicle driving. Stress and fatigue are well-established as risk factors for crashes. There is also a wider occupational literature linking stress to cognitive failures and errors (e.g., Wallace & Vodanovich, 2003). In an occupational sample, Rowden, Matthews, Watson, and Biggs (2011) showed that both job stress and more general life stress were associated with errors and violations during driving. Studies of simulated driving show that manipulations such as reduction of driver control over the vehicle and prolonged workload and monotony may elicit differing patterns of change in subjective state (Matthews, Saxby, Funke, Emo, & Desmond, 2011). Comparable state change patterns are seen in field studies of professional and nonprofessional drivers.

Finally, Matthews and Falconer (2000, 2002) explored the interplay between stress state and performance in two studies of customer service agents. Simulated tasks were developed in collaboration with the employing organizations (a supermarket chain and a telecommunications company). Findings confirmed that answering customer queries and handling complaints may be intrinsically stressful, in that substantial distress responses were observed, irrespective of the level of experience of the agent. Matthews and Falconer (2002) also found evidence that distress and worry correlated with some aspects of performance impairment.

Thus, stress state assessment has two potential benefits for the occupational health psychologist. First, it allows a direct evaluation of the extent to which performance of a specific task produces distress, worry, or loss of task engagement, with potential consequences for the employee's health, morale, and organizational commitment. Second, it allows an assessment of the employee's vulnerability to objective cognitive performance deficits, to the extent that the task makes demands on sustained attention or working memory.

Use of subjective measures should be qualified by the potential vulnerability of such scales to unconscious impression management or deliberate faking, although mood measures appear to be largely free of social desirability bias (Matthews, Jones, & Chamberlain, 1990). Thus, they may not be suitable for high-stakes assessment, such as deciding whether to terminate a person's employment because of excessive stress. Their value is more in situations in which the employee is actively cooperating, for example, in choosing between alternate task assignments or career guidance.

Psychophysiological Assessments of Cognition

The review of psychophysiological and brain-imaging measures as related to cognition would be a formidable task in itself. Placing such a body of literature within the context of occupational health psychology in a few pages is equally challenging. We will look at two general issues that may limit the utility of psychophysiological and imaging techniques, before providing examples of recent advancements that may have potential for workplace studies.

The first limitation is the restrictions imposed by the technological challenges and the equipment required, so that the use of psychophysiological and other imaging techniques in field settings is more the exception than the rule. Imaging techniques such as PET and fMRI may have their greatest utility in exploring basic processes rather than in environments with greater ecological validity.

Ideally, ambulatory equipment should be used to record data as the person engages in normal activities. Fahrenberg, Myrtek, Pawlik, and Perrez (2007) have argued that technological improvements in ambulatory monitoring should make researchers look again at the advantages of these techniques, including ecological validity and objectivity of assessment. Ambulatory assessments may be especially useful for the study of movement and activity (electromyogram and actigraphy; e.g., Bussmann, Ebner-Priemer, & Fahrenberg, 2009), and for responses that are relatively insensitive to ambient electrical interference and body movement, such as the electrocardiogram (ECG). However, even with highly portable recording systems, there are inevitable restrictions on recording, and technical debates regarding type, size, location, and methods for attaching electrodes (and associated artifacts) appear to have no end in sight (see Cacioppo, Tassinary, & Berntson, 2000).

The second limitation is making inferences about cognitive functioning from psychophysiological data. Traditionally, measures such as ECG and electrodermal activity (EDA) were interpreted as indices of a general arousal state, which was deemed central to stress and emotion. Indeed, much of the current interest in ambulatory monitoring relates to naturalistic study of clinical emotion disorders (e.g., Ebner-Priemer & Trull, 2009). In performance research, a further inferential step is needed to link indices of arousal to cognitive processes, and this step has proved problematic. The popular Yerkes-Dodson law, which states that there is an inverted-U relationship between arousal and performance, has proved highly fallible (Hancock & Ganey, 2003; Matthews & Gilliland, 1999). Specific autonomic and central arousal indices have also been shown to be of limited utility as performance indicators (Matthews, 2001).

Workload Monitoring

One application of psychophysiological assessment is the continuous monitoring of task workload, for which cardiovascular measures have proved useful. Vogt, Hagemann, and Kastner (2006) investigated workload in German air traffic controllers by recording heart rate and systolic blood pressure during actual work, and in simulations in which task demands could be manipulated systematically. Factors such as traffic density and potential conflicts influenced these indices. Modeling of workload allowed 50% of the variance in self-reported strain to be predicted. The authors report that the model was used in negotiating new collective work agreements for German controllers. Specifically, it helped in setting staffing levels and ensuring acceptable workload levels for individual controllers. However, the study, like other comparable ones, was primarily geared toward mitigating operator stress, rather than directly linking workload indices to performance and cognitive functioning.

A more sophisticated development of this approach is to develop algorithms that integrate information from multiple recording systems to monitor operator workload. Wilson and Russell (2003) manipulated workload on a multicomponent laboratory task reproducing some of the cognitive demands of piloting an airplane. They recorded ECG, EEG, electrooculographic (EOG), and respiration indices. Neural network modeling allowed operator workload to be predicted with a high degree of accuracy (around 90% classification accuracy). Modeling was done separately for each individual participant, accommodating individual variation in sensitivity to workload of the various indices. Although Wilson and Russell (2003) were concerned with aviation applications, a similar approach might be used to monitor for excessive workload in a variety of occupational settings. Studies of this kind belong within the emerging field of neuroergonomics which seeks to use indices of brain functioning to monitor and augment performance (Parasuraman & Wilson, 2008).

EEG Methods

Electroencephalography (EEG) was for many years one of the few methods researchers had for assessing brain function in relationship to cognition. Clearly, EEG is potentially an effective tool for assessing matters of concern to occupational health psychology, including fatigue and sleepiness. Fatigue, especially sleepiness, produces distinct EEG patterns, characterized especially by slow-wave activity (delta and theta waves), which may be used to monitor alertness during work (see Lal & Craig, 2001). These authors caution that there is some interindividual variability in EEG patterns. Another limitation is that EEG may be more effective in determining sleepiness rather than stress that is induced by task overload or underload in the wakeful operator (Matthews, Saxby, Funke, Emo, & Desmond, 2011). For example, Grandjean (1979) demonstrated that EEG activation levels during both boredom and fatigue are typically low, thus rendering the states fairly indistinguishable, although they may potentially have differing performance consequences.

The digital recording of EEG (as opposed to analogue paper recording) has supported recent advances in mathematical processing of the digital waveforms. Analysis using sophisticated processing algorithms provides a quantitative EEG (qEEG), which may be more diagnostic of cognitive processes than standard spectral analyses of power densities in different frequency bands; qEEG techniques consolidate, typically, EEG amplitude measures across the brain, forming an "EEG brain map." These brain maps should not be confused with brain mapping by other imaging techniques to which EEG brain maps may or may not have direct functional relationships (Nuwer, 1997). Applications of qEEG relevant to occupational psychology are in their infancy, but some recent studies demonstrate their potential. For example, qEEG has been shown to correlate with deficits associated with toxigenic mold exposure (Crago & Nelson, 2004), suggesting the utility of the technique in studies of environmental health and cognition.

Transcranial Doppler Sonography

Another recent innovation in psychophysiology is the use of transcranial Doppler sonography (TCD) to measure cerebral bloodflow velocity (CBFV) during sustained performance. CBFV is typically measured from the medial cerebral arteries, which perfuse large areas of the cortex. Increases in the metabolic activity of the neurons concerned elevate CO_2 concentration, which results in increased blood flow to remove metabolic waste products. Measurement is based on analysis of the Doppler shift of ultrasound pulses reflected back to the TCD transceiver from the blood corpuscles moving within the artery. TCD may provide a global measure of metabolic activity during task performance. Indeed, short, high workload tasks typically elicit increases in TCD (Stroobant & Vingerhoets, 2000).

Studies of vigilance (reviewed by Warm, Matthews & Parasuraman, 2009) have shown that on longer (typically 30–40 min) signal detection tasks, CBFV typically declines. Two key observations show that this decline represents more than some general loss of arousal. First, if the person simply performs the monotonous activity of watching task stimuli without attempting to detect signals, there is no change in CBFV. Second, workload parameters that control vigilance decrement, such as memory load and cueing, also produce corresponding changes in CBFV decline. Thus, by contrast with autonomic arousal measures, CBFV appears to relate directly to information-processing. Warm et al. (2009) suggest that CBFV may reflect the depletion of attentional resources that results from prolonged high-intensity mental work.

TCD is less invasive than brain-imaging techniques and may have applications in occupational health psychology, especially for work where the person is seated. Monitoring for loss of CBFV during prolonged work may provide an index of vulnerability to cognitive fatigue and loss of alertness in contexts that include vehicle driving (Reinerman, Warm, Matthews, & Langheim, 2008). In addition, CBFV may provide a predictive index of readiness to perform. Matthews, Warm, Reinerman, Langheim, Washburn, and Tripp (2010) showed that the short-term increase in CBFV elicited by a battery of short high-workload tasks predicted performance on subsequent sensory and cognitive vigilance tasks. Structural modeling of data demonstrated that CBFV relates to subjective task engagement, but the two measures contributed independently to prediction of performance.

Conclusions

We conclude with some general remarks and a summary of applications. We have highlighted the availability of contrasting methods for assessment of cognitive function, which suggest the utility of multivariate assessment in many occupational settings. It seems that the domain of occupational health psychology lies in a region bound by a number of applied psychology specialties, and further that the content of those specialties will inevitably be drawn into the activities of occupational health psychologists to one degree or another. Specifically, the rather separate traditions represented by basic research on human performance and cognition and by clinical neuropsychology appear to be converging. Thus, a challenge for practitioners is to recognize the appropriate domain of expertise necessary to transition basic research into occupational applications.

We emphasize also the absence of any gold standard for assessment of cognitive function. Operator cognitive efficiency cannot be adequately captured because of the multifaceted nature of information-processing which means that no single task, scale, or physiological index can make this determination. Furthermore, key constructs, including cognitive workload, stress, and neural activity are interrelated, but should be discriminated psychometrically and conceptually. Thus, occupational health psychologists should adopt a true scientist–practitioner perspective, selecting assessment tools on the basis of a theoretical understanding of task demands, use of CTA, and the larger psychosocial context for performance (Matthews, 2001). Some of these techniques may be unfamiliar to many practitioners and may, indeed, require validation in the occupational context. We discern a need to investigate the practical utility of the various new advancements in measurement of cognitive functioning that are being generated by basic research. With these general principles in mind, we summarize specific applications thus:

- *Long-term performance prediction*. The traditional goal of personnel selection is to assess aptitude for performance of work activities over an extended future period. A similar aim attaches to evaluating the performance strengths and weakness of individuals with disabilities. Conventional psychometric tests of ability (and, to a lesser degree, personality) remain essential for prediction of trait performance. However, the newer generation of psychometrically sound tests of cognitive functions and systems (e.g., executive processes, working memory) may be more informative about processes contributing to unsatisfactory performance, or, indeed, excellence in performance. If configured thoughtfully, test batteries (e.g., ANAM) can provide a systematic approach that also discriminates underlying processing components. Subjective and psychophysiological assessments of stress responses to short work-sample tests may also have some utility for long-term prediction.

- *Readiness-to-perform.* The concept of using cognitive testing as a method to ensure employees were fully prepared for work requires prediction of performance over short durations of a few hours. Gilliland and Schlegel (1993) completed a report for the FAA which covers the conceptual basis of cognitive testing for risk factor screening, as well as a range of practical issue. The systematic approach advocated in this report has helped to shape testing standards in aviation, trucking, and other industries with safety-sensitive jobs. We have seen here how multivariate approaches may optimize prediction. In the case of sustained performance (e.g., Matthews, Warm, Reinerman, Langheim, Washburn, & Tripp, 2010), unique information is potentially provided by cognitive tests (short vigilance task), by subjective measures (task engagement), and by psychophysiological measures (TCD, EEG).
- *Continuous performance monitoring.* Especially in safety-critical work, such as vehicle operation, it may be important to monitor cognitive functioning continually, in order to detect imminent error or loss of alertness. Performance may be directly monitored, but, by the time impairment is detected, it may be too late to intervene (e.g., a sleepy driver closing her eyes). Early warning may be most effectively provided by psychophysiological monitoring, and we discussed the utility of TCD and EEG for this purpose. Detecting loss of alertness in the wakeful-but-fatigued operator is more challenging than detecting sleepiness.
- *Workload monitoring.* We distinguish workload monitoring from performance monitoring because associations between workload and performance may be complex (Shultz et al., 2010). Again, psychophysiology may offer the most effective means for detecting spikes in workload during performance that may impose excessive strain on the operator, and *may* indicate vulnerability to performance defi-

cit. We discussed the utility of ECG and multivariate modeling of the individual's psychophysiological response to workload for this purpose.

Finally, important to any discussion of assessing cognition within the context of occupational health psychology is the critical need to amalgamate current advances in cognitive psychology, neuroscience, psychophysiology, and assessment of individual differences in traits and states. It is only in this crucible that we will advance a more accurate understanding of cognitive functions and processes and their role in the workplace.

Disclosure

The University of Oklahoma (OU) holds the exclusive license from the US Army for the Automated Neuropsychological Assessment Metrics (ANAM). Kirby Gilliland has a standard university royalty agreement with OU regarding ANAM sales. VistaLifeSciences Inc. manages all commercial sales and distribution of ANAM through an exclusive licensing agreement with OU.

References

Ackerman, P. L., Beier, M. E., & Boyle, M. O. (2005). Working memory and intelligence: The same or different constructs? *Psychological Bulletin*, *131*, 30–60.

Bartram, D., & Bayliss, R. (1984). Automated testing: Past, present and future. *Journal of Occupational Psychology*, *57*, 221–237.

Beckmann, B., Holling, H., & Kuhn, J. (2007). Reliability of verbal-numerical working memory tasks. *Personality and Individual Differences*, *43*, 703–714.

Boone, K. B. (Ed.). (2007). *Assessment of feigned cognitive impairment*. New York: Guilford.

Bussmann, J. B. J., Ebner-Priemer, U. W., & Fahrenberg, J. (2009). Ambulatory activity monitoring: Progress in measurement of activity, posture, and specific motion patterns in daily life. *European Psychologist*, *14*, 142–152.

Cacioppo, J. T., Tassinary, L. G., & Berntson, G. G. (2000). *Handbook of psychophysiology*. New York: Cambridge University Press.

Carroll, J. B. (1993). *Human cognitive abilities: A survey of factor-analytic studies*. New York: Cambridge University Press.

Crago, B. R., & Nelson, A. (2004). QEEG and neuropsychological consequences of exposure to toxic molds. *Journal of Neurotherapy*, *8*, 142–143.

Ebner-Priemer, U. W., & Trull, J. (2009). Ambulatory assessment: An innovative and promising approach for clinical psychology. *European Psychologist, 14*, 109–119.

Echemendia, R. J. (2006). *Sports neuropsychology: Assessment and management of traumatic brain injury*. New York: Guilford.

Egeland, J., & Kovalik-Gran, I. (2010). Measuring several aspects of attention in one test: The factor structure of Conners's Continuous Performance Test. *Journal of Attention Disorders, 13*, 339–346.

Englund, C. E., Reeves, D. L., Shingledecker, C. A., Thorne, D. R., Wilson, K. P., & Hegge, F. W. (1987). *The Unified Tri-Service Cognitive Performance Assessment Battery (UTC-PAB): I. Design and specification of the battery* (No. NHRC-TR-87-10). San Diego, CA: Naval Health Research Center.

Ericsson, K. A., & Ward, P. (2007). Capturing the naturally occurring superior performance of experts in the laboratory: Toward a science of expert and exceptional performance. *Current Directions in Psychological Science, 16*, 346–350.

Eriksen, B. A., & Eriksen, C. W. (1974). Effects of noise letters upon the identification of a target letter in a non-search task. *Perception & Psychophysics, 16*, 143–149.

Fahrenberg, J., Myrtek, M., Pawlik, K., & Perrez, M. (2007). Ambulatory assessment—Monitoring behavior in daily life settings: A behavioral-scientific challenge for psychology. *European Journal of Psychological Assessment, 23*, 206–213.

Fan, J., McCandliss, B. D., Sommer, T., Raz, A., & Posner, M. I. (2002). Testing the efficiency and independence of attentional networks. *Journal of Cognitive Neuroscience, 14*, 340–347.

Gilliland, K., & Schlegel, R. E. (1993). *Readiness-to-perform testing: A critical analysis of the concept and current practices* (FAA report no. FAA-AM-93-13). Washington, DC: Department of Transportation, Federal Aviation Administration, Office of Aviation Medicine.

Gilliland, K. & Schlegel, R. E. (1995, January). Readiness–to-perform testing and the worker. *Ergonomics in Design*, 14–19.

Grandjean, E. (1979). Fatigue in industry. *British Journal of International Medicine, 36*, 175–186.

Hancock, P. A., & Ganey, N. (2003). From the inverted-U to the extended-U: The evolution of a law of psychology. *Journal of Human Performance in Extreme Environments, 7*, 5–14.

Heilman, K. M., & Valenstein, E. (2003). *Clinical neuropsychology* (4th ed). New York: Oxford University Press.

Hurrell, J. J., & Murphy, L. R. (1992). An overview of occupational stress and health. In W. M. Rom (Ed.), *Environment and occupational medicine* (2nd ed., pp. 675–684). Boston, MA: Little, Brown.

Ilkowska, M., & Engle, R. W. (2010). Trait and state differences in working memory capacity. In A. Gruszka, G. Matthews, & B. Szymura (Eds.), *Handbook of individual differences in cognition: Attention, memory and executive control* (pp. 295–320). New York: Springer.

Kabat, M., Kane, R., Jefferson, A., & DiPino, R. (2001). Construct validity of selected Automated Neuropsychological Assessment Metrics (ANAM) battery measures. *Clinical Neuropsychologist, 15*, 498.

Kandel, E. R., Schwartz, J. H., & Jessell, T. M. (2000). *Principles of neural science.* New York: McGraw-Hill.

Kane, R., & Kay, G. (1992). Computerized assessment in neuropsychology: A review of tests and test batteries. *Neuropsychology Review, 3,* 1–117.

Karasek, R. (1979). Job demands, job decision latitude, and mental strain: Implications for job re-design. *Administrative Science Quarterly, 24,* 285–306.

Kay, G. G., & Spector, J. (1991). Applications of computerized performance assessment batteries in neuropsychology [Abstract]. *Archives of Clinical Neuropsychology, 6,* 198–199.

Lal, S., & Craig, A. (2001). A critical review of the psychophysiology of driver fatigue. *Biological Psychology, 55,* 173-194.

Larrabee, G. J. (Ed.). (2007). *Assessment of malingered neuropsychological deficits.* New York: Oxford University Press.

Lazarus, R. S. (1999). *Stress and emotion: A new synthesis.* New York: Springer.

Matthews, G. (2001). Levels of transaction: A cognitive science framework for operator stress. In P. A. Hancock & P. A. Desmond (Eds.), *Stress, workload and fatigue* (pp. 5–33). Mahwah, NJ: Erlbaum.

Matthews, G., & Campbell, E. (2010). Dynamic relationships between stress states and working memory. *Cognition and Emotion, 24,* 357–373.

Matthews, G., Campbell, S. E., Falconer, S., Joyner, L., Huggins, J., Gilliland, K., ... Warm, J. S. (2002). Fundamental dimensions of subjective state in performance settings: Task engagement, distress and worry. *Emotion, 2,* 315–340.

Matthews, G., Davies, D. R., Westerman, S. J., & Stammers, R. B. (2000). *Human performance: Cognition, stress and individual differences.* London: Psychology Press.

Matthews, G., Emo, A. K., Funke, G., Zeidner, M., Roberts, R. D., Costa, P. T., Jr., & Schulze, R. (2006). Emotional intelligence, personality, and task-induced stress. *Journal of Experimental Psychology: Applied, 12,* 96–107.

Matthews, G., & Falconer, S. (2000). Individual differences in task-induced stress in customer service personnel. *Proceedings of the Human Factors and Ergonomics Society, 44,* 145–148.

Matthews, G., & Falconer, S. (2002). Personality, coping and task-induced stress in customer service personnel. *Proceedings of the Human Factors and Ergonomics Society, 46,* 963–967.

Matthews, G., & Gilliland, K. (1999). The personality theories of H. J. Eysenck and J. A. Gray: A comparative review. *Personality and Individual Differences, 26,* 583–626.

Matthews, G., Gruszka, A., & Szymura, G. (2010). Conclusion: The state of the art in research on individual differences in executive control and cognition. In A. Gruszka, G. Matthews, & B. Szymura (Eds.), *Handbook of individual differences in cognition: Attention, memory and executive control* (pp. 437–462). New York: Springer.

Matthews, G., Jones, D. M., & Chamberlain, A .G. (1990). Refining the measurement of mood: The UWIST Mood Adjective Checklist. *British Journal of Psychology, 81,* 17–42.

Matthews, G., Saxby, D. J., Funke, G. J., Emo, A. K., & Desmond, P. A. 2011). Driving in states of fatigue or stress. In D. Fisher, M. Rizzo, J. Caird, & J. Lee (Eds.), *Handbook of driving simulation for engineering, medicine and psychology* (pp. 29-1–29-11). Boca Raton, FL: Taylor & Francis.

Matthews, G., Warm, J. S., Reinerman, L. E., Langheim, L. K., & Saxby, D. J. (2010). Task engagement, attention and executive control. In A. Gruszka, G. Matthews, & B. Szymura (Eds.), *Handbook of individual differences in cognition: Attention, memory and executive control* (pp. 205–230). New York: Springer.

Matthews, G., Warm, J. S., Reinerman, L. E., Langheim, L, Washburn, D. A., & Tripp, L. (2010). Task engagement, cerebral blood flow velocity, and diagnostic monitoring for sustained attention. *Journal of Experimental Psychology: Applied, 16,* 187–203.

Matthews, G., Warm, J. S., Shaw, T. H., & Finomore, V. (2010). A multivariate test battery for predicting vigilance. *Proceedings of the Human Factors and Ergonomics Society, 54,* 1072–1076.

Miyake, A., Friedman, N. P., Emerson, M. J., Witzki, A. H., & Howerter, A. (2000). The unity and diversity of executive functions and their contributions to complex "frontal lobe" tasks: A latent variable analysis. *Cognitive Psychology, 41,* 49–100.

Nuwer, M. (1997). Assessment of digital EEG, quantitative EEG, and EEG brain mapping: Report to the American Academy of Neurology and the American Clinical Neurophysiology Society. *Neurology, 49,* 277–292.

Ochsner, K. N., & Gross, J. J. (2008). Cognitive emotion regulation: Insights from social cognitive and affective neuroscience. *Current Directions in Psychological Science, 17,* 153–158.

Olson, E. M., Roth, W. T., & Melin, L. (2010). Psychophysiological characteristics of women suffering from stress-related fatigue. *Stress and Health, 26,* 113–126.

Parasuraman, R., & Wilson, G. F. (2008). Putting the brain to work: Neuroergonomics past, present, and future. *Human Factors, 50,* 468–474.

Pashler, H. (1998). *The psychology of attention*. Cambridge, MA: MIT Press.

Reeves, D. L., Winter, K. P., Bleiberg, J., & Kane, R. L. (2007). ANAM Genogram: Historical perspectives, description, and current endeavors. *Archives of Clinical Neuropsychology, 22 Supplement 1,* S15–S37.

Reeves, D. L, Winter, K., LaCour, S., Raynsford, K., Kay, G., Elsmore, T., & Hegge, F. W. (1992). *Automated Neuropsychological Assessment Metrics documentation: Volume I Test administration guide*. Silver Springs, MD: Office of Military Performance Assessment Technology.

Reinerman, L. E., Warm, J. S., Matthews, G., & Langheim, L. K. (2008). Cerebral blood flow velocity and subjective state as indices of resource utilization during sustained driving. *Proceedings of the Human Factors and Ergonomics Society, 52,* 1252–1256.

Roth, E. M. (2008). Uncovering the requirements of cognitive work. *Human Factors, 50,* 475–480.

Rowden, P., Matthews, G., Watson, B., & Biggs, H. (2011). The relative impact of occupational stress, life stress, and driving environment stress on driving outcomes. *Accident Analysis and Prevention, 4,* 1332–1340.

Schatz, P., & Browndyke, J. (2002). Applications of computer-based neuropsychological assessment. *Journal of Head Trauma Rehabilitation, 17,* 395–410.

Shaw, T. H., Matthews, G., Warm, J. S., Finomore, V., Silverman, L., & Costa, P. T., Jr. (2010). Individual differences in vigilance: Personality, ability and states of stress. *Journal of Research in Personality, 44,* 297–308.

Shultz, K. S., Wang, M., & Olson, D. A. (2010). Role overload and underload in relation to occupational stress and health, *Stress and Health,* 26, 99–111.

Spielberger, C. D., & Reheiser, E. C. (2004). Measuring anxiety, anger, depression, and curiosity as emotional states and personality traits with the STPI, STAXI and STPI. In M. J. Hilsenroth & D. L. Segal (Eds.), *Comprehensive handbook of psychological assessment: Vol. 2. Personality assessment* (pp. 70–86). Hoboken, NJ: Wiley.

Stroobant, N., & Vingerhoets, G. (2000). Transcranial Doppler ultrasonography monitoring of cerebral hemodynamics during performance of cognitive tasks: A review. *Neuropsychology Review, 10,* 213–231.

Thayer, R. E. (1989). *The biopsychology of mood and arousal.* Oxford, England: Oxford University Press.

Vasterling, J., Proctor, S., Amoroso, P., Kane, R., Heeren, T., & White, R. (2006). Neuropsychological outcomes of Army personnel following deployment to the Iraq war. *JAMA: Journal of the American Medical Association, 296,* 519–529.

Vogt, J., Hagemann, T., & Kastner, M. (2006). The impact of workload on heart rate and blood pressure in en-route and tower air traffic control. *Journal of Psychophysiology, 20,* 297–314.

Wallace, J. C., & Vodanovich, S. J. (2003). Workplace safety performance: Conscientiousness, cognitive failure and their interaction. *Journal of Occupational Health Psychology, 8,* 316–327.

Warm, J. S., Matthews, G., & Parasuraman, R. (2009). Cerebral hemodynamics and vigilance performance. *Military Psychology, 21,* S75–S100.

Watson, D., & Clark, A. (1997). Measurement and mismeasurement of mood: Recurrent and emergent issues. *Journal of Personality Assessment, 68,* 267–296.

Wilson, G. F., & Russell, C. A. (2003). Real-time assessment of mental workload using psychophysiological measures and artificial neural networks. *Human Factors, 45,* 635–643.

Worley, J. A., Vassar, M., Wheeler, D. L., & Barnes, L. L. B. (2008). Factor structure of scores from the Maslach Burnout Inventory: A review and meta-analysis of 45 exploratory and confirmatory factor-analytic studies. *Educational and Psychological Measurement, 68,* 797–823.

Zeidner, M., & Matthews, G. (2005). Evaluation anxiety. In A. J. Elliot & C. S. Dweck (Eds.), *Handbook of competence and motivation* (pp. 141–163). New York: Guilford.

12

Experimental and Quasi-Experimental Designs in Occupational Health Psychology[1]

Peter Y. Chen, Konstantin P. Cigularov, and Lauren M. Menger

Various research methods such as observation, interview, survey, experiment, or quasi-experiment have been utilized to advance the literature in occupational health psychology (OHP) with the goal of improving the quality of work life and promoting healthy workplaces (Chen, DeArmond, & Huang, 2007). Among these methods, experiments are likely considered the most robust method to investigate causal relationships between variables of interest. The unique strength of experiments is attributed to the fact that "causes" are manipulated (either deliberately or not) and their "effects" or "impacts" are systematically assessed in a controlled setting.

Experiments are generally classified into two major categories (Shadish, Cook, & Campbell, 2002): randomized experiments (e.g., participants are *randomly* assigned to two stress management intervention programs and a waitlist control condition; Searle, 2008) and quasi-experiments (e.g., participants are exposed either to nothing, or they are exposed to an office workstation ergonomics intervention program implemented by municipal offices; May, Reed, Schwoerer, & Porter, 2004). The major difference between randomized experiments and quasi-experiments is whether participants are exposed to

1. Preparation for this chapter is supported by Occupational Health Psychology Training, U. S. NIOSH (1T42 OH009229-01) awarded to Peter Chen. Its contents are solely the responsibility of the authors and do not necessarily represent the official views of NIOSH.

treatments randomly or not. When participants are exposed to treatments randomly, all of them have an equal chance of being selected to or placed into the experimental condition(s). In contrast, participants in quasi-experiments do not have an equal chance of being in the experimental condition(s) because their participation may be attributed to self-selection, naturally occurring circumstances (e.g., merger or shooting), or administrative decisions.

Survey of Experimental Studies in OHP Literature

To understand characteristics of experiments conducted as reported in the OHP literature, we reviewed two main OHP journals, *Journal of Occupational Health Psychology* (JOHP) and *Work & Stress* (WS), between January 1, 2000 and September 30, 2010. Among them, 36 empirical articles applied either randomized experimental or quasi-experimental designs. These studies included at least a treatment, an outcome measure, units of assignment, and some comparison from which change could be inferred and might be attributed to the treatment (Shadish et al., 2002). Frequencies and percentage use indices (PUIs; Stone-Romero, Weaver, & Glenar, 2002) were calculated for each year by type of experimental design and journal and are presented in Table 12.1. PUIs were calculated by dividing the number of relevant studies for a given year by the total number of studies in that year and then multiplying the resultant number by 100. Between 2000 and 2010, 20 articles applying experimental designs were published in JOHP, ranging from 0 to 4 articles per year (mean = 1.82), and 16 articles applying experimental designs were published in WS, ranging from 0 to 4 articles per year (mean = 1.46).

Two types of quasi-experiments are described in Table 12.1: generally interpretable quasi-experimental designs (i.e., quasi-experiment I), and uninterpretable quasi-experimental designs (i.e., quasi-experiment II), the latter including one group posttest-only design, one group pretest–posttest design, and posttest-only design with nonequivalent groups (Shadish et al., 2002). Reasons why these designs are considered uninterpretable will be presented in a later section.

Overall, randomized experimental studies appeared more frequently in JOHP than in WS over the 11-year period (4% vs. 2%, respectively), whereas the reverse trend was observed for the two types of quasi-experimental studies (2% vs. 3%, respectively for quasi-experiments I,

Table 12.1 Frequencies and Percentage Use Indices of Randomized Experimental and Quasi-Experimental Studies Published in JOHP and WS between 2000 and 2010

	2000	2001	2002	2003	2004	2005	2006	2007	2008	2009	2010	2000–2010
JOHP												
Randomized experiments	4 (11%)	0 (0%)	3 (12%)	0 (0%)	0 (0%)	0 (0%)	1 (4%)	3 (10%)	2 (7%)	0 (0%)	0 (0%)	12 (4%)
Quasi-experiments I	0 (0%)	2 (7%)	0 (0%)	0 (0%)	1 (4%)	0 (0%)	1 (3%)	0 (0%)	1 (4%)	1 (3%)	0 (0%)	7 (2%)
Quasi-experiments II	0 (0%)	0 (0%)	0 (0%)	0 (0%)	0 (0%)	0 (0%)	0 (0%)	0 (0%)	1 (4%)	0 (0%)	0 (0%)	1 (0.3%)
Overall experiments	4 (11%)	2 (7%)	3 (12%)	0 (0%)	1 (4%)	0 (0%)	2 (7%)	3 (10%)	4 (14%)	1 (3%)	0 (0%)	20 (6.3%)
All published articles	36	28	25	22	24	36	29	29	28	34	26	317
WS												
Randomized experiments	1 (5%)	1 (4%)	0 (0%)	0 (0%)	1 (4%)	1 (4%)	1 (4%)	0 (0%)	0 (0%)	0 (0%)	0 (0%)	5 (2%)
Quasi-experiments I	0 (0%)	0 (0%)	2 (7%)	2 (10%)	0 (0%)	0 (0%)	2 (9%)	0 (0%)	0 (0%)	0 (0%)	1 (10%)	7 (3%)
Quasi-experiments II	0 (0%)	0 (0%)	2 (7%)	0 (0%)	0 (0%)	1 (4%)	1 (4%)	0 (0%)	0 (0%)	0 (0%)	0 (0%)	4 (2%)
Overall experiments	1 (5%)	1 (4%)	4 (15%)	2 (10%)	1 (4%)	2 (8%)	4 (17%)	0 (0%)	0 (0%)	0 (0%)	1 (10%)	16 (7%)
All published articles	22	25	27	21	26	26	23	20	19	20	10	239
JOHP & WS												
Randomized experiments	5 (9%)	1 (2%)	2 (4%)	0 (0%)	1 (2%)	1 (2%)	2 (4%)	3 (6%)	2 (4%)	0 (0%)	0 (0%)	17 (3%)
Quasi-experiments I	0 (0%)	2 (4%)	3 (6%)	2 (5%)	1 (2%)	0 (0%)	3 (6%)	0 (0%)	1 (2%)	1 (2%)	1 (3%)	14 (3%)
Quasi-experiments II	0 (0%)	0 (0%)	2 (4%)	0 (0%)	0 (0%)	1 (2%)	1 (2%)	0 (0%)	1 (2%)	0 (0%)	0 (0%)	5 (1%)
Overall experiments	5 (9%)	3 (6%)	7 (15%)	2 (5%)	2 (4%)	2 (3%)	6 (12%)	3 (6%)	4 (9%)	1 (2%)	1 (3%)	36 (7%)
All published articles	58	53	52	43	50	62	52	49	47	54	36	556

Note. All percentages reported are based on rounding from three places after the decimal point. JOHP=*Journal of Occupational Health Psychology*; WS=*Work & Stress*.

and 0.3% vs. 2%, respectively for quasi-experiments II). PUIs of randomized experiments published in JOHP and WS generally decreased across years. In contrast, the use of quasi-experiments I fluctuated across years. Due to the low PUIs for quasi-experiments II, no yearly trends can be meaningfully interpreted.

A summary of the 36 studies is presented in Table 12.2, which describes research topics, settings where the studies were conducted, purpose of the studies, outcome variables, characteristics of samples, characteristics of experimental design, as well as key findings. Based on the OHP literature (e.g., Quick & Tetrick, 2010), six broad OHP topical areas were used to categorize each study: health promotion (e.g., health coaching, or substance prevention programs), occupational stress (e.g., unemployment, burnout), occupational safety, work organization (e.g., work arrangements/scheduling, job characteristics/design), workplace aggression (e.g., harassment, violence, bullying, counterproductive behaviors), and work–nonwork interface (e.g., work–family conflict/balance/spillover). As shown in Table 12.2, most studies focused on occupational stress (72%), with fewer studies focusing on health promotion (17%), work organization (14%), and occupational safety (8%). It should be noted that randomized experiments tended to be used more in the area of occupational stress, health promotion, and occupational safety. In contrast, no studies examining work–nonwork interface and workplace aggression with either randomized experimental or quasi-experimental designs were published in either JOHP or WS between 2000 and 2010.

Settings were classified either as (a) field, if the manipulation and data collection took place in participants' natural setting (e.g., at work), or as (b) laboratory, if the manipulation and data collection occurred in a laboratory or a simulated setting created with the specific purpose of conducting research (Stone-Romero, 2002). Our review indicated that most (81%) of the studies were conducted in a field setting, regardless of type of experimental design or journal. Furthermore, three-quarters of the articles aimed to evaluate the effectiveness of a program or an intervention by means of randomized experimental and quasi-experimental designs.

Basic Components of Experiments

While the topics summarized in Table 12.2 are diverse, the basic structure of these experiments is similar, consisting of three critical elements: variables of interest, control, and measurement.

Table 12.2 Summary of Randomized Experimental and Quasi-Experimental Studies Published in JOHP and WS between 2000 and 2010

Study	Topic/Setting	Design Characteristics	Study Objective/Findings
Randomized Experiments			
Bond & Bunce (2000)	Stress/field	2 treatments/control; pre & follow-up (27 wk)	Evaluated the effectiveness of two stress management interventions. Improvements in mental health and propensity to innovate were found with both interventions.
Cigrang et al. (2000)	Stress/field	Treatment/usual care pre; personnel records	Evaluated the utility of stress inoculation training in improving graduation from basic military training. Treatment and usual care groups did not differ significantly in graduation rates.
de Jong & Emmelkamp (2000)	Stress/field	2 treatments/control; pre, post, & follow-up (6 mo)	Examined the effectiveness of stress management training. Participants in both treatment groups reported lower levels of distress, unassertiveness, trait anxiety, and psychosomatic complaints, and more problem-focused coping. No differences between treatment groups.
Mikkelsen et al. (2000)	Stress/field	Treatment/control; pre & 2 follow-ups (1 wk & 1 yr)	Investigated the effects of a participatory intervention in health care institutions. The intervention had a positive effect on work-related stress, job demands, social support, role harmony, learning climate, and management quality.
Vinokur et al. (2000)	Stress/field	Treatment/control; pre & follow-up (24 mo)	Examined the effectiveness of a job-search workshop. The treatment group reported higher levels of reemployment and monthly income, less depressive symptoms, fewer major depressive episodes, and better role and emotional functioning.
Searle et al. (2001)	Stress/lab	Studies 1, 2, & 3 had 4, 8, & 6 exp. conditions, respectively; pre & post	Investigated the role of social support as an intervening variable in the job strain model using a computer simulated mail-sorting work environment. Stress was highest in high demand and low control conditions; no significant demand/control interaction effects; social support increased arousal, satisfaction, and perceived performance.

Eckerman et al. (2002)	Safety/lab	4 exp. conditions; pre, post, & 2 follow-ups (1 wk & 2 mo)	Tested the effect of training delivery on knowledge about respiratory protection. PI-active (interactive training) produced significantly more correct test answers than the other three conditions at posttest and follow-ups.
Probst (2002)	Stress & safety/lab	Treatment/control; post only	Assessed the effects of layoff threat, which was related to increased productivity, increased violations of safety rules, and decreased quality of outputs.
Vuori et al. (2002)	Stress/field	Treatment/control; pre & 2 follow-ups (2 wk & 6 mo)	Examined the effectiveness of the Työhön Job Search Program, which had positive effects after 6 months on the quality of reemployment and levels of distress.
Salemea-Aro et al. (2004)	Stress/field	2 treatments/control; pre, midintervention, follow-up (1 mo)	Examined the effects of two group psychological interventions on burnout. Participants in both interventions showed decreased levels of burnout.
Gardner et al. (2005)	Stress/field	2 treatments/control; pre, post, & follow-up (3 mo)	Determined the effects of cognitive and behavioral coping interventions on job stress. Participants in both interventions showed reductions in stress symptoms.
Atlantis et al. (2006)	Health/field	Treatment/control; pre & follow-up (24 wk)	Evaluated the effectiveness of a behavioral/exercise intervention, which was found to improve sleep quality compared to a control condition.
Blonk et al. (2006)	Stress/field	2 treatments/control; pre & 2 follow-ups (4 & 10 mo)	Assessed the effects of two interventions on return to work and psychological complaints. Significant effects on return to work were found for combined intervention but not CBT. No effects on psychological complaints.
Armitage (2007)	Health/field	Treatment/control; pre & follow-up (2 mo)	Evaluated the effects of a worksite intervention on smoking. Significantly more people quit smoking in the treatment condition than in the control condition.
Bono & Vey (2007)	Stress/Lab	2 exp. conditions; pre & post; DV recorded throughout	Examined the association between personality and emotional performance, stress, and heart rate using two role-play tasks. High self-monitors reported less stress, more deep acting, and lower heart rate during emotional performance.

(*continued*)

Table 12.2 Continued

Study	Topic/Setting	Design Characteristics	Study Objective/Findings
Goldberg & Grandey (2007)	Stress/lab	4 exp. conditions; post only	Examined the effects of display rules and customer hostility on emotional exhaustion and task errors using a call center simulation. Positive display rules and customer hostility were linked to higher emotional exhaustion and number of errors.
de Vente et al. (2008)	Stress/field	2 treatments/usual care; pre & 3 follow-ups (4, 7, & 10 mo)	Assessed the effectiveness of stress management training in reducing burnout, distress, and absenteeism. All groups demonstrated similar decreases in sickness absences and complaints of burnout and distress as a result of treatment.
Searle (2008)	Stress/lab	2 treatments/control; pre & 2 follow-ups (7 & 13 wk)	Evaluated the effects of two problem-focused stress management interventions on proactive behavior and strain. Both interventions were similarly effective at reducing strain, while no changes were observed for the control group.
Quasi-Experimental I			
Bennett & Lehman (2001)	Health/field	2 treatments/control; pre, post (2–4 wk), & follow-up (6 mo)	Evaluated the effectiveness of two workplace substance abuse prevention programs. Significant increases in group privacy regulation, trust, help seeking, knowledge, and peer encouragement were observed in team training. Stigma decreased for information training.
Bond & Bunce (2001)	Stress/field	Treatment/matched control; pre & follow-up (8 mo)	Tested the effectiveness of a work reorganization intervention. Improvements in mental health, sickness absence rates, and job performance were observed.
Lundberg et al. (2002)	Stress/lab	Treatment; measurements taken throughout	Investigated the effect of mental stress on muscle activity and heart rate. Mental stress induced significant increases in muscle activity and heart rate, similar to physical demands.
Shakespeare-Finch et al. (2002)	Stress/field	Case/control; cross-sectional	Examined the impact of trauma in the workplace on family functioning and coping. Social support predicted family functioning across groups. Cognitive/rational coping predicted family conflict only in the case group.

Berkhout et al. (2003)	Work org/ field	Treatment/matched control; pre & 2 follow-ups (6 & 16 mo)	Evaluated the effects of a resident-oriented care model on job design characteristics. The intervention was found partly successful: improvements in resident assignment, use of the nursing process, and resident-oriented tasks.
Hansen et al. (2003)	Stress & work org/ field	Case/control; cross-sectional	Examined the effects of occupation, repetitive work, and adverse work environment on endocrinological indicators of stress. Case group, as well as individuals reporting more repetitive work and more adverse work environment, showed higher endocrinological indicators of stress.
May et al. (2004)	Work org/ field	Treatment/control; pre & follow-up (4 mo)	Investigated the effectiveness of an office workstation ergonomics intervention program. Workstation ergonomic improvements were associated with enhanced perceptions of the workstation and less upper back pain.
Butterworth et al. (2006)	Health/field	Treatment/control; pre & post	Examined the effects of motivational interviewing health coaching program on physical and mental health. The treatment group showed improvements in both physical and mental health.
Moriana & Herruzo (2006)	Stress/field	Case/matched control; DVs were collected over 2 mo	Examined the relationships of sociodemographic, organizational, and personality variables with psychiatric sick leave. Teachers who were emotionally exhausted, highly competitive and hostile, and had low job satisfaction were at greater risk of taking psychiatric sick leave.
Nielsen et al. (2006)	Health/field	Treatment/control; pre & follow-up (20 mo)	Evaluated the effectiveness of a health promotion and empowerment intervention across four industrial canteens in Denmark. Improvements in one treatment canteen and one control canteen; no changes in the remaining two canteens.
Chen et al. (2009)	Stress/field	Treatment/control; pre & 2 follow-ups (2 wk & 2 mo)	Investigated the effectiveness of an IT resource workshop. Increase in users' mean efficacy for treatment group; decrease in efficacy for control group; only control group had an increase in dissatisfaction and exhaustion from the new IT.
Rau et al. (2010)	Stress/field	Case/control; cross-sectional	Examined the association of job demands and control with depression. Both expert-rated and self-rated job demands were related to major depression; only self-rated job control was related to major depression. Demand-control interaction was not significant.

(*continued*)

Table 12.2 Continued

Study	Topic/Setting	Design Characteristics	Study Objective/Findings
Quasi-Experimental II			
Rissen et al. (2002)	Work org/ field	Treatment; pre & post	Evaluated the effectiveness of a job rotation intervention. Reductions in muscle activity and blood pressure and improvements in work attitudes and perceptions and positive arousal were observed after job rotation. No changes in perceived stress and hurry, workload, and pain.
Saksvik et al. (2002)	Stress & health/field	Treatment; pre & post	Identified implementation barriers for stress and health interventions in different work environments.
Eklof & Torner (2005)	Safety/field	Treatment; pre, post, & follow-up (2 mo)	Tested the effects of a group-based intervention on perceived manageability of risks, risk acceptance, safety work, and accidents/incidents. Results suggested some increases in safety work during and after the intervention, and decreases in perceived manageability of risks.
Petterson et al. (2006)	Stress & work org/ field	Treatment; pre & post	Evaluated the impact of an intervention program on workload, staff resources, staff health and well-being, and quality of care. Only limited improvements were observed, mainly in perceptions of quality of care.
Elo et al. (2008)	Stress/field	Treatment; pre & post	Investigated the effect of level of participation in a stress management program on well-being, psychosocial, and organizational context factors. Level of participation had only a positive effect on feedback from supervisor and flow of information.
Probst et al. (2008)	Stress/field	Treatment; pre & post	Assessed the effects of a workplace intervention on knowledge about methods of HIV/AIDS transmission, workplace violence, drug use, and job stress. Improvements were found on only 5 of the 12 knowledge questions.

Note. Stress = occupational stress; health = health promotion; safety = occupational safety; work org = work organization; pre = pretest; post = posttest.

Variables of Interest

Variables are attributes or properties of an object or events of interest, which can be classified qualitatively or measured quantitatively. In the context of OHP, objects or events can be employees, work organizations, family, management, intervention, or almost anything else. Attributes can be types of programs/interventions (e.g., Armitage, 2007; de Vente, Kamphuis, Emmelkamp, & Blonk, 2008; Mikkelsen et al., 2000; Vuori, Silvonen, Vinokur, & Price, 2002), perceived job stability, or job satisfaction (de Jong & Emmelkamp, 2000).

In a typical experiment, three types of variables are involved: independent variables, dependent variables, and confounding variables. However, any variable can be conceptualized as an independent variable, dependent variable, intervening variable (i.e., mediator), modifying variable (i.e., moderator), or confounding (or extraneous) variable at any given time, depending on conceptual models proposed by researchers (see examples illustrated in Spector, Zapf, Chen, & Frese, 2000) .

Independent variables (IVs) are presumed causes that are deliberately manipulated by researchers or occur naturally without being manipulated by researchers. Examples of IVs reported in Table 12.2, which are deliberately manipulated by researchers, include stress management interventions (Blonk, Brenninkmeijer, Lagerveld, & Houtman, 2006; Bond & Bunce, 2000; Gardner, Rose, Mason, Tyler, & Cushway, 2005; Searle, 2008); job search workshop (Vinokur, Schul, Vuori, & Price, 2000); task demands (Searle, Bright, & Bochner, 2001); layoff threat (Probst, 2002); exercise intervention (Atlantis, Chow, Kirby, & Singh, 2006); or customer hostility (Goldberg & Grandey, 2007). An example of a naturally occurring IV is traumatic incidents experienced by police, nurses, and fire fighters in their work (e.g., Shakespeare-Finch, Smith, & Obst, 2002).

In contrast to IVs, dependent variables (DVs) are referred to as effects or outcomes that are affected (i.e., influenced, impacted, or caused) by IVs. DVs in Table 12.2 include well-being (Elo, Ervasti, Kuosma, & Mattila, 2008); safety (Eklof & Törner, 2005); blood pressure and heart rate (Rissén, Melin, Sandsjö, Dohns, & Lundberg, 2002); eyestrain and back pain (May et al., 2004); nicotine dependence (Armitage, 2007); and burnout (de Vente et al., 2008). It is important to note that both IVs and DVs are equally important, and neither can be overlooked in scientific inquiries. However, there has been a tendency to emphasize IVs in order to predict DVs in the organizational behavior literature (see Staw & Oldham, 1978).

Both IVs and DVs in an experiment assist us to organize and make sense of correspondent constructs or concepts (Pedhazur & Schmelkin, 1991). One could conceptualize causal relationships between IVs and DVs in an experiment as snapshots of what causal relationships between the correspondent constructs may look like. As illustrated in Figure 12.1, an IV (i.e., number of workers present) and a DV (i.e., frequencies of arguments) capture small fractions of their correspondent constructs, group pressure, and interpersonal conflict, as shown in paths *1* and *2*. By observing changes in the DV after the IV is manipulated by researchers, a causal relationship between number of workers present and frequencies of argument is substantiated (path *3*), which infers the causal relationship between group pressure and interpersonal conflict (path *4*), assuming all confounding variables are controlled and construct validities of IV and DV measures are demonstrated.

Confounding variables are uncontrolled variables remaining in an experiment that distort inferences of a causal relationship between an IV and a DV. For instance, one may argue that the number of coworkers present may decrease the level of anxiety, which in turn reduces frequencies of arguments, as shown in paths *5* and *6* in Figure 12.1. Furthermore, social desirability or other unknown factors might be affected by the number of coworkers present or affect whether participants argue with others. The above illustration points out that the failure to eliminate confounding variables in an experiment may hamper researchers in interpreting the findings with confidence because other plausible explanations why IVs relate to DVs may exist. When this happens, researchers would have little faith in the internal validity of the study (i.e., inferences about the observed causal relationship between an IV and a DV, path *3* and *4* in Figure 12.1).

It is important to recognize that it is path *4* rather than path *3* that is interesting to scientific inquiries, although path *4* can only be inferred. In other words, verification of the existence of path *4* has to be built upon the evidence of path *3*, yet path *3* can only provide necessary but not sufficient evidence for path *4*.

There are various factors that challenge the interpretation of evidence about path *3*. These factors, also referred to as threats to internal validity, include testing, selection, history, and maturation (Shadish et al., 2002). For instance, testing refers to the phenomenon that a pretest measure influences participants' subsequent responses on the posttest measure. On the other hand, selection bias occurs when participants attend different experimental conditions in a non-random fashion, which could lead participants in these conditions to be different on

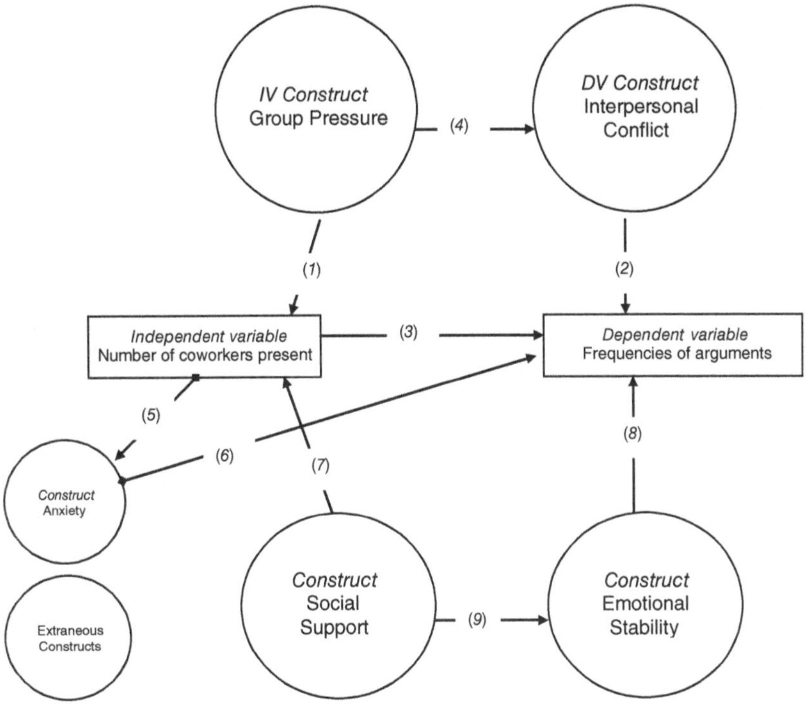

Figure 12.1 Hypothesized causal relationship between social support and job satisfaction in an experiment

variables of interest at the beginning of the experiment. History refers to an external event (e.g., tragedy of September 11, 2001, Hurricane Katrina in 2005, Deepwater Horizon oil spill in 2010) that occurs during the period of experimentation and may potentially alter the DVs. In contrast to history, maturation threatens the internal validity of an experiment because some internal changes (e.g., adjustment experience through years) occur during the period of experimentation. The above examples clearly show that DVs may *fluctuate across time* even in the absence of IVs of interest. To reduce threats to the internal validity of an experiment, it is imperative that researchers exercise rigorous experimental controls and demonstrate construct validity of IVs and DVs in research designs.

Control

Three types of strategies (i.e., manipulation, elimination or inclusion, and randomization) are often utilized in experimentation to control confounding variables. The former two strategies can be applied in either randomized experiments or quasi-experiments. However, the latter control is the very factor that distinguishes between randomized experiments and quasi-experiments.

Although statistical control is widely practiced in data analysis, it should not be confused with the concept of experimental control (i.e., control implemented in the design of the study). Statistical control is an application of statistical techniques (e.g., analysis of covariance or hierarchical regression analysis) to artificially keep the effects of presumed confounding variables constant for participants. A detailed discussion of its uses and implications can be found in Spector et al. (2000).

Control by Manipulation. Control by manipulation refers to the procedure when the attributes of IVs are systematically changed (e.g., participants receive either a stress management program or no program), and the attributes are *identical* for participants within each experimental condition.

Control by Elimination or Inclusion. In some cases, researchers can control confounding variables by eliminating them from or including them in an experiment. For instance, if we suspect that the gender of facilitators might influence the effects of two types of health promotion programs, the same facilitator may conduct both programs. As a result, any potential effect caused by gender is held constant (or controlled) in both conditions. In contrast to control by elimination, we can include the suspected confounding variables in an experiment. Modifying the previous example, we can include gender as an additional IV by having male and female facilitators.

Control by Random Assignment. Practically, not all confounding variables can be fully controlled by elimination or inclusion. It is random assignment that indirectly controls effects of confounding variables. Random assignment is the process of assigning participants to each of experimental conditions with an equal chance. Through the random process of assigning participants to experimental conditions, confounding variables would likely be evenly distributed across experimental conditions. As a result, effects of confounding variables are held constant. Most importantly, randomized experiments tend to show unbiased estimates of effect, compared to nonrandomized experiments (Shadish et al., 2002).

However, researchers should be cautiously optimistic about the effect of random assignment on controlling confounding variables. One of the underappreciated concerns with random assignment is the belief that the random assignment process would distribute confounding variables evenly across all experimental conditions. In reality, this process does not guarantee that all experimental conditions are equal before any treatment. Imagine that 100 people are randomly assigned to two groups. It is unlikely to expect the same or even similar average height between the two groups. The average height, however, would be approximately the same if we repeat the random assignment process repeatedly, say 1,000 trials. Average height of 1,000 trials would likely be the same or very similar between these two groups. Similarly, one would not expect to get five heads after tossing 10 coins. However, after 1,000 times of tossing 10 coins, the average of getting heads would be five approximately.

Random assignment should not be confused with random selection. In contrast to random assignment, which facilitates causal inferences by equating participants in all experimental conditions, random selection is a process of randomly selecting a sample of participants from a population of interest as a way to ensure that findings from the sample can be generalized to the population. Although researchers can use various probability sampling strategies such as systematic sampling to select a sample from a population, the experimental studies published in the field of OHP as well as in other organizational behavior literatures tend to rely on convenience samples (e.g., researchers select participants based on the convenience). Therefore, results of an experiment using random assignment based on a convenience sample may or may not be similar to those found in another randomized experiment should participants be selected randomly from the population.

Measurement

Evidence of construct validity (i.e., inferences about the extent to which operations or measures of variables are similar to their correspondent constructs) for both IVs and DVs is vital. In the context of conducting experimental studies, construct validity of IVs is inferred based on their operationalization (e.g., job insecurity was manipulated by telling some participants that laying off 50% of the workers is necessary due to low sales of paintings and declining company profits; Probst, 2002) and verified through some sort of manipulation check.

In some cases, manipulation procedures can be different for the same IVs, and, ironically, they can be similar for different IVs. For instance, Twenge, Baumeister, Tice, and Stucke (2001) manipulated social exclusion in the first experiment of their study by providing participants with bogus feedback on a personality test, telling them that they would end up alone in life. In a second experiment they first asked participants to choose two people they would most like to work with from a group of four to six people. Participants in the social exclusion condition were told "no one chose you as someone they wanted to work with" (p. 1063). This example demonstrates that the same construct can be manipulated by two different manipulations, which provides additional convergent evidence.

Similar operationalization, however, could be used to manipulate two different constructs. For instance, "Number of coworkers present" may reflect group pressure and may also capture the degree of social support (path *7*), and "frequencies of arguments" may capture aggression and may also capture the level of emotional stability (path *8*). Therefore, even if there is strong evidence supporting path *3*, researchers would have less confidence to determine if path *3* represents the causal relationship of path *4* rather path *9*.

To substantiate the construct validity of variables in an experiment, researchers might consider examining both the intended as well as unintended effects of the manipulation (Pedhazur & Schmelkin, 1991). In addition, researchers should make efforts to rule out possible threats to construct validity such as inadequate explication of the construct, construct confounds, mono-operation bias, mono-method bias, reactivity to the experimental situation, experimenter expectancies, and resentful demoralization (see Shadish et al., 2002 for further information on threats to construct validity). For instance, mono-operation bias and mono-method bias are biases that arise when only one operationalization and one method are used to manipulate or assess IVs and DVs, respectively. If results are similar when the same construct is operationalized differently or measured by different methods, we would have greater confidence about construct validity of IVs and DVs.

Experimental Designs

There are a wide variety of experimental designs illustrated in the literature (Shadish et al., 2002), six of which are presented below. These designs can be expanded or modified, and can be applied either in the laboratory or in the field, although most experimental studies published in JOHP and WS between 2000 and 2010 were conducted in the field. This trend may be attributed to the nature of the OHP discipline that may prefer mundane realism (i.e., the extent to which IVs manipulated in an experiment resemble reality) over experimental realism (i.e., the extent to which the participants both believe and are affected by the experimental manipulation).

To illustrate these designs, we use the following notations: *O* stands for one or more DV measures, *X* represents one or more IVs, T refers to conditions where participants are exposed to a treatment, C refers to the condition where participants are not exposed to a treatment, and W/*R* or WO/R indicates if the random assignment approach is employed.

One-Group Design.

This design consists of one group, which has different variations including the one-group posttest only, X O_{post} (T) or one-group pretest–posttest, O_{pre} X O_{post} (T). This design is often used for practical reasons such as when a company wants all of their employees to receive an intervention. Thus, a random assignment procedure cannot be applied because there is only one group available. Results of this design are generally considered uninterpretable because the design suffers from many threats to internal validity. For instance, change or improvement cannot be substantiated when there is no pretest in the one-group posttest-only design. In the one-group pretest–posttest design, the researcher cannot be certain if changes between O_{pre} and O_{post} are attributed to the IV if evidence about no change between O_{pre} and O_{post} in a control group is lacking.

Nevertheless, one-group designs remain commonly employed in field evaluation studies because they are relatively easy to conduct and are in many instances the only option for evaluators when practical and ethical constraints prevent the use of multiple groups. For example, Petterson and colleagues (2006) employed a one-group pretest–posttest design to assess the effectiveness of an 18-month intervention program, which aimed to improve the work and health conditions among eldercare staff (i.e., nurses and nursing assistants) and the perceived quality of care. The researchers used questionnaires to measure the evaluation outcomes immediately before and after the intervention and found only limited improvements, mostly in perceived quality of care. They discussed several challenges that might have compromised the internal validity of their findings, including inability to use a control group due to practical and ethical reasons, as well as high employee turnover (44%), temporary employment, and relocation. The extended period (i.e., 18 months) of the intervention evaluated in Petterson et al. (2006) presents a further challenge to meaningfully interpret their findings, due to possible co-occurring events (e.g., history, maturation) and significant attrition of participants.

The above example represents a common dilemma encountered by OHP researchers and practitioners who set out to evaluate the effectiveness of various interventions or training programs in the field; namely, how to optimize the validity of inferences under suboptimal conditions (e.g., when organizational constraints make one-group designs the only feasible option). So, what can evaluators do to overcome the inherent methodological weaknesses of one-group designs when the use of multiple groups is not an option? In other words, how can alternative explanations of the observed intervention/program effects be eliminated under such circumstances?

A shorter time interval between pretest and posttest, as well as multiple pretest and posttest assessments, can mitigate certain internal validity threats, such as history, maturation, and attrition. In addition, two supplementary approaches may strengthen the weakness of this design. The first one is suggested by Cook and Campbell (1979), the nonequivalent dependent variables design, which was later termed the Internal Referencing Strategy (IRS) by Haccoun and Hamtiaux (1994). Rather than use a control *group*, the IRS design extends the one group pretest–posttest design through the use of control *items* in the pretest and posttest measures. In the context of evaluating an intervention program, the pretest and posttest measures contain *trained* items that reflect treatment content, and also include *untrained* items based on similar content yet not included in the intervention program. If participants demonstrate significant improvement on the trained items, and no significant change on the untrained items, the researcher would have more confidence about the results derived from the one-group pretest–posttest design.

The second approach concerns whether participants have achieved a minimum competency of skill, knowledge, or responses on DVs (Sackett & Mullen, 1993). For instance, researchers can determine in advance the level of minimum performance on trained behaviors or the minimum attitudinal responses (e.g., 80% of participants agree that the program is useful to deal with difficult situations at work) or beliefs upon completion of an intervention program (see the demonstration by Cigularov, Chen, Thurber, & Stallones, 2008). Furthermore researchers may determine the level of minimum change or growth. It is important to acknowledge that significant results between pretest and posttest scores may appear, yet the change or the posttest score does not necessarily meet the criterion. Thus, this approach adds an additional safeguard in evaluating OHP interventions.

Treatment-Control Posttest Design

This design consists of two groups: a treatment group and control group. Only participants in the treatment group receive the treatment (i.e., a manipulation). Although the control group generally receives a placebo or receives no manipulation, participants in the control group could subsequently receive treatment for either ethical or practical reasons. An assessment of DVs from both groups is conducted after an IV is manipulated. The structure of this design can be depicted as the following:

WR (WO/R)	X	O	(T)
WR (WO/R)		O	(C)

There are other variations of this basic design. For instance, the above design may consist of multiple treatment groups as well as a control group, and participants in each group receive different or no treatments.

Our review revealed that none of the *field* experiments published in JOHP and WS between 2000 and 2010, which employed treatment and control groups to assess the effectiveness of various OHP-focused interventions, utilized the treatment-control posttest design. This design appeared more applicable to lab experiments, which aimed to manipulate a key job stressor (e.g., job insecurity) and investigate its effects on outcomes of concern (e.g., employee productivity, product quality, and adherence to safety policies; Probst, 2002).

Observing changes from pretest to posttest is a key strategy in evaluation efforts (along with comparing these changes between treatment and control groups), which aim to infer that changes in DVs are attributable to a focal intervention (Sackett & Mullen, 1993). Since the treatment-control posttest design lacks pretest measurement, the effect of an intervention or treatment is inferred only from differences in DVs between the treatment and control groups, which makes random assignment an important requirement to strengthen this design.

Factorial Design

When describing manipulation by inclusion, we pointed out that more than one IV could be included in an experiment. This design is often referred to as a factorial design, where IVs are labeled as factors (e.g., treatment and length of treatment), and each factor has at least two levels/types (e.g., treatment A vs. treatment B vs. control; or 2-week vs. 4-week vs. 8-week treatment). A major advantage of this design is to investigate interactive effects of IVs on DVs.

The simplest version of this design consists of four experimental conditions created by two factors (A and B) and two levels (1 and 2) within each factor. Based on the combination of factors and levels, participants in the experiment receive one of four treatments: X_{A1B1}, X_{A1B2}, X_{A2B1}, and X_{A2B2}. Similar to previous cases, a pretest measure can be included in this design.

WR (WO/R)	X_{A1B1}	O
WR (WO/R)	X_{A1B2}	O
WR (WO/R)	X_{A2B1}	O
WR (WO/R)	X_{A2B2}	O

For example, Goldberg and Grandey (2007) manipulated emotion display rules (positive display rule vs. display autonomy) and customer hostility (hostile vs. polite) to examine their main and interaction effects on emotional exhaustion and task errors using a call center simulation. They randomly assigned 86 psychology and business undergraduate students to the four conditions and measured the DVs of interest with a posttest survey. Their findings revealed that participants who had to follow positive emotion display rules (i.e., "service with a smile") reported higher emotional exhaustion and number of errors at the end of the simulation compared to those given autonomy about how to display their emotions. Further, customer hostility positively predicted emotional exhaustion, as well as errors during the call from the hostile customer. No significant interaction effects were found.

Pretest–Posttest Control Group Design

Without having a pretest in the above designs, positive findings may be inconclusive if participants drop out of an experiment or fail to complete outcome measures in the posttests. With the pretest information, we can verify if those who drop out are similar to those who complete the experiment. We can also determine if participants are similar prior to the experiment, and if there are changes between O_{pre} to O_{post}. In general, the measures at pretest (O_{pre}) and posttest (O_{post}) should be identical. In some cases due to practical constraints, parallel or alternate forms of measures may be used with the assumption that both measures assess the same underlying construct. The structure of this design can be depicted as below. Similar to the above design, multiple treatment groups as well as a control group could be integrated into this design so that participants in each group receive different or no treatments.

WR (WO/R)	O_{pre}	X	O_{post}	(T)
WR (WO/R)	O_{pre}		O_{post}	(C)

The above design is the most strongly recommended design due to its ability to control for all major threats to internal validity. However, this design is susceptible to various threats to external validity (Shadish et al., 2002). One such threat concerns the interaction between pretest measurements and the intervention (or treatment), and how this interaction may affect the results for the treatment group. In other words, it might be possible that pretest measures affected participants' responses toward dependent variables in the treatment group, rather than the treatment itself.

The importance of considering drop-out rates, as well as randomization, in the pretest–posttest control group design is exemplified in the field experiment conducted by Butterworth, Linden, McClay, and Leo (2006) to assess the effectiveness of a 3-month motivational interviewing based health coaching intervention in improving employees' physical and mental health. Although Butterworth et al. showed improvements from pretest to posttest in the treatment group compared to no improvements in the control group, the internal validity of these results was tainted by lack of randomization and differential drop rates (17% and 10% of participants in the treatment and control groups, respectively, dropped out of the study during the 3-month intervention). The intervention was incorporated into a larger ongoing employee wellness program at the participating organization, which made randomization unfeasible. As a result, participants self-selected into the treatment and control groups, raising a concern for possible selection bias (Shadish et al., 2002). This concern was reinforced when participants in the treatment group indicated significantly higher mental and physical health issues than the participants in the control group at the onset of the study.

To examine the possibility of selection bias, Butterworth et al. (2006) matched subsamples of treatment and control group participants on baseline characteristics using propensity scores and conducted a sensitivity analysis (Linden, Adams, & Roberts, 2006). They also compared the characteristics of the individuals who left the study prematurely with those who completed it. However, the above analyses would not preclude the possibility of selection and drop-out interaction effects (Shadish et al., 2002), especially considering the unequal drop-out rates and differences between groups at pretest found in that study.

Solomon Four-Group Design

The Solomon four-group design is especially useful to control for the main effects of testing, as well the interaction of testing with the intervention (or treatment) as described above. In that sense, this design, whose structure is depicted below, represents an upgrade on the pretest–posttest control group design:

WR (WO/R)	O_{pre}	X	O_{post}	(T_1)
WR (WO/R)		X	O_{post}	(T_2)
WR (WO/R)	O_{pre}		O_{post}	(C_1)
WR (WO/R)			O_{post}	(C_2)

According to Solomon and Lessac (1968), the means of group T_1 and group C_1 on the pretest measure, assuming they are equal, enable researchers to estimate the pretest performance of group T_2 and group C_2. Furthermore, researchers can examine the change of the posttest performance of group T_2 by comparing it to the means of the pretest measures obtained from groups T_1 and C_1. Comparisons on the posttest measures between group T_1 and group T_2 as well as between group C_1 and group C_2 allow researchers to examine the main effect of the pretest measure. Interactions between the pretest measure and the experimental treatment can be investigated by the divergence between two sets of differences on the posttest measure (group T_1 vs. group T_2 and group C_1 vs. group C_2). Assuming there is a positive effect from the treatment (e.g., the posttest measure increases after the treatment), further evidence for the treatment effect exists if (a) the posttest measure of group T_1 is higher than the pretest measure of group T_1, (b) the posttest measure of group T_1 is higher than that of group C_1, (c) the posttest measure of group T_2 is higher than that of group C_2, and (d) the posttest measure of group T_2 is higher than that of group C_1.

The main drawback of the Solomon four-group design lies within its name; that is, its rigor comes at the expense of practicality as the number of groups doubles compared to the traditional pretest–posttest control group design. Thus, implementing a Solomon four-group design, especially in the field, could be challenging, as evidenced by the lack of such studies in our review of the recent OHP literature. An utilization of the features of this design in an OHP-related topic was demonstrated in a field experiment by Jackson in 1983, who assessed the effects of an intervention aiming to improve participation in decision making on job stressors (e.g., role conflict and ambiguity) and outcomes (e.g., job satisfaction, absenteeism, and turnover intentions) with one pretest (2 months before intervention) and two posttests (3 and 6 months following the intervention). However, one methodological limitation of that study was that participants in the pretest and posttest samples were not identical, albeit overlapping.

Time Series Design

Compared to the prior designs, a time series design provides stronger evidence for internal validity. Rohsenow, Monti, Martin, Michalec, and Abrams (2000) reported that participants receiving a treatment reported using substances for the first 6 months significantly fewer days than those who received the placebo treatment; however, the treatment effect diminished when the participants were retested 12 months after the interventions. Without several observations after the treatment, the "lack of" a long-term benefit of the treatment would not have been known. This design consists of multiple pretests and posttests over a period of time. The numbers of pretests and posttests do not need to be the same, and generally there are more posttests than pretests. As yet, there is no definitive rule when and how often pre and post measures should be assessed, and the choices may vary depending on theoretical mechanisms or patterns of the causal relationships (e.g., when the outcome becomes stable or when the outcome would likely occur), as well as practical constraints. An example of the structure of the time series design is as follows:

WR (WO/R)	O_{pre}	O_{pre}	X	O_{post}	O_{post}	O_{post}	O_{post}	(T)
WR (WO/R)	O_{pre}	O_{pre}		O_{post}	O_{post}	O_{post}	O_{post}	(C)

For example, de Vente and colleagues (2008) used one pretest and three posttests (4, 7, and 10 months) to assess the long-term effectiveness of a cognitive-behavior-based stress management training (SMT). A total of 82 patients on sickness leave with complaints of work-related stress were randomly assigned to three treatment groups (a control group was deemed unethical): (a) individual SMT, (b) group SMT, and (c) care as usual (CAU). Complaints of burnout and distress, as well as self-reported sickness absences, were reduced across all treatment groups in the first 4 months following the treatments. After that the complaints leveled out but sick absences continued to decrease. However, no support was found in favor of any of the treatment conditions. In their discussion, the researchers pointed out three possible threats to the validity of their findings: (a) inability to use a control group, resulting in less contrast between conditions, (b) substantially unequal drop-out rate among groups, and (c) a reactive arrangement effect stemming from participants receiving the CAU treatment who expressed strong disappointment about not obtaining SMT, which might explain the increased drop-out rate in the CAU group.

Three intervention studies reported by Zohar and Luria (2003), which were designed to improve supervisory practices in three companies in an effort to increase safety behaviors, provide an even better illustration of the key features of the time series design (i.e., multiple pretests and posttests). These researchers collected IV (i.e., safety-related supervisory interactions) and DV (i.e., worker safety behaviors) data through brief questionnaires and onsite observations, respectively, for a total of 40 weeks, twice a week: 9 weeks prior to the intervention, 12 weeks during the intervention, and 19 weeks after the intervention.

Conclusion

Experimental designs are useful tools to advance our knowledge of the causal process between IVs and DVs under reasonable conditions. As arguably laid out in *Logic* by J. S. Mill in 1843 (Boring, 1954), and further elaborated by others (e.g., Shadish et al., 2002), three prerequisite conditions should be met for confirmatory analysis. There are (a) observed associations or covariation between a proposed cause and its effect, (b) a temporal sequence between a cause and its effect, and (c) unequivocal isolation or elimination between a cause and its effect (i.e., the association between IV and DV is not due to another variable or process).

However, OHP research rarely meets required conditions 2 (i.e., temporal sequence) and 3 (i.e., elimination of alternative variables) in experimental studies. Suppose that the causal relation between social support and interpersonal conflict is tenable as depicted in Figure 12.1. Researchers can manipulate social support today and measure interpersonal conflict a month later. The observed relationship as well as the order of manipulating an IV in an experiment doesn't guarantee that we can draw a convincing causal conclusion about what the true temporal sequence is. In reality, it is possible that interpersonal conflict influences levels of social support received. In other words, it is reasonable to argue that one's belief about *a* causal process between an IV and a DV may not be the same as *the* actual process, whatever it may be.

In sum, it is rare that any one experiment can possess all optimal characteristics. In the case of internal and external validity, it may be necessary to compromise the generalizability of an experiment's results to the world outside the laboratory or an organization (external validity) in order to ensure causation with a strong manipulation (internal validity). When conducting an experiment, it is appropriate to acknowledge the stronger features of the design (e.g., manipulation and random assignment) and the weaker aspects (e.g., weak experimental realism and mundane realism) so that later experiments can build on the solid components and ameliorate the limitations. Randomized experiments and quasi-experiments should be conceptualized as small pieces of a larger puzzle where each study provides a small understanding of a larger overarching OHP phenomenon.

References

*Articles reviewed in Table 12.2.

*Armitage, C. J. (2007). Efficacy of a brief worksite intervention to reduce smoking: The roles of behavioral and implementation intentions. *Journal of Occupational Health Psychology, 12*(4), 376–390.

*Atlantis, E., Chow, C. M., Kirby, A., & Singh, M. A. F. (2006). Worksite intervention effects on sleep quality: A randomized controlled trial. *Journal of Occupational Health Psychology, 11*(4), 291–304.

*Bennett, J. B., & Lehman, W. E. K. (2001). Workplace substance abuse prevention and help seeking: Comparing team-oriented and informational training. *Journal of Occupational Health Psychology, 6*(3), 243–254.

*Berkhout, A. J. M. B., Boumans, N. P. G., Nijhuis, F. J. N., Van Breukelen, G. P. J., & Abu-Saad, H. H. (2003). Effects of resident-oriented care on job characteristics of nursing caregivers. *Work & Stress, 17*(4), 337–353.

*Blonk, R. W. B., Brenninkmeijer, V., Lagerveld, S. E., & Houtman, I. L. D. (2006). Return to work: A comparison of two cognitive behavioural interventions in cases of work-related psychological complaints among the self-employed. *Work & Stress, 20*(2), 129–144.

*Bond, F. W., & Bunce, D. (2000). Mediators of change in emotion-focused and problem-focused worksite stress management interventions. *Journal of Occupational Health Psychology, 5*(1), 156–163.

*Bond, F. W., & Bunce, D. (2001). Job control mediates change in a work reorganization intervention for stress reduction. *Journal of Occupational Health Psychology, 6*(4), 290–302.

*Bono, J. E., & Vey, M. A. (2007). Personality and emotional performance: Extraversion, neuroticism, and self-monitoring. *Journal of Occupational Health Psychology, 12*(2), 177–192.

Boring, E. G. (1954). The nature and history of experimental control. *The American Journal of Psychology, 67*, 573–589.

*Butterworth, S., Linden, A., McClay, W., & Leo, M. C. (2006). Effect of motivational interviewing-based health coaching on employees' physical and mental health status. *Journal of Occupational Health Psychology, 11*(4), 358–365.

Chen, P. Y., DeArmond, S., & Huang, Y. H. (2007). Occupational health psychology. In S. G. Rogelberg (Ed.), *The encyclopedia of industrial and organizational psychology* (pp. 525–528). Thousand Oaks, CA: Sage.

*Chen, S., Westman, M., & Eden, D. (2009). Impact of enhanced resources on anticipatory stress and adjustment to new information technology: A field-experimental test of conservation of resources theory. *Journal of Occupational Health Psychology, 14*(3), 219–230.

Cigularov, K., Chen, P. Y., Thurber, B. W., & Stallones, L. (2008). Investigation of the effectiveness of a school-based suicide education program using three methodological approaches. *Psychological Service, 5*, 262–274.

*Cigrang, J. A., Todd, S. L., & Carbone, E. G. (2000). Stress management training for military trainees returned to duty after a mental health evaluation: Effect on graduation rates. *Journal of Occupational Health Psychology, 5*(1), 48–55.

Cook, T. D., & Campbell, D. T. (1979). *Quasi-experimentation: Design and analysis issues for field settings*. Skokie, IL: Rand McNally.

*de Jong, G. M., & Emmelkamp, P. M. G. (2000). Implementing a stress management training: Comparative trainer effectiveness. *Journal of Occupational Health Psychology, 5*(2), 309–320.

*de Vente, W., Kamphuis, J. H., Emmelkamp, P. M. G., & Blonk, R. W. B. (2008). Individual and group cognitive-behavioral treatment for work-related stress complaints and sickness absence: A randomized controlled trial. *Journal of Occupational Health Psychology, 13*(3), 214–231.

*Eckerman, D. A., Lundeen, C. A., Steele, A., Fercho, H. L., Ammerman, T. A., & Anger, W. K. (2002). Interactive training versus reading to teach respiratory protection. *Journal of Occupational Health Psychology, 7*(4), 313–323.

*Eklöf, M., & Törner, M. (2005). Participatory analysis of accidents and incidents as a tool for increasing safety behaviour in fishermen: A pilot intervention study. *Work & Stress, 19*(4), 360–369.

*Elo, A. L., Ervasti, J., Kuosma, E., & Mattila, P. (2008). Evaluation of an organizational stress management program in a municipal public works organization. *Journal of Occupational Health Psychology, 13*(1), 10–23.

*Gardner, B., Rose, J., Mason, O., Tyler, P., & Cushway, D. (2005). Cognitive therapy and behavioural coping in the management of work-related stress: An intervention study. *Work & Stress, 19*(2), 137–152.

*Goldberg, L. S., & Grandey, A. A. (2007). Display rules versus display autonomy: Emotional regulation, emotional exhaustion, and task performance in a call center simulation. *Journal of Occupational Health Psychology, 12*(3), 301–318.

Haccoun, R. R., & Hamtiaux, T. (1994). Optimizing knowledge tests for inferring learning acquisition levels in single group training evaluation designs: The internal referencing strategy. *Personnel Psychology, 47,* 593–603.

*Hansen, A. M., Kaergaard, A., Andersen, J. H., & Netterstrom, B. (2003). Associations between repetitive work and endocrinological indicators of stress. *Work & Stress, 17*(3), 264–276.

Jackson, S. E. (1983). Participation in decision making as a strategy for reducing job-related strain. *Journal of Applied Psychology, 68,* 3-19.

Kray, L. J., Thompson, L., & Galinsky, A. (2001). Battle of the sexes: Gender stereotype confirmation and reactance in negotiations. *Journal of Personality and Social Psychology, 80,* 942–958.

Linden, A., Adams, J., & Roberts, N. (2006). Strengthening the case for disease management effectiveness: Unhiding the hidden bias. *Journal of Evaluation in Clinical Practice,12,* 140–147.

*Lundberg, U., Forsman, M., Zachau, G., Eklöf, M., Palmerud, G., Melin, B., & Kadefors, R. (2002). Effects of experimentally induced mental and physical stress on motor unit recruitment in the trapezius muscle. *Work & Stress, 16*(2), 166–178.

*May, D. R., Reed, K., Schwoerer, C. E., & Potter, P. (2004). Ergonomic office design and aging: A quasi-experimental field study of employee reactions to an ergonomics intervention program. *Journal of Occupational Health Psychology, 9*(2), 123–135.

*Mikkelsen, A., Saksvic, P. O., & Landsbergis, P. (2000). The impact of participatory organizational intervention on job stress in community health care institutions. *Work & Stress, 14*(2), 156–170.

*Moriana, J. A., & Herruzo, J. (2006). Variables related to psychiatric sick leave taken by Spanish secondary school teachers. *Work & Stress, 20*(3), 259–271.

*Nielsen, K., Fredslund, H., Christensen, K. B., & Albertsen, K. (2006). Success or failure? Interpreting and understanding the impact of interventions in four similar worksites. *Work & Stress, 20*(3), 272–287.

Pedhazur, E. J., & Schmelkin, L. P. (1991). *Measurement, design, and analysis: An integrated approach.* Hillsdale, NJ: Erlbaum.

Petterson, I. L., Donnersvärd, H. A., Lagerström, M., & Toomingas, A. (2006). Evaluation of an intervention programme based on empowerment for eldercare nursing staff. *Work & Stress, 20*(4), 353–369.

*Probst, T. M. (2002). Layoffs and tradeoffs: Production, quality, and safety demands under the threat of job loss. *Journal of Occupational Health Psychology, 7*(3), 211–220.

*Probst, T. M., Gold, D., & Caborn, J. (2008). A preliminary evaluation of SOLVE: Addressing psychosocial problems at work. *Journal of Occupational Health Psychology, 13*(1), 32–42.

Quick, J. C., & Tetrick, L. E. (Eds.). (2010). *Handbook of occupational health psychology* (2nd ed.). Washington, DC: American Psychological Association.

*Rau, R., Morling, K., & Rösler, U. (2010). Is there a relationship between major depression and both objectively assessed and perceived demands and control? *Work & Stress, 24*(1), 88–106.

*Rissén, D., Melin, B., Sandsjö, L., Dohns, I., & Lundberg, U. (2002). Psychophysiological stress reactions, trapezius muscle activity, and neck and shoulder pain among female cashiers before and after introduction of job rotation. *Work & Stress, 16*(2), 127–137.

Rohsenow, D. J., Monti, P. M., Martin, R. A., Michalec, E., & Abrams, D. B. (2000). Brief coping skills treatment for cocaine abuse: 12-month substance use outcomes. *Journal of Consulting and Clinical Psychology, 68,* 515–520.

Sackett, P. R., & Mullen, E. J. (1993). Beyond formal experimental design: Towards an expanded view of the training evaluation process. *Personnel Psychology, 46,* 613-627.

*Saksvik, P. O., Nytro, K., Dahl-Jorgensen, C., & Mikkelsen, A. (2002). A process evaluation of individual and organizational stress and health interventions. *Work & Stress, 16*(1), 37–57.

*Salema-Aro, K., Näätänen, P., & Nurmi, J. E. (2004). The role of work-related personal projects during two burnout interventions: A longitudinal study. *Work & Stress, 18*(3), 208–230.

*Searle, B. J. (2008). Does personal initiative training work as a stress management intervention? *Journal of Occupational Health Psychology, 13*(3), 259–270.

*Searle, B., Bright, J. E. H., & Bochner, S. (2001). Helping people to sort it out: The role of social support in the job strain model. *Work & Stress, 15*(4), 328–346.

Shadish, W. R., Cook, T. D., & Campbell, D. T. (2002). *Experimental and quasi-experimental designs for generalized causal inference.* Boston, MA: Houghton Mifflin.

*Shakespeare-Finch, J., Smith, S., & Obst, P. (2002). Trauma, coping resources, and family functioning in emergency services personnel: A comparative study. *Work & Stress, 16*(3), 275–282.

Solomon, R. L., & Lessac, M. S. (1968). A control group design for experimental studies of developmental processes. *Psychological Bulletin, 70,* 145–150.

Spector, P. E., Zapf, D., Chen, P. Y., & Frese, M. (2000). Why negative affectivity should not be controlled in job stress research: Don't throw out the baby with the bath water. *Journal of Organizational Behavior, 21,* 79–95.

Staw, B. M., & Oldham, G. R. (1978). Reconsidering our dependent variables: A critique and empirical study. *Academy of Management Journal, 21,* 539–559.

Stone-Romero, E. F. (2002). The relative validity and usefulness of various empirical research designs. In S. G. Rogelberg (Ed.), *Handbook of research methods in industrial and organizational psychology* (pp. 77–98). Malden, MA: Blackwell.

Stone-Romero, E. F., Weaver, A. E., & Glenar, J. L. (2002). Trends in research design and data analytic strategies in organizational research. *Journal of Management, 21*, 141–157.

Twenge, J. M., Baumeister, R. F., Tice, D. M., & Stucke, T. S. (2001). If you can't join them, beat them: Effects of social exclusion on aggressive behavior. *Journal of Personality and Social Psychology, 81*, 1058–1069.

*Vinokur, A. D., Schul, Y., Vuori, J., & Price, R. H. (2000). Two years after a job loss: Long-term impact of the JOBS program on reemployment and mental health. *Journal of Occupational Health Psychology,* 5(1), 32–47.

*Vuori, J., Silvonen, J., Vinokur, A. D., & Price, R. H. (2002). The Työhön Job Search program in Finland: Benefits for the unemployed with risk of depression or discouragement. *Journal of Occupational Health Psychology, 7*(1), 5–19.

Zohar, D., & Luria, G. (2003). The use of supervisory practices as leverage to improve safety behavior: A cross-level intervention model. *Journal of Safety Research, 34*, 567–577.

13

Event-Sampling Methods in Occupational Health Psychology

Sabine Sonnentag, Carmen Binnewies, and Sandra Ohly

Over the last decade, researchers have increasingly used the experience sampling method (ESM; cf. Hektner, Schmidt, & Csikszentmihalyi, 2007) as a methodological research strategy to examine a wide variety of research questions in occupational health psychology (OHP) and related fields. This methodological approach is also referred to as event-sampling method (Reis & Gable, 2000), ecological momentary assessment (Weiss, Nicholas, & Daus, 1999), or diary method (Bolger, Davis, & Rafaeli, 2003). Basically, the ESM approach aims at assessing people's experiences and behaviors as well as everyday events and situational conditions in situ, or as Bolger at al. (2003) stated it: "capturing life as it is lived" (p. 579). In this chapter, we will use the term *ESM* for this broader range of approaches that are comprised of event-sampling studies, ecological momentary assessments, and diary studies.

This chapter presents an overview of the ESM approach. In the first section, we describe core features of ESM studies. In the second section, we characterize typical research questions to be answered with ESM studies. We provide empirical examples from OHP in order to illustrate how this methodological approach can be used to examine research questions within this specific field. In the third section, we discuss how the ESM approach can advance theory in OHP. We will also address issues where theory development is needed. In the fourth and fifth sections we focus on the more practical side of ESM studies and provide information about how to conduct an ESM study and how to deal with specific challenges inherent in this approach.

Core Features and Types of ESM Studies

In an ESM study a person is typically asked to respond to survey questions multiple times either within a day or over several days; participants provide data over multiple days, sometimes even over several weeks. By asking a person about his or her feelings and certain situational factors at the very moment, ESM studies provide the advantage that memory biases (e.g., recency effects), different salience of experiences, and the problem that ratings may be influenced by momentary affective states are minimized. In addition, the ecological validity of ESM studies is considered to be very high because participants can be asked questions in their natural environments; for example, at their workplace or at home. As multiple measures are collected from the participants of ESM studies, researchers are able to investigate variation between persons as well as variation within persons. The most common statistical method to examine both within- and between-variation is hierarchical linear modeling (also often referred to as multilevel modeling). But ESM data can also be analyzed with other approaches (e.g., time series analysis).

Although all ESM studies aim at collecting multiple measures per person, there are three different types of data collecting protocols (Wheeler & Reis, 1991): (a) interval-contingent protocols, (b) signal-contingent protocols, and (c) event-contingent protocols. In interval-contingent protocols participants are instructed to report to the surveys at regular, predetermined intervals (e.g., every 4 hr). The length of the intervals should represent "theoretically or logically meaningful units of time" (Reis & Gable, 2000, p. 198). For example, in a study of Totterdell and Parkinson (1999), participants responded every 2 hr to surveys assessing their current mood and the affect regulation strategies they had used over the last 2 hr. In signal-contingent protocols, participants respond to surveys when a signal (e.g., the alarm of a handheld computer or cell-phone) is delivered. Signals can follow fixed or random schedules, or a combination of the two (e.g., random within a certain time interval, such as every 2 hr). For example, in a classical study of Csikszentmihalyi and LeFevre (1989), participants received signals randomly within 2-hr periods within a time interval from 7:30 a.m. to 10:30 p.m. to answer questions about their quality of experiences and performed activities. Event-contingent protocols require participants to respond to questions whenever a certain event (which has to be defined for participants) occurs. For example, in the study of Wheeler and Nezlek (1977), participants were instructed to fill in a survey whenever they encountered a social interaction. Wheeler and Nezlek (1977) defined a social interaction (the event) as "any encounter of 10 minutes or longer with another person(s) in which the participants attended to one another and adjusted their behavior in response to one another" (p. 743). Reis and Gable (2000) provide a comparison of the different data collection protocols and situations favoring a certain protocol: Interval-contingent protocols are recommended when the time interval is inherently meaningful (e.g., 1 working day), when susceptibility to retrospection bias is low and when participants' burden has to be minimized. This approach enables researchers "to conduct time-series analyses and evaluate cyclical patterns of variation and co-variation" (Reis & Gable, 2000, p. 200). Signal-contingent protocols are advised when researchers aim at describing and comparing "different domains of activity or mental states during different activities" (Reis & Gable, 2000, p. 200). This protocol should be used when susceptibility to retrospection bias is high and it is important to verify the time of recording. Event-contingent protocols should be used when researchers aim at investigating "a specific class of events or states, especially rare, clearly defined states" and comparing "relatively infrequent variations within a class of events" (Reis & Gable, 2000, p. 200). Examples for such events could be an argument with one's supervisor or the experience of flow at work. This protocol is also advised when susceptibility to retrospection bias is high and when it is important to obtain many episodes of the event or state in question. The data collection tools to conduct ESM studies advanced over the last years as new technology was introduced into everyday life. In addition to the more traditional paper-and-pencil surveys (Sonnentag, 2003), researchers now collect data by handheld computers (Totterdell & Parkinson, 1999), by web-based surveys (Ilies et al., 2007), as well as by cell phones or smartphones (Courvoisier, Eid, Lischetzke, & Schreiber, 2010; Song, Foo, & Uy, 2008).

Research Questions Answered by ESM Studies

ESM data can be used to examine several types of research questions. Bolger et al. (2003) suggested a typology of research questions to be answered with ESM studies, namely questions referring to (a) aggregation over time, (b) modeling the time course, and (c) modeling within-person processes. This typology, which is comprised of three types of research questions, is also a useful framework for summarizing typical research questions addressed by ESM studies within OHP. We start our description of the various approaches by describing research that refers to within-person processes, because this is the most popular approach when using ESM data within OHP; then we move to research questions related to time courses and finally address aggregation approaches. For all three types of research questions we provide specific examples of empirical studies.

Modeling Within-Person Processes

ESM studies can directly address within-person variability and examine the antecedents, correlates, and consequences of states that fluctuate within persons. For example, the level of time pressure an employee has to deal with may differ from day to day, and also a person's level of fatigue at the end of the working day varies from day to day (Grech, Neal, Yeo, Humphreys, & Smith, 2009). When examining the association between time pressure and fatigue from a within-person perspective, one tests if on days when an employee faces a high level of time pressure he or she experiences higher levels of fatigue at the end of the working day than on days when time pressure is low. Moreover, ESM data can be used to examine if persons differ with regard to within-person processes. Here, a typical research question would be: are there differences between persons in the strength of the association between time pressure and fatigue at the end of the working day? A next step then is to search for factors that can explain the between-person differences in within-person processes. For example, employees high on negative affectivity might react with more fatigue to time pressure than employees low on negative affectivity.

Within-person processes are addressed in several areas of OHP. For example, researchers are interested in how job stressors are related to strain symptoms (Ilies et al., 2007; Kammeyer-Mueller, Judge, & Scott, 2009; Zohar, 1999); how unfavorable social interaction processes that include social conflicts and the perception of unfair treatment are related to negative affect or depressive mood (Ilies, Johnson, Judge, & Keeney, 2011; Meier, Semmer, & Hupfeld, 2009); how experiences during work breaks and leisure time are associated with subsequent affective experiences and performance at work (Binnewies, Sonnentag, &

Mojza, 2009; Sonnentag, Binnewies, & Mojza, 2008; Trougakos, Beal, Green, & Weiss, 2008); or how emotion regulation strategies enacted at certain moments during the working day are related to the concurrent and subsequent experience of stress and exhaustion (Bono, Foldes, Vinson, & Muros, 2007; Judge, Woolf, & Hurst, 2009). Although most ESM studies solely rely on self-report data, some studies included physiological measures (Ilies, Dimotakis, & De Pater, 2010; Klumb, Hoppmann, & Staats, 2006) or assessments provided by significant others (Ilies, Wilson, & Wagner, 2009) to overcome problems associated with same-source bias.

To illustrate these kinds of studies, we will describe one study on the stressor–strain relation in more detail. Kammeyer-Mueller and his coworkers (2009) analyzed the association between day-specific job stressors, coping, and strain (emotional and physical exhaustion) in an ESM study, with a particular emphasis on the role of core self-evaluations and emotional stability as individual-difference variables that might buffer the negative effects of stressors on strain. A total of 252 persons were asked to complete a daily survey over a period of 2 weeks. Data from a total of 1,718 days were available for data analysis with the mean number of responses per person being 6.8. Between-person variation in job stressors was 63% of the total variance, and within-person variation was 37%. For strain, between-person variation was 42% of the total variance, and within-person variation was 58%. These figures demonstrate that both stressors and strains fluctuate within-person from day to day. Also, coping measures varied substantially between and within persons. For further data analysis, day-level predictor variables were centered at the person mean, implying that all between-person variance was removed from these variables. Hierarchical linear modeling showed that day-specific job stressors were a strong predictor of the day-specific strain level, also when taking day-specific coping processes into account. Interestingly, emotional stability moderated the relation between job stressors and strain: persons low on emotional stability had not only higher overall levels of strain, but showed also a stronger association between stressors and strain. Core self-evaluations did not moderate the stressor–strain relation. Overall, this study relying on day-specific measures demonstrates that elevated stressor levels go hand in hand with elevated strain levels on a day-to-day basis. Studies like this one have the potential to disentangle between-person differences from within-person processes. They make it possible for OHP research to be more process-oriented.

Modeling the Time Course

ESM data are also well suited for testing temporal dynamics. For example, the data can be used to examine how strain levels of an employee change over time (e.g., over the course of the working day or over the course of one or several work weeks). Moreover, it can be tested if employees differ in these changes (e.g., if older employees have a steeper increase in fatigue over the course of a working day than younger employees). Similar questions can also be addressed with more traditional longitudinal studies; but such more traditional longitudinal research designs often only comprise a small number of measurement occasions (ranging from two to four or five) and have longer time intervals between each measurement occasions (e.g., ranging from months to years), whereas ESM studies comprise a much higher number of measurement points (Beal & Ghandour, 2011; Fuller et al., 2003).

In typical studies addressing the time course, researchers examined how accountants experienced time pressure and strain over the course of a month (Teuchmann, Totterdell, & Parker, 1999) or how strain developed in university employees over the course of a semester (Fuller et al., 2003). A more recent study captured daily ratings of affect over the course of 21 days and examined the impact of fluctuations of daily affective events (i.e., intrinsic task motivation) and of a unique major negative event (i.e., hurricane Ike) on positive and negative affect (Beal & Ghandour, 2011). Again, there was substantial variation of positive (43%) and of negative (44%) affect within persons. The authors identified a cyclical pattern of affect over the week, a negative impact of hurricane Ike on positive affect (but no relations with negative affect), and an increase in positive affect over time. These findings imply that positive and negative affect follow a systematic cyclical pattern over the week (further analyses showed that positive affect had the lowest level on Wednesday and increased toward the weekend; negative affect showed a reversed pattern with high levels on Wednesday and low levels toward the weekend). Furthermore, the impact of the hurricane as a major disturbing event was reflected in a decrease in positive affect; the increase in positive affect over the study period, however, was unexpected. A reason for this unexpected increase in positive affect may be the unforeseen event of hurricane Ike and the specific timing of affect measurement with most measurement points occurring after the hurricane. Thus the increase in positive affect may reflect affective recovery from the hurricane and may be therefore very specific for this study. Complex study designs such as the one chosen by Beal and Ghandour (2011) enable researchers to simultaneously examine several intertwined processes and thereby disentangle the complex interplay between various factors that impact on affective experiences over time.

Aggregating Data over Time

Finally, ESM data can also be used to generate person-level data by aggregating information assessed at multiple measurement occasions at the person level. For example, multiple assessments of fatigue over several working days may be summed up in order to reflect an employee's average fatigue level during work. Analyses, then, are performed at the person level. Aggregating ESM data is advantageous over the use of single summary statement ("How fatigued are you on a typical working day?") that may suffer from retrospection bias. Aggregation processes may not only be used for the computation of average scores, but also for the computation of variability scores. For instance, variability scores would indicate the degree of fluctuation in specific job stressors (e.g., situational constraints, stressful customer interactions) and resources (e.g., social support, feedback) from day to day. Such scores would be needed to answer research questions such as: "Are more constantly present job stressors (as opposed to fluctuating job stressors) more detrimental for employees' well-being? Are constantly available resources (as opposed to fluctuating resources) more helpful in buffering the effects of job stressors on strains?"

In ESM studies within OHP, researchers have not often relied on aggregation of data. One reason might be that because gathering ESM data is time-consuming, researchers may want to use all the data for microlevel analyses and may eventually be reluctant to loose information when aggregating the data. Exceptions from this reluctance to aggregate ESM data can be found in a study on leadership and employee emotions (Bono et al., 2007). In this study, the researchers analyzed employees' emotions at work, depending on the type of interaction partner. Emotion data aggregated by interaction partners showed that the level of positive emotions was lower when participants interacted with their supervisors than when they interacted with coworkers or clients. Another study aiming at the predicting of task conflicts also used aggregated data (Spychala & Sonnentag, 2011). Specifically, situational constraints and proactive behavior as potential predictors of task conflicts were gathered with daily surveys over the period of four working days and subsequently aggregated at the person level; task conflicts as the outcome variable were assessed once at the person level (cf., also Weiss et al., 1999, for a study using aggregation).

Relevance of ESM Studies for Advancing Theories in OHP

Empirical studies using an ESM approach have clearly shown that affect and other experiences (e.g., the perceptions of stressors) fluctuate over time. This knowledge about the fluctuation of affect and other constructs has been very influential in some of the more recent theoretical developments in work and organizational psychology, including OHP. Probably the most prominent example of a theoretical approach that explicitly incorporates the fluctuation of affect over time is affective events theory (AET; Weiss & Cropanzano, 1996). Basically, AET states that persons show affective reactions to events encountered at the workplace that, in turn, result in affect-driven behaviors and work attitudes. By relying on empirical work that used ESM and similar methods, Weiss and Cropanzano (1996) argued that affect levels fluctuate over time and that the "patterns of affective reactions influence both overall feelings about one's job and discrete behaviors at work" (p. 11). Thus, basic assumptions of AET are closely linked to the availability of ESM methodology.

In addition to the relevance of ESM studies for testing assumptions of AET, ESM data are useful when testing more traditional theories in OHP. For example, the basic assumptions of the transactional stress model (Lazarus & Folkman, 1984) refer to rather short-term cognitive processes that make an ESM study a suitable approach for testing such processes. Using an ESM procedure, researchers may want to assess events at work, acts of primary and secondary appraisal, coping responses, and subsequent affective and other outcomes. Importantly, because appraisals may fluctuate within persons, ESM is a fruitful approach for tracking these within-person fluctuations. Empirical studies within and outside the area of OHP demonstrate that ESM provides new insights into appraisal and coping processes (Marco, Neale, Schwartz, Shiffman, & Stone, 1999; Ohly & Fritz, 2010; Weinstein, Brown, & Ryan, 2009). One of the major challenges with this approach refers to the difficulty of using a data-collection protocol that is fine grained enough to reflect subtle changes in appraisals and coping.

An important extension of the transactional stress model refers to the inclusion of person-level and situation-level variables as moderators in the appraisal and coping process. The associations between stress appraisals, coping, and affective outcomes may not be uniform for all persons and across all situations. For instance, the question whether one appraises an event as primarily harmful or primarily challenging may depend on more stable person factors and job characteristics. Furthermore, employees enjoying job control and other resources on the job may react differently to momentary stress appraisals than employees who do not have such resources available. Similarly, the effective use of coping strategies may be influenced by personality factors, for instance emotional stability. By modeling such cross-level interactions, ESM studies provide the opportunity to test this extension of the transactional stress model (for a similar approach, see Judge et al., 2009).

ESM studies are also important for developing and refining OHP theories in the future, particularly when it comes to theories referring to within-person processes and theories describing cumulative effects over time and other processes evolving over time.

Within-Person Processes

With respect to personality structure, Cervone (2005) has argued that within-person processes have to be distinguished from phenomena associated with between-person variation. Within-person processes refer to intraindividual processes "that underlie people's distinctive patterns of experience and action" (p. 424), whereas between-person approaches focus on "interindividual variations by identifying between-person categories or dimensions of population variation" (p. 424). A similar distinction also applies to theories in OHP and related fields. The within-person perspective aims at explaining the experience as well as the action (and often also reaction) of employees in an occupational context; for example, when facing a specific configuration of working conditions, which in itself is also dynamic. For example, a study may aim at describing how the level of sleeping problems fluctuates within persons, depending on what happened at work during the preceding day. The between-person perspective, on the other hand, aims at describing between-person variability that is associated with between-person differences. For example, studies adopting this perspective have examined person and situation factors that are associated with between-person differences in burnout (Demerouti, Bakker, Nachreiner, & Schaufeli, 2001; Halbesleben, 2006; Schaufeli & Bakker, 2004).

Until now, the majority of empirical studies in OHP have used study designs that allow conclusions about between-person variability, but not within-person processes. Also, the dominant approaches used for meta-analyses addressed between-person differences. Thus, this empirical work is helpful for testing and developing theories on between-person variability, but not for theories on within-person experiences and actions/reactions.

When addressing within-person fluctuation in meaningful variables, one core issue refers to theoretical models concerning the degree of within-person variation. Thus, to gain a better understanding of psychological processes relevant for work (and beyond), it is not only important to know *if* phenomena such as affect or attention fluctuate over time. It is also important to know their *degree* of fluctuation and to understand the factors that influence this degree of fluctuation. For example, it has been argued that the degree to which persons differ in their core affect trajectories is meaningfully related to other between-person differences (Kuppens, Van Mechelen, Nezlek, Dossche, & Timmermans, 2007). From the perspective of OHP, it is important to recognize that not only processes within persons may fluctuate over time, but that also features of the work situation might show some degree of fluctuation. For example, workload may differ from day to day or from week to week (Teuchmann et al., 1999). To achieve a better understanding about how aspects of the work situation impact on individuals' health it is important to keep in mind that not only the mean level of job stressors and other job-related features might be important, but also the degree to which they fluctuate over time. Theoretical models about the role of fluctuating situational features are still in their infancy. On the one hand, one might argue that a high degree of fluctuation implies a high level of unpredictability and therefore might lead to impaired well-being; on the other hand one could argue that a high level of fluctuation implies periods of low demands that allow for some recuperation between periods of high demands, what might result in better well-being. Thus, high levels of fluctuations in situational features might thus have positive or negative consequences for individuals' well-being and health.

In the context of performance research, Beal et al. (2005) have developed a model on episodic performance. By adopting a within-person perspective, they argued that job performance within persons is not stable over time, but largely fluctuates, also within working days. They proposed that within-person variability of performance can largely be attributed to fluctuations in self-regulation of attention, which in turn is assumed to be influenced, among other factors, by task-relevant affect, regulatory resources, and off-task attentional demands. This model focuses on performance and might therefore appear not to be of core relevance for OHP at first sight. However, we suggest that theoretical models on health-related processes within persons may be inspired by this model of episodic performance. Although major health impairments (besides impairments resulting from accidents) most often do not develop during single days, such a model might be highly fruitful for understanding health-relevant behaviors. Examples here include behaviors referring to unsafe acts and risky decision making (Daniels, Beesley, Cheyne, & Wimalasiri, 2008) or behaviors including alcohol use (Wang, Liu, Zhan, & Shi, 2010) and unhealthy eating (Jones, O'Connor, Conner, McMillan, & Ferguson, 2007). Theoretical models would need to specify the proximal and more distal person and situational processes that result in such behaviors. ESM studies are needed to provide the data for testing (parts of) such models; between-person data (even if gathered in longitudinal studies) would not be useful here. For example, when adopting theoretical approaches such as the ego-depletion model (Baumeister, Bratslavsky, Muraven, & Tice, 1998) specifically for the job context, it would be helpful to draw on ESM data.

Time-Related Processes

Time is an important aspect of theorizing about psychological processes, although the notion of time is often not explicitly addressed in psychological theories (George & Jones, 2000; Roe, 2008). For example, with respect to OHP it is important to know how long it takes for employees to develop signs of health impairment after they have been exposed to unfavorable working conditions or how long it takes for employees to calm down after they have been involved in a social conflict incident. Of course, many of the time-related questions ask for longitudinal studies with repeated measurements over longer periods of time (with time lags ranging between months and years).

However, theoretical models that aim at describing more short-term processes can benefit from the availability of ESM data, which are indispensable, particularly when modeling change trajectories over days and weeks. Here, one might think of developing novel approaches that describe change trajectories in affect and behavior as responses to experiences at work. Probably, it would be fruitful to integrate basic assumptions from affective events theory into this line of research (Weiss & Cropanzano, 1996).

How to Conduct an ESM Study in OHP

When conducting an experience sampling study, a number of decisions have to be made. In this section, we address rather practical issues such as the sample size required in experience-sampling studies, questionnaire design, measures, devices, and approaches to data analysis.

Sample Size

When planning a study, the question arises as to how many participants are required. The answer to this question depends on the primary aim of the study (temporal patterns, predictors on the daily level, or aggregation). It is also important to note that there are at least two sample sizes in experience sampling studies: number of participants (N) and total number of incidents (k) on the lower level (e.g., days, events). Often, researchers are interested in the relationship between daily stressors and daily well-being. Here, the total number of days (events) k (= N * days sampled per person) is relevant to determine the statistical power. Studies of this kind have typically used 100 to 200 participants, but only 4 to 10 daily assessments (e.g., Fritz & Sonnentag, 2009; Judge, Woolf, & Hurst, 2009; for a review see Ohly, Sonnentag, Niessen, & Zapf, 2010).

If researchers are interested in temporal patterns of stressors or well-being, an even smaller number of participants might be sufficient, but more daily (or event-related) assessments are needed. Studies of this kind have typically sampled less than 20 participants but used up to 75 daily assessments (Fuller et al., 2003; Teuchmann et al., 1999). When ESM data is aggregated to the person-level, sample size requirements are the same as in conventional survey studies, and N is the relevant sample size. Studies of this kind have sampled at least 100 participants (e.g., Fisher, 2002). In ESM studies, researchers are often interested in interaction effects between person-level and day-level (or event-level) variables (cross-level interaction). When testing such cross-level interactions, power issues need specific attention (Culpepper, Mathieu,

Aguinis, & Chen, 2010). More specifically, in multilevel studies cross-level interactions are often not detected because of insufficient power. Culpepper et al. identified the sample size at the higher level (in the case of ESM studies mostly the person level), the sample size of the lower level (in the case of ESM studies mostly event or day level), the variability of the lower level slopes, and the magnitude of the cross-level interaction as the most influential determinants of the power to detect cross-level interactions. Choosing larger sample sizes at the lower and the higher level of analysis is the most obvious (and feasible consequence) that should help in identifying cross-level interaction that are present in the specific data set.

Questionnaire Design and Measures

To reduce participants' burden and to enhance collaboration, issues related to questionnaire design also need to be considered. Reis and Gable (2000) recommended that filling out the questionnaire should not exceed 5 to 7 min each time, but others recommend even shorter durations (Sonnentag & Geurts, 2009; Uy, Foo, & Aguinis, 2010). The frequency of ESM assessment also needs to be determined in advance when planning the study. Multiple assessments per day provide a more accurate picture of individuals' experiences. Furthermore, lagged effects of stressors on outcomes such as well-being can be tested, and retrospective bias can be minimized when arranging multiple assessments. For example, the rating of sleep quality might be more accurate when participants are asked to report their sleep quality early in the morning instead of in the afternoon when returning home from work.

The requirement that ESM assessments should not take up too much time often requires the use of abbreviated scales (e.g., Van Hooff, Geurts, Taris, & Kompier, 2007), but these might not always be available so that established scales need to be adapted. Outcomes of interest to OHP researchers (affect, fatigue) are frequently conceptualized on a daily level (and instruments are available), but the same is not true for stressors or resources, so that adapting established scales is more problematic. For example, frequently having to deal with demanding customers (emotion work requirement) is conceptually different from having a single episode dealing with a demanding customer. Care should thus be taken when adapting scales to a specific ESM time frame and it may become necessary to collect new validation data.

Devices

Different devices are available for data collection: Paper questionnaires, web-based surveys, handheld computers, or smartphones (see Ohly et al., 2010; Sonnentag & Geurts, 2009; Uy et al., 2010 for detailed discussions of (dis-)advantages). The advantage of electronic devices is that compliance with the study protocol (e.g., predetermined times to answer the questions) can be tracked via an electronic time stamp. Furthermore, electronic devices can be programmed to remind participants of the questionnaire. Paper and pencil questionnaires have the advantage of being readily available. The use of smartphones is an attractive alternative since they became more common (see Song et al., 2008; Uy et al., 2010 for details; for a detailed discussion on paper versus electronic devices see Green, Rafaeli, Bolger, Shrout, & Reis, 2006).

Data Analysis

Data gathered with ESM methods need specific attention during data analysis. Because day-level (or event-level) data are nested within persons, multilevel approaches are often needed; they take the interdependence of data into account (Bryk & Raudenbush, 1992; Snijders & Bosker, 1999). When interested in trajectories of changes over time, time series analysis (cf. Fuller et al., 2003) or latent growth models (cf. Crosby et al., 2009) can be used.

Specific Challenges of ESM Studies

In the following, we will point out some of the specific challenges inherent in an ESM study. First, ESM studies require participants to devote a substantial amount of time and effort to constantly respond to all measurements occasions. Therefore, it is important to keep participants motivated over the time of data collection and to ensure that they comply with study instructions (e.g., answering a survey at the correct time or immediately after an event). In addition to strengthening participants' commitment by building a personal relationship, researchers can encourage participation by rewarding participants based on their compliance (e.g., the compensation is dependent on the number of surveys answered, participants get a higher chance of winning a prize if they answer all surveys). However, Green et al. (2006) warn against the use of incentives contingent on number of surveys because it might provoke faked responses.

A second issue prevalent amongst researchers in OHP refers to the generalizability of the findings and selectivity of the samples: Individuals who experience high levels of stressors or suffer from more strain symptoms might be less likely to be willing to participate at all, and might be more likely to drop out during the study process (see Ohly et al., 2010); here, establishing a good relationship with potential participants is even more important—see also Green et al. (2006) for a discussion of issues with noncompliance. Research with burnout patients, however, demonstrates that it is possible to conduct ESM studies with persons showing high strain levels (Sonnenschein, Sorbi, van Doornen, Schaufeli, & Maas, 2007).

Third, although new technologies facilitate conducting and monitoring ESM studies, conducting ESM studies is very time consuming also for researchers and requires substantial resources. Responding to surveys on a handheld computer or cell phone needs a minimum amount of instruction and training for participants, often requiring a personal meeting at the beginning of the study. As a consequence, sampling will be restricted to persons that can be personally approached by the research team; for instance, those persons living in a certain local area. Using web-based surveys requires that participants can easily access a computer with an Internet connection at a specific time of the day or immediately after a certain event. While this design may be a good choice when addressing research questions in the work context it may be difficult to implement when focusing on the interplay between work and private life because participants may be less motivated to check their (work-related) e-mails at home and to start their private computer in the evening to respond to a daily survey. In sum, it is most crucial that researchers design their ESM study in such a way that it requires a minimum amount of time and effort from participants and that it is most convenient, interesting, and pleasant.

Fourth, participants may miss several surveys or may not give answers at the exact times as instructed. Researchers should anticipate missing data in advance and design their ESM study in such a way that they have a sufficient amount of data points from each participant even if one or more surveys are missing or if some persons have to be eliminated from analyses (depending on the necessary statistical power, etc.). Therefore, the majority of researchers sample data over a period of 1 or

2 weeks resulting in data sets from 5 to 10 working days (Ilies, Scott, & Judge, 2006; Sonnentag, Binnewies, & Mojza, 2008). In addition, it is important that researchers examine whether missing data results in a selective sample of persons and measurement occasions (see Schafer & Graham, 2002 for an overview of how to deal with missing data). Determining if an answer was given at the right or wrong time (e.g., answering the lunch-break survey in the evening) is always dependent on study instructions (i.e., were participants required to answer the survey at an exact time or after an event?). To validate compliance with timing issues one should collect the necessary information by asking participants or using other data sources (e.g., recorded working times).

Furthermore, conducting an ESM study may cause participants' reactivity and influence the behavior of interest. For example, if one is interested in examining health behaviors, such as physical activities, with an ESM study, the repeated questions as to whether one engages in physical activities, or how long one engages in which activities may cause participants to reflect about their behavior and to change it (e.g., to engage in more physical activities than usual). If researchers think that reactivity might be a problem in their study, they should aim at testing such an effect by analyzing patterns of behavior over time (e.g., do participants increase or decrease certain behaviors over the time of data collection? Is there a day-of-the-week effect?), validating participants' self-report data with other-source data (e.g., observation, archival data, ratings of other persons), or using a control group design to test reactivity effects of certain questions. Another option is to rotate the specific items that are used at the single measurement occasions so that it becomes difficult for participants to track and to anticipate what will be asked at the various measurement occasions (Miner, Glomb, & Hulin, 2005).

Conclusion

Taken together, the ESM approach is a useful method for addressing a broad range of research questions in OHP that could not be examined with other methods. It is particularly important when researchers want to take a close look at processes. Strictly speaking, most ESM designs do not allow drawing conclusions about causality. But the ESM approach can be easily combined with field experiments. Moreover, ESM has an enormous potential to inspire and inform further theory development within OHP. As the availability of affordable electronic devices and software solution is increasing, we can expect that the use of ESM will also increase.

References

Baumeister, R. F., Bratslavsky, E., Muraven, M., & Tice, D. M. (1998). Ego depletion: Is the active self a limited resource? *Journal of Personality and Social Psychology, 74*, 1252–1265.

Beal, D. J., & Ghandour, L. (2011). Stability, change, and the stability of change in daily workplace affect. *Journal of Organizational Behavior, 32*, 526–546.

Binnewies, C., Sonnentag, S., & Mojza, E. J. (2009). Daily performance at work: Feeling recovered in the morning as a predictor of day-level job performance. *Journal of Organizational Behavior, 30*, 67–93.

Bolger, N., Davis, A., & Rafaeli, E. (2003). Diary methods: Capturing life as it is lived. *Annual Review of Psychology, 54*, 579–616.

Bono, J. E., Foldes, H. J., Vinson, G., & Muros, J. P. (2007). Workplace emotions: The role of supervision and leadership. *Journal of Applied Psychology, 92*, 1357–1367.

Bryk, A. S., & Raudenbush, S. W. (1992). *Hierarchical linear models: Application and data analysis methods*. Newbury Park, CA: Sage.

Cervone, D. (2005). Personality architecture: Within-person structures and processes. *Annual Review of Psychology, 56*, 423–452.

Courvoisier, D. S., Eid, M., Lischetzke, T., & Schreiber, W. H. (2010). Psychometric properties of a computerized mobile phone method for assessing mood in daily life. *Emotion, 10*, 115–124.

Crosby, R. D., Wonderlich, S. A., Engel, S. G., Simonich, H., Smyth, J., & Mitchell, J. E. (2009). Daily mood patterns and bulimic behaviors in the natural environment. *Behaviour Research and Therapy, 47*, 181–188.

Csikszentmihalyi, M., & LeFevre, J. (1989). Optimal experience in work and leisure. *Journal of Personality and Social Psychology, 56*, 815–822.

Culpepper, S. A., Mathieu, J. E., Aguinis, H., & Chen, G. (2010). *Searching for needles in haystacks: Determining the power of testing for cross-level interactions in random coefficient modeling.* Paper presented at the Conference of the Society of Industrial and Organizational Psychology, Atlanta, GA.

Daniels, K., Beesley, N., Cheyne, A., & Wimalasiri, V. (2008). Coping processes linking the demands-control-support model, affect and risky decisions at work. *Human Relations, 61*, 845–874.

Demerouti, E., Bakker, A. B., Nachreiner, F., & Schaufeli, W. B. (2001). Job demands-resources model of burnout. *Journal of Applied Psychology, 86*, 499–512.

Fisher, C. D. (2002). Antecedents and consequences of real-time affective reactions at work. *Motivation and Emotion, 26*, 3–30.

Fritz, C., & Sonnentag, S. (2009). Antecedents of day-level proactive behavior at work: A look at job stressors and positive affect during the workday. *Journal of Management, 35*, 94–111.

Fuller, J. A., Stanton, J. M., Fisher, G. G., Spitzmüller, C., Russell, S. S., & Smith, P. C. (2003). A lengthy look at the daily grind: Time series analyses of events, mood, stress, and satisfaction. *Journal of Applied Psychology, 88*, 1019–1033.

George, J. M., & Jones, G. R. (2000). The role of time in theory and theory building. *Journal of Management, 26*, 657–684.

Grech, M. R., Neal, A., Yeo, G., Humphreys, M., & Smith, S. (2009). An examination of the relationship between workload and fatigue within and across consecutive days of work: Is the relationship static of dynamic? *Journal of Occupational Health Psychology, 14*, 231–242.

Green, A. S., Rafaeli, E., Bolger, N., Shrout, P. E., & Reis, H. T. (2006). Paper or plastic? Data equivalence in paper and electronic diaries. *Psychological Methods, 11*, 87–105.

Halbesleben, J. R. B. (2006). Sources of social support and burnout: A meta-analytic test of the conservation of resources model. *Journal of Applied Psychology, 91*, 1134-1145.

Hektner, J. M., Schmidt, J. A., & Csikszentmihalyi, M. (2007). *Experience sampling method: Measuring the quality of everyday life*: Thousand Oaks, CA: Sage.

Ilies, R., Dimotakis, N., & De Pater, I. E. (2010). Psychological and physiological reactions to high workloads: Implications for well-being. *Personnel Psychology, 63*, 407–436.

Ilies, R., Johnson, M. D., Judge, T. A., & Keeney, J. (2011). A within-individual study of interpersonal conflict as a work stressor: Dispositional and situational moderators. *Journal of Organizational Behavior, 32*, 44–64.

Ilies, R., Schwind, K. M., Wagner, D. T., Johnson, M. D., DeRue, D. S., & Ilgen, D. R. (2007). When can employees have a family life? The effects of daily workload and affect on work-family conflict and social behavior at work. *Journal of Applied Psychology, 92*, 1368–1379.

Ilies, R., Scott, B. A., & Judge, T. A. (2006). The interactive effects of personal traits and experienced states on intraindividual patterns of citizenship behavior. *Academy of Management Journal, 49*, 561–575.

Ilies, R., Wilson, K. S., & Wagner, D. T. (2009). The spillover of daily job satisfaction onto employees' family lives: The facilitating role of work–family integration. *Academy of Management Journal, 52*, 87–102.

Jones, F., O'Connor, D. B., Conner, M., McMillan, B., & Ferguson, E. (2007). Impact of daily mood, work hours, and iso-strain variables on self-reported health behaviors. *Journal of Applied Psychology, 92*, 1731–1740.

Judge, T. A., Woolf, E. F., & Hurst, C. (2009). Is emotional labor more difficult for some than for others? A multilevel, experience-sampling study. *Personnel Psychology, 62*, 57–88.

Kammeyer-Mueller, J. D., Judge, T. A., & Scott, B. A. (2009). The role of core self-evaluations in the coping process. *Journal of Applied Psychology, 94*, 117–195.

Klumb, P., Hoppmann, C., & Staats, M. (2006). Work hours affect spouse's cortisol secretion: For better and for worse. *Psychosomatic Medicine, 68*, 742–746.

Kuppens, P., Van Mechelen, I., Nezlek, J. B., Dossche, D., & Timmermans, T. (2007). Individual differences in core affect variability and their relationship to personality and psychological adjustment. *Emotion, 7*, 262–274.

Lazarus, R. S., & Folkman, S. (1984). *Stress, appraisal, and coping.* New York: Springer.

Marco, C. A., Neale, J. M., Schwartz, J. E., Shiffman, S., & Stone, A. A. (1999). Coping with daily events and short-term mood changes: An unexpected failure to observe effects of coping. *Journal of Consulting and Clinical Psychology, 67*, 755–764.

Meier, L. L., Semmer, N. K., & Hupfeld, J. (2009). The impact of unfair treatment on depressive mood: The moderating role of self-esteem level and self-esteem instability. *Personality and Social Psychology Bulletin, 35*, 643–655.

Miner, A. G., Glomb, T. M., & Hulin, C. (2005). Experience sampling mood and its correlates. *Journal of Occupational and Organizational Psychology, 78*, 171–193.

Ohly, S., & Fritz, C. (2010). Work characteristics, challenge appraisal, creativity and proactive behavior: A multi-level study. *Journal of Organizational Behavior, 31*, 543–565.

Ohly, S., Sonnentag, S., Niessen, C., & Zapf, D. (2010). Diary studies in organizational research: An introduction and some practical recommendations. *Journal of Personnel Psychology, 9*, 79–93.

Reis, H. T., & Gable, S. L. (2000). Event-sampling and other methods for studying everyday experience. In T. H. Reis & M. C. Judd (Eds.), *Handbook of research methods in social and personality psychology* (pp. 190–222). New York: Cambridge University Press.

Roe, R. A. (2008). Time in applied psychology: A study of "what happens" rather than "what is." *European Psychologist, 13*, 37–52.

Schafer, J. L., & Graham, J. W. (2002). Missing data: Our view of the state of the art. *Psychological Methods, 7*, 147–177.

Schaufeli, W. B., & Bakker, A. B. (2004). Job demands, job resources, and their relationship with burnout and engagement: a multi-sample study. *Journal of Organizational Behavior, 25*, 293–315.

Snijders, T. A. B., & Bosker, R. J. (1999). *Multilevel analysis: An introduction to basic and advanced multilevel modeling.* London: Sage.

Song, Z., Foo, M.-D., & Uy, M. A. (2008). Mood spillover and crossover among dual-earner couples: A cell phone event sampling study. *Journal of Applied Psychology, 93*, 443–452.

Sonnenschein, M., Sorbi, M. J., van Doornen, L. J. P., Schaufeli, W. B., & Maas, C. J. M. (2007). Electronic diary evidence on energy erosion in clinical burnout. *Journal of Occupational Health Psychology, 12*, 402–413.

Sonnentag, S. (2003). Recovery, work engagement, and proactive behavior: A new look at the interface between nonwork and work. *Journal of Applied Psychology, 88*, 518–528.

Sonnentag, S., Binnewies, C., & Mojza, E. J. (2008). "Did you have a nice evening?" A day-level study on recovery experiences, sleep and affect. *Journal of Applied Psychology, 93*, 674–684.

Sonnentag, S., & Geurts, S. (2009). Methodological issues in recovery research. In S. Sonnentag, P. L. Perrewé, & D. C. Ganster (Eds.), *Current perspectives in job-stress recovery. Research in ooccupational stress and well-being* (pp. 1–36). Bingley, UK: Emerald.

Spychala, A., & Sonnentag, S. (2011). The dark and the bright sides of proactive work behavior and situational constraints: Longitudinal relationships with task conflicts. *European Journal of Work and Organizational Psychology, 20*, 564–580.

Teuchmann, K., Totterdell, P., & Parker, S. K. (1999). Rushed, unhappy, and drained: An experience sampling study of relations between time pressure, perceived control, mood, and emotional exhaustion in a group of accountants. *Journal of Occupational Health Psychology, 4*, 37–54.

Totterdell, P., & Parkinson, B. (1999). Use and effectiveness of self-regulation strategies for improving mood in a group of trainee teachers. *Journal of Occupational Health Psychology, 4*, 219–232.

Trougakos, J. P., Beal, D. J., Green, S. G., & Weiss, H. M. (2008). Making the break count: An episodic examination of recovery activities, emotional experiences, and positive affective displays. *Academy of Management Journal, 51*, 131–146.

Uy, M. A., Foo, M.-D., & Aguinis, H. (2010). Using experience sampling methodology to advance entrepreneurship theory and research. *Organizational Research Methods, 13*, 31–54.

Van Hooff, M. L. M., Geurts, S., Taris, T. W., & Kompier, M. A. J. (2007). How fatigued do you currently feel? Convergent and discriminant validity of a single-item fatigue measure. *Journal of Occupational Health Psychology, 49*, 224–234.

Wang, M., Liu, S., Zhan, Y., & Shi, J. (2010). Daily work–family conflict and alcohol use: Testing the cross-level moderation effects of peer drinking norms and social support. *Journal of Applied Psychology, 95*, 377–386.

Weinstein, N., Brown, K. W., & Ryan, R. M. (2009). A multi-method examination of the effects of mindfulness on stress attribution, coping, and emotional well-being. *Journal of Research in Personality, 43*, 374–383.

Weiss, H. M., & Cropanzano, R. (1996). Affective events theory: A theoretical discussion of the structure, causes and consequences of affective experiences at work. In B. M. Staw & L. L. Cummings (Eds.), *Research in organizational behavior* (Vol. 18, pp. 1–74). Stamford, CT: JAI Press.

Weiss, H. M., Nicholas, J. P., & Daus, C. S. (1999). An examination of the joint effects of affective experiences and job beliefs on job satisfaction and variations in affective experiences over time. *Organizational Behavior and Human Decision Processes, 78*, 1–24.

Wheeler, L., & Nezlek, J. (1977). Sex differences in social participation. *Journal of Personality and Social Psychology, 35*, 742–754.

Wheeler, L., & Reis, H. T. (1991). Self-recording of everyday life events: Origins, types, and uses. *Journal of Personality, 59*, 339–354.

Zohar, D. (1999). When things go wrong: The effect of daily work hassles on effort, exertion and negative mood. *Journal of Occupational and Organizational Psychology, 72*, 265–283.

14

Sampling in Occupational Health Psychology

An Epidemiological Perspective

Sue Ann Sarpy, Felicia Rabito, and Nancy Goldstein

As a discipline, occupational health psychology (OHP) relies on a multidisciplinary approach from diverse specialties (Leka & Houdmont, 2010; Quick & Tetrick, 2003; Sauter, Hurrel, Fox, Tetrick, & Barling; 1999). Traditional public health fields, such as epidemiology, have received increased attention in their potential contribution to advance the field of OHP through scientific research and methodology (Macik-Frey, Quick, & Nelson, 2007). Epidemiology is "the study of the distribution and determinants of health-related events and the application of this study to control health problems" (Last, 2000, p. 84). Epidemiology has several specific objectives that are consistent with those found in OHP (Gordis, 2009). One objective of epidemiology is to identify the etiology of disease through the identification and quantification of relevant risk factors. This objective directly contributes to the field of OHP in that identification of risk factors is a necessary first step in the development of intervention and prevention programs. Another objective of epidemiology is to study the natural history of disease to determine baseline rates and to identify trends in disease occurrence. A final objective is the design and evaluation of intervention programs developed in response to risk factors found to be associated with disease. These objectives are consistent with occupational health psychology research efforts such as those found in intervention effectiveness research.

When one considers the role that the science of epidemiology may play in advancing occupational health psychology, it is apparent that previous efforts have tended to focus on research design and methodological issues (Kasl & Jones, 2003; Marmot & Madge, 1987). In fact, research methods, in general, and methodological issues surrounding workplace interventions and their evaluations have received substantial attention in designing and conducting occupational health psychology research (Taris, deLange, & Kompier, 2010). For example, in a review of longitudinal research examining workplace stress, 45 studies were assessed to identify high quality studies and label these studies worthy of the title as the "very best of the millennium" (deLange, Taris, Kompier, Houtman, & Bongers, 2003). Several critical criteria of optimal methodological quality were provided. These characteristics included type of longitudinal design, length of time lags between the waves of the study, quality of measures, method of statistical analysis, and nonresponse analysis. Interesting, although nonresponse analysis was among these criteria, sampling itself was not specified as a critical factor in the decision-making process toward assessing quality. In fact, sampling and related design issues have been relatively ignored by the field of occupational health psychology. Epidemiology, which emphasizes sampling and related methodological issues, can be used to bridge this gap. Therefore, the present chapter will provide an epidemiological perspective to emphasize the key concepts in sampling, the importance of survey sampling, applications of sampling to the field of occupational health psychology, and the practical implications for future research.

Basic Concepts

In order to provide an overview of survey sampling, it is necessary to provide some background information regarding basic concepts in sampling, the process of identifying a population of interest and selecting a portion of that population for inclusion in a study. The intent is to generalize findings from the sample to the total population. Sampling has two major advantages: gains in study efficiency and gains in study validity. Efficient samples are those that collect high quality data with the lowest cost. By using a sample of the population rather than the total population, it is readily apparent that time and resources are kept to a minimum with subsequent cost savings. With respect to sampling, a valid study is one where participants are selected in an unbiased manner (resulting in internal validity) and one in which study findings can be generalized beyond the study participants (external validity). Sampling can increase validity on both of these parameters as discussed in detail below.

A key component of sampling is distinguishing between various types of populations (Levy & Lemshow, 2008). The population to which one would like to apply the results is known as the *theoretical* or *target population*. For example, a researcher may be interested in studying the effects of exercise on employee stress and conducts a study at a given health care organization. In this case, the theoretical population may be all employed individuals in the United States. It is rarely possible or even desirable to study or even sample the theoretical or target population due to financial constraints and other practical considerations. However, through careful selection of study participants, results can be generalized beyond the study participants and the particular company for which they work.

More readily available is the *accessible or source population*. The source population refers to the population to which the researcher has access. This access may be attained through documentation such as membership lists, national registers, or telephone directories. This listing of the accessible population is called the *sampling frame*. The source population for a researcher studying the effects of exercise on employee stress may be all employees working for the health care organization.

Researchers need not include the entire source population in their studies. Sampling is employed to obtain an unbiased sample of the source population. In other words, the individuals who meet the inclusion criteria set by the researcher are eligible to participate and may be selected for the study (i.e., the *eligible population*). For example, the inclusion criteria may include only those employees who are full-time workers and receive company benefits. From this eligible population, the study participants are selected. This is referred to as the *selected sample* (Gliner, Morgan, & Leech, 2009). The selection can be performed in a number of different ways as will be described below. To continue the previous example, a researcher may randomly select 100 full-time employees from the health care organization to participate in their study designed to assess the effects of exercise on employee stress. Of the 100 employees asked to participate, only a percentage of them may actually agree and provide data for analysis. The data from these participants comprise the study population. The ratio of the study population to the selected population is known as the study's *response rate*. If 75 participants completed the study (study population) whose data are analyzed and reported from the 100 individuals who were selected (selected population), then the response is 75% (75/100). A low response rate reduces the quality of the sample, especially if the individuals who comprise the study population differ from those who did not respond. For this reason, it is important to report the response rate and to analyze the similarities and differences between responders and nonresponders (i.e., to assess for *nonresponse bias*). That is, when the response rate is less than 100%, one must consider the possibility that those who do not respond differ in some important way from those who do respond introducing bias into study findings (Thomsen, 2000). Therefore, the lower the response rate, the lower the internal validity of the study.

In order to make valid inferences from the study population to the target population, the sample must be representative of the target population. The ability to generalize findings from a study to a population is referred to as the *population external validity*, and it is based on the representativeness of the accessible population to the theoretical population, as well as the adequacy of the sampling method utilized and the response rate. External validity, however, involves more than generalizing the sample and includes the ability to generalize the study settings and variables.

Sampling Design

The procedure or process used to select the sample is referred to as the sampling method. As previously discussed, sampling is employed as an efficient way to capture data on a target population. There are two major types of sampling: probability sampling and nonprobability sampling. The difference between the two techniques lies in the way in which participants are selected. In probability sampling, every participant has a known, nonzero probability of being selected. The probability of any *element* or *unit* (for example, a study participant) appearing in the sample must be known. To do this, the sampling frame must be identified. Recall that the sampling frame is essentially a list including all elements or units in the population. Probability sampling involves some form of random selection (e.g., simple random sampling; stratified random sampling) from the sampling frame. Sampling is necessary to capture the inherent heterogeneity of the population of interest and probability sampling is an approach to reflect this variation with limited bias. Random selection of study participants coupled with high response rates will result in an unbiased study population.

Nonprobability sampling, on the other hand, does not involve random or systematic selection and is subject to bias. Studies that rely on nonprobability sampling are subject to selection bias because individuals who choose to participate in the study are typically not representative of the total population that they are meant to represent (selection bias). Therefore, the major limitation of nonprobability sampling is that it precludes generalizing study findings to the larger target population. This sampling technique is often used when there is no reliable roster of study participants from which to sample, when convenience is a major consideration, and when there are cost and time restrictions. Probability and nonprobability sampling techniques will be discussed in more detail below, with examples of research studies using each sampling technique to illustrate their application in OHP research.

Probability Sampling

Random Sampling

Random sampling involves selecting study participants in such a way that all participants in the study population have an equal and independent chance of being included in the sample. Two most commonly used techniques are simple random sampling (i.e., using a table of random numbers, a computer random number generator, or a mechanical device) and systematic random sampling (i.e., every Nth unit is selected from the list of population members). Random sampling is considered the purest form of probability sampling and ensures that a representative sample will be selected given a large enough sample and high response rates. Probability sampling requires that the source/accessible population is defined by the researcher, the population is enumerated, and that the researcher has access to the population from which to draw the sample. In the case of very large populations, it may be difficult to identify *all* members of the population and a roster from which to sample is often unavailable. Because of these requirements, nonprobability sampling is often used.

Although not numerous, there are examples of random sampling within the OHP literature. Bamberger, Sonnenstauhl, and Vashdi (2006) used random sampling to examine the performance of the Drinking Problems Index (DPI), a screening instrument for drinking problems and problematic alcohol consumption patterns among older, blue collar workers. Bamberger et al., (2006) expanded upon the previous research base, which included study participants selected from hospitalized, emergency room/ambulatory, or primary care patients being treated for disorders other than alcoholism. The reliance on study participants obtained from clinical settings limited the generalizability of the findings to the broader population. To address this limitation, Bamberger et al. (2006) included a population-based sample. The subjects were identified from membership lists of nine national and local unions employed in three blue collar sectors in the United States (i.e., manufacturing, transportation, and construction). Previous research had identified that blue collar workers in these sectors have a high risk of becoming heavy drinkers and developing alcohol problems. All eligible local union members were included in the sampling frame and a random sample of 1,279 national union members was selected, producing an overall response rate of 46%. Although the study findings suggested that the DPI was an effective screening instrument, several limitations were identified based on the study sample and follow-up analyses that were conducted. Because the study included primarily White, male workers, the possibility exists that the DPI may not be as effective in screening for older members of ethnic minorities or women. The study does, however, demonstrate a significant step forward in sampling in this research domain beyond the previous reliance on clinical studies.

Another example of the use of random sampling in OHP research is presented in examining the severity of work psychosocial stressors and occurrence of musculoskeletal discomfort and the relationship between musculoskeletal discomfort and work-related stress. Palliser, Firth, Feyer, and Paulin (2005) conducted a cross-sectional survey of 524 dentists randomly selected from 1,562 dentists listed in the New Zealand Dental Register. Of these, 505 dentists were found to be eligible for inclusion in the study and 413 completed the surveys for a response rate of 82%. The authors concluded that the sample was representative of the population of working dentists in New Zealand with respect to age, gender, and ethnicity. Because the study obtained a high response rate (82%), it is unlikely that response bias occurred. However, because only dentists currently practicing were eligible for inclusion in the study, selection bias may have been introduced into the study sample. Musculoskeletal disorders and stress-related illness are the leading causes for early retirement among dentists; therefore, limiting inclusion to currently practicing dentists excludes those who may have retired due to their illness, which would result in an underestimate of the association between stress and musculoskeletal disorders. To capture this population and increase the validity of the study, a more expansive sampling frame could have been constructed to include both currently employed and retired dentists. This is an example of the challenges of survey sampling. High response rates, coupled with a representative sampling frame, are needed for an unbiased effect estimate.

Stratified Random Sampling

Stratified random sampling (also referred to as proportional or quota random sampling) is employed to ensure that all segments of a population, particularly those not in the majority, are represented in the sample. It is used when the researcher has knowledge of common characteristics of the accessible population or study frame that are important in obtaining a representative sample. To conduct stratified random sampling, the population is placed into strata. A stratum is a subset of the population that shares a common characteristic. A stratum could be males and females, managers and nonmanagers, or geographical regions. After the researcher identifies the relevant strata, their actual representation in the population is enumerated. Subjects are then randomly sampled from each stratum so that the sample contains the proportion of subjects as in the overall population. Thus, the overall population is represented as well as the key subgroups of interest in the population.

There are two types of stratified random sampling: *proportionate and disproportionate.* Proportionate stratified random sampling uses the same sampling fraction within each stratum. Disproportionate stratified random sampling uses different sampling fractions in the strata and it is used when the subgroup of interest is extremely small (e.g., minority groups). In this case, the researcher may oversample the small group to ensure a large enough sample for meaningful analysis. Effect estimates are statistically adjusted (i.e., weighted on the within-group estimates) to account for the weighted sample.

Stratified random sampling generally has greater statistical precision than simple random sampling but only to the extent that the strata are homogeneous. If the groups within the strata are homogeneous, the variability within-groups is expected to be lower than the variability for the population as a whole. Stratified sampling capitalizes on that fact; sampling variation is reduced and representativeness of the sample to the population is increased.

Stratified random sampling has been used in OHP research. Lacey, Lewis, and Sim (2007) provided a noteworthy example of its use in a cross-sectional study investigating the relationship between piecework and musculoskeletal pain and general health, and the extent to which this is influenced by perceived workplace psychosocial factors in a general population in the United Kingdom (UK). In order to select a representative sample of the general population in the UK, the sampling frame consisted of the general practice (GP) database of the North Staffordshire District Health Authority in the UK (the GP database as a whole contains approximately 98% of the population in the UK). A random sample was selected in which 10,000 adults, drawn from a broader epidemiological study (Sim, Lacey, & Lewis, 2006), were chosen in equal numbers from four age groups: 18 to 44; 45 to 54; 55 to 64; 65 to 75. This is an example of proportionate stratified random sampling. Nonrespondents were contacted twice, at two weekly time intervals in an effort to bolster response rates. A total of 5,133 people returned the questionnaire (54%). A total of 1,193 respondents currently employed in their main job completed the question regarding piecework. Of these individuals, 201 reported being paid according to their rate of work (i.e., piecework) in their current main job, whereas 992 reported that their main job was nonpiecework.

The sampling technique used in this study advanced this area of research in several ways. First, the study was population-based rather than industry-specific, the latter of which previously tended to dominate this research stream. Similarly, whereas previously most studies in this domain focused on female-specific samples, the current study included both male and female piecework employees.

Another noteworthy example of a stratified random sample is presented in an interdisciplinary approach to examine voice use, vocal disorders, and psychosocial working conditions among teachers (Bermudez de Alvear, Martinez-Arquero, Baron, and Hernandez-Mendo, 2010). This study included a representative sample of 282 teachers from 51 kindergartens and elementary schools selected from currently employed, full-time kindergarten and primary school teachers. The source population included 2,174 teachers employed in 82 public schools, in the Malaga City District of Andalusia, Spain. This population was divided into six geographic strata with similar numbers of teachers to create a proportionally distributed sample. In addition, the variety of socioeconomic levels was equally represented in every stratum and thus social interactions between teachers and students were homogeneous and comparable.

Surveys were mailed to 879 teachers. Two hundred eighty-two teachers (32% response rate) from 51 schools completed the surveys, representing 13% of the total teaching population in the Malaga City District. Given the low response rate, the ability to generalize these findings to the total population of teachers is limited. However, the number of female and male teachers in the sample was equal to that of the reference population as well as the age of the subjects in the sample was comparable to the general population of teachers increasing the confidence that the sample is representative and that findings can be generalized beyond the study sample.

Cluster Sampling

Cluster sampling is a probability sampling technique in which the population of interest is divided into groups or clusters. Examples of clusters include towns, workplaces, schools, and hospitals. Each cluster must be mutually exclusive and together the clusters must include the entire population. To select the study sample, a *random sample* of clusters is selected from the population (e.g., a random sample of high schools is selected from the population of all high schools in the United States). After clusters are selected, individual units within the clusters are then selected. When all the units within a cluster are selected (e.g., all students from each of the selected high schools), the technique is referred to as *one-stage cluster sampling*. Alternatively, a random sample of participants from each of the clusters can be selected (e.g., a random sample of students from each of the selected high schools) for inclusion in the study. If a subset of participants is selected from each included cluster, this technique is called *two-stage cluster sampling*.

In cluster sampling, the clusters (e.g., high schools) are the *primary sampling unit* and the participants within the clusters are the *secondary sampling units*. It is important to keep these two levels in mind when evaluating the representativeness of the sample.

Cluster sampling is primarily used in situations where constructing a *sampling frame* that identifies every individual in the population is either impossible or cost prohibitive. Therefore, cluster sampling is an efficient sampling approach and can result in significant cost savings. Cluster sampling, although efficient, may reduce the representativeness of the sample because participants from a specific cluster tend to be more similar to each other than participants who are selected at random from the total population. Cluster sampling also may result in greater sampling error because sampling error may occur at each stage of the sample. For example, in the previous example provided, errors in the initial selection of high schools may occur in addition to those that may occur when selecting students within each high school. In simple random sampling, sampling errors occur only at a single point of subject selection.

Cluster sampling is cost effective when the individuals in the population are scattered over a geographically dispersed area. For example, if a researcher wanted a representative sample of all high school students in the United States and drew a *simple random sample* of high school students, he would likely have to visit many schools throughout the United States to collect data. With cluster sampling, high schools would first be clustered by geographic region, then students within each of the selected schools would be sampled, thus saving both time and research dollars.

Cluster sampling is well illustrated in a study conducted by Halkos and Bousinakis (2010) investigating the effects of stress and job satisfaction on productivity. The study population consisted of 425 randomly selected individuals working in private enterprises and public organizations throughout Greece. The target population was defined as all employees in private and public sectors in middle and high positions. In this cluster sample, job sectors served as the clusters. Seventeen job sectors were identified (e.g., agriculture, fish farming, manufacturing, and telecommunication/internet). A total of 94 companies were randomly selected for inclusion in the study. Next, employees from each of the 94 companies were randomly chosen to obtain the sample of 425 employees.

Briefly, results of this study showed that productivity is seriously affected by stress and satisfaction. As expected, increased stress was found to lead to reduced productivity and increased satisfaction was found to lead to increased productivity. Due to the cluster sampling methodology, results of this study are generalizable across job sectors among middle and high-level employees in Greece. The authors note that a limitation of their study is the lack of generalizability to other populations where the relation between stress and productivity may be reasonably expected to differ (e.g., high and middle level employees in America).

Nonprobability Sampling

Quota Sampling

In quota sampling, the population is first segmented into *mutually exclusive* subgroups based on a particular characteristic of the target population thought to be important to the research question (e.g., race, geographical region, age, gender). The proportion of the target population with each characteristic is enumerated and participants are then selected in a nonrandom manner until the quota for each subgroup is filled. With quota sampling, because the sample selection is non*random* (unlike probability sampling) the sample is often an unreliable representation of the target population. For example, in studies where participants are recruited in neighborhoods, interviewers might be tempted to interview only those people in the street who appear most helpful or willing to participate. The flaw in this approach is that the study population may be biased by an overrepresentation of a particular type of person and a systematic underrepresentation of others resulting in a biased sample because not everyone had an equal chance of selection. This nonrandom element is the greatest weakness of quota sampling. However, quota sampling is useful when time is limited, a *sampling frame* is not available, the research budget is very tight, or when detailed accuracy or generalizability is not critical.

A study conducted by Wadsworth et al. (2007) is an example of quota sampling in the field of OHP. The study aimed to establish rates of perceived work stress in three ethnic groups (i.e., Black-African-Caribbean, Bangladeshi, and UK born Whites) and focused on the contribution of the reported experience of racial discrimination on perceived stress, and assessed the association between perceived work stress and well-being.

Using quota sampling, interviews with 200 respondents from each of the three ethnic groups were intended. Altogether, 3,181 households were selected from the electoral registers and post office address files for five council wards in Hackney and Tower Hamlets, East London. The total population of the five wards was 47,722, of which 10% were Black African-Caribbean, 19% Bangladeshi, and 55% Whites. Interviews were attempted at 3,176 households. If the interview was refused, or the resident did not fit the research criteria, interviewers tried up to six households on either side of the identified one until a participant was recruited. Inclusion criteria were employment in paid work, aged between 18 and 65 years, and self-reported ethnicity of Black African-Caribbean, Bangladeshi, or White (UK born).

In total, 626 people took part in the survey, for a response rate of 55%. Results indicate the perceived work stress may be caused by racial discrimination at work among Black African-Caribbean women, and this may affect their psychological well-being.

The use of quota sampling allowed for the study of differing experiences of work stress, racial discrimination, and psychological distress among the three ethnic groups investigated in a convenient, relatively low cost, and less time-consuming fashion. However, it is not unreasonable to assume that those who had suffered discrimination and workplace stress were more likely to participate in the survey than less affected groups. Therefore, due to the nonrandom nature of subject selection, the ability to generalize findings from this sample to the overall population of people in these three ethnic groups is limited.

Convenience Sampling

Convenience sampling is the most commonly used nonprobability based sampling method and is commonly used in occupational health psychology research. Its popularity is due to its ease of use and limited cost. As with all nonprobability sampling, it does not involve *random* selection. Thus, convenience samples may or may not represent the target population, to whom research inferences are to be made. As the name implies, participants are selected on the basis of convenience, rather than relying on the theory of probability to represent the target or theoretical population. Convenience sampling is also known as accidental or haphazard sampling (Kish, 1995). The use of college students, passersby, and employees in a company willing to participate are all examples of convenience sampling.

Many examples of convenience sampling exist in OHP. One notable example is a study conducted by Mohr and Wolfram (2010) that investigated the importance of dynamic tasks, predictability, and social support in predicting stress reactions of managers. Study participants included 142 managers selected from 46 readily accessible German companies. The managers' response rate was 76%. In this study, the researchers did not define their theoretical or target population. They simply chose participants to whom they had easy access and who were willing to partake in the study. Study findings indicate that dynamic tasks can lead to increased stress reactions, and that this relation is moderated by the predictability of task and social support. However, the use of a convenience sample prohibits the generalization of these results to other workers, occupations, or work settings.

Another occupational health study that used convenience sampling was conducted by Grant and Langan-Fox (2007) to investigate the impact of the big five personality dimensions on the occupational stressor–strain relationship. The sample consisted of 211 managers from an Australian department store. The sample was drawn from 41 stores out of a total of 50 stores (selected at the organization's discretion). Findings suggested that three of the Big Five personality dimensions, Extraversion, Neuroticism, and Conscientiousness, play an important role in occupational health and well-being and that Agreeableness and Openness may be less important in this context. However, because a convenience sample was used, one cannot be sure who the study population represents, and, therefore, results cannot be generalized to other employees.

The most obvious criticism regarding convenience sampling in these studies is that the samples used are not representative of the population. Thus, study findings cannot be generalized to any other sample or population of interest (i.e., low external validity).

Respondent Driven Sampling

Certain populations are particularly difficult to sample. These include populations for which there is no enumerated sampling frame, populations where the sample characteristic is rare, or populations that are hard to reach because the sample characteristic is considered undesirable (e.g., injection drug users; Kral et al., 2010). In these situations probability sampling is difficult if not impossible. In these cases, one approach to sample selection is to obtain the study sample via convenience sampling as described above. Another approach that may be used is to identify the most accessible populations (e.g. injection drug users from needle exchange programs), and randomly recruit participants from these groups. However, the resulting sample, while statistically valid (if response rates are adequate), precludes generalizing these results to the larger population of drug users because those attending needle exchange programs are not representative of the total population of injection drug users.

An alternative approach is to use a sampling method that relies upon referrals by study participants. Such methods are commonly referred to as network-based samples or chain referral samples. Two commonly used chain referral methods are snowball sampling and respondent driven sampling. In both methods, initial participants (seeds) are incentivized to recruit other study members from among their social network. An important principle of snowball sampling is the need for the seeds to be chosen at random from the target population. Given the lack of a sampling frame for hidden and hard-to-reach populations, random sampling of the initial population is not possible. The nonrandom nature of the study sample, as described above, introduces bias. Therefore, in snowball sampling, the resulting study population is comprised of volunteers, those most willing to participate, those who are most accessible, and those who are not isolated or fear social stigma (Heckathorn, 1997). Snowball sampling is an expedient way to select a convenience sample and, while snowball sampling increases sample coverage, the sample lacks statistical validity.

In response to the limitations of snowball sampling, respondent driven sampling (RDS) was introduced to compensate for the inability to select a random sample of study participants (Heckathorn, 1997). RDS combines elements of snowball sampling (chain referral) with a mathematical model that weights the sample in an effort to create population-based estimates. Briefly, as in snowball sampling, initial seeds recruit additional study members from within their social network. These new recruits in turn recruit other participants creating a chain referral system. RDS involves carefully tracking these recruitment chains and recording the size of each chain and their corresponding characteristics in order to quantify and describe each participant's social network. Survey results are then adjusted to compensate for clustering effects and varying inclusion probabilities of the sampling process (Heckathorn, 2002; Salganik & Heckathorn, 2004). Theoretically, with adequate length chains, the selected sample will be a statistical representation of the target population.

RDS has primarily been used to study injection drug users, HIV positive populations, prostitutes, and gay men. It also has been used extensively in the developing world where access to enumerated sampling frames is limited. However, there are examples of its use in other populations and specifically to occupational cohorts. RDS was used in a longitudinal study of undocumented Latino migrant workers in the Greater New Orleans area (Rabito et al., 2011). The purpose of the study was to assess the occupational and health experience of a cohort of Latino immigrants who recently migrated to the New Orleans area in search of day labor work and to assess the extent to which their occupation is related to select health outcomes. Immigrants, particularly those who are undocumented, may be disproportionately exposed to workplace health and safety hazards and may suffer from high rates of occupational injury. However, due to their undocumented status and legal concerns as well as language barriers, migrant workers are a hidden population. Given these constraints, they are not only hard to reach, but also a roster of workers, which could be used for sampling purposes, does not exist.

Therefore, studying Latino migrant workers and selecting unbiased samples is a challenge and RDS was employed. To begin enrollment of the Latino migrant worker cohort, eight seeds were identified in collaboration with organizations in the New Orleans area that serve Latinos. Country of origin was chosen as the demographic characteristic of interest in order to enroll a representative sample of Latino workers, and because of the degree of heterogeneity within Latino worker networks, interethnic mixing was unknown. Analysis of the final sample indicated that network ties were formed randomly, indicating heterogeneity of the study sample and overall representation of the sample to the larger population of Latino migrant workers.

Another example of the use of RDS was in a study of the work experience of jazz musicians in four U.S. cities (Heckathorn & Jeffri, 2001; Jeffri, 2003). Jazz musicians can also be considered a hidden population because a large portion of them do not belong to an organized group, and national labor statistics (often used as a source for occupational survey research) are not stratified by musical type. Therefore, constructing a sampling frame from which to sample jazz musicians is not possible and RDS was employed. In each of the four participating cities, a coordinator invited six to eight musicians as the initial seeds. Each of them was then asked to contact four additional study participants. These participants then recruited additional musicians to form recruitment chains. The use of RDS allowed for analysis of the social network of jazz musicians. The authors concluded that the RDS method highlighted the strength of social cohesion of the jazz community and identified specific patterns of affiliation. Data from the study are being used to devise strategies to promote the work and well-being of jazz musicians.

Sampling Issues and Occupational Health Psychology

The previous section focused on exemplary illustrations of the use of various sampling techniques in OHP research. However, it should be noted that in the vast majority of cases, nonprobability methods in general and convenience sampling in particular are used. As described above, convenience sampling is commonly used because it is not only time and cost efficient, but also may be the only sampling technique available to the researcher when studying particular populations. In an attempt to examine the extent to which convenience sampling is used in OHP research, the present authors conducted a review of sampling techniques in a particular topic area. More specifically, sampling techniques were identified in studies focusing on workplace stressors published in the *Journal of Occupational Health Psychology*, from January 2005 to the present. Of the 36 articles identified, 28 (78%) used convenience sampling to assemble the study population. Thus, although convenience sampling is least preferred because of the biases introduced by this technique and the inability to determine if the sample is truly representative of the target population, it remains a popular technique, particularly for exploratory research.

To completely ensure that the findings of a research study are representative of the population of interest, the entire target population should be investigated. Unfortunately, examining the entire target population is often not feasible. Selecting less than the total population (i.e., sampling) can be advantageous to researchers for several reasons. Most importantly, it is less expensive and it takes less time to study fewer subjects. Additionally, studying fewer subjects can allow a researcher to have better quality control over participants.

In order to reap the benefits of sampling and ensure that the study findings can be generalized to the target population, a representative sample must be utilized. If this does not occur, even the best designed studies can produce meaningless findings. It should be noted that many of the aforementioned studies and examples have discussed representativeness relative to demographic variables. However, oftentimes in OHP, representativeness refers to populations that are defined by other phenomena of interest including cognitive abilities, occupational characteristics, and risk factors. For example, a researcher may be interested in examining how the effects of an intervention differ on workers who are higher in conscientiousness relative to those workers who are lower in conscientiousness. However, the researcher only has access to workers who are higher in conscientiousness.

With this in mind, several steps must be taken to obtain a representative sample. First, the target or theoretical population must be clearly defined. Second, a source population should be identified and the sampling frame constructed. Finally, a sample of subjects should be chosen in a nonbiased manner (i.e., probability sampling) from the sampling frame. Each possible population participant should have a known, nonzero probability of being chosen as a member of the selected sample. This will result in a study sample that represents the target population.

In reality, the study sample, that is, the participants that actually complete the study and whose data are used in data analysis and reporting, is not likely to include the entire selected sample. The key issue to maintain validity of the study is to calculate the study's response rate. Lower response rates will lower the quality of the sample. The best that researchers can do in situations of low response rates is to determine if there are differences between study respondents and nonrespondents on variables relevant to the study. This information allows the researcher and consumers of such research to judge the representativeness of the actual sample and the validity of the study findings.

There are several features of research studies that lead to samples that are not representative of the target population. As noted throughout this chapter, researchers tend to choose their study sample based on convenience sampling. Probability sampling from an enumerated source population (as opposed to the target population) is seldom utilized. The problem with this strategy is that those who self-select to be in a study are seldom comparable to the total population that they are meant to represent. Finally, response rates are often not identified and no comparisons between respondents and nonrespondents are made.

Once a representative actual sample is obtained, there are a variety of methods of distributing subjects into the different study arms. Allocations of participants to a research group will impact the internal validity of the study. The internal validity of the study refers to lack of study bias and the subsequent ability to attribute study findings to the intervention under study. This topic is beyond the scope of this chapter; however, it is imperative to understand that internal validity will never be established even with the best methods of group assignment if the actual study sample does not represent the target population. Thus, the field of OHP would benefit from incorporation of sampling principles discussed in the present chapter.

References

Bamberger, P., Sonnenstuhl, W., & Vashdi, D. (2006). Screening older workers for drinking problems: Comparing CAGE and the Drinking Problem Index using a sample of retirement-eligible workers. *Journal of Occupational Health Psychology, 11*(1), 119–134.

Bermudez de Alvear, R. M., Martinez-Arquero, G., Baron, F. J., & Hernandez-Mendo, A. (2010). An interdisciplinary approach to teachers' voice disorders and psychosocial working conditions. *Folio Phoniatrica et Logopaedica, 62*, 24–34.

deLange, A. H., Taris, T. W., Kompier, M. A., Houtman, I. L. D., & Bongers, P. M. (2003). "The very best of the millennium": Longitudinal research and the demand-control-(support) model. *Journal of Occupational Health Psychology, 8*(4), 282–305.

Gliner, J. A., Morgan, G. A., & Leech, N. L. (2009). *Research methods in applied settings: An integrated approach to design and analysis* (2nd ed.). New York: Routledge.

Gordis, L. (2009). *Epidemiology* (4th ed.). Philadelphia, PA: Saunders.

Grant, S. L., & Langan-Fox, J. (2007). Personality and the occupational stressor–strain relationship: The role of the big five. *Journal of Occupational Health Psychology, 12*, 20–33.

Halkos, G., & Bousinakis, D. (2010). The effect of stress and satisfaction on productivity. *International Journal of Productivity and Performance Management, 59*(5), 415–431.

Heckathorn, D. D. (1997). Respondent driven sampling: A new approach to the study of hidden populations. *Social Problems, 44*(2), 174–199.

Heckathorn, D. D. (2002). Respondent driven sampling II: Deriving valid population estimates from chain referral samples of hidden populations. *Social Problems, 49*(2), 11–34.

Heckathorn, D. D., & Jeffri, J. (2001). Finding the beat: Using respondent-driven sampling to study jazz musicians. *Poetics, 28*, 307–329.

Jeffri, J. (2003). *Changing the beat: A study of the worklife of jazz musicians: Vol. 3. Respondent-driven sampling: Survey results by the Research Center for Arts and Culture* (Report No. 43). Washington, DC: National Endowment for the Arts Research Division.

Kasl, S. V., & Jones, B. A. (2003). A epidemiological perspective on research design, measurement, and surveillance strategies. In J. C. Quick & L. E. Tetrick (Eds.), *Handbook of occupational health psychology* (pp. 379–398). Washington, DC: American Psychological Association.

Kish, L. 1995. *Survey sampling.* New York: Wiley.

Kral, A. H., Malekinejad, M. Vaudrey, J., Martinez, A. N., Lorvick, J., McFarland, W., & Raymond, H. F. (2010). Comparing respondent-driven sampling and targeted sampling methods of recruiting injection drug users in San Francisco. *Journal of Urban Health, 87*(5), 839–50.

Lacey, R. J., Lewis, M., & Sim, J. (2007). Piecework, musculoskeletal pain and the impact of workplace psychosocial factors. *Occupational Medicine, 57*, 430–437.

Last, J. M. (2000). *A dictionary of epidemiology* (4th ed.). New York: Oxford University Press.

Leka, S., & Houdmont, J. (2010). *Occupational health psychology.* Malden, MA: Wiley-Blackwell.

Levy, P. S., & Lemeshow, S. (2008). *Sampling of populations: Methods and applications* (4th ed.). Hoboken, NJ: Wiley.

Macik-Frey, M., Quick, J. C., & Nelson, D. L. (2007). Advances in occupational health: From a stressful beginning to a positive future. *Journal of Management, 33*(6), 809–840.

Marmot, M. G., & Madge, N. (1987). An epidemiological perspective on stress and health. In S. V. Kasl & C. L. Cooper (Eds.), *Research methods in stress and health psychology* (pp. 4–26). New York: Wiley.

Mohr, G., & Wolfram, H. J. 2010. Stress among managers: The importance of dynamic tasks, predictability, and social support in unpredictable times. *Journal of Occupational Health Psychology, 15*(2), 167–179.

Palliser, C. R., Firth, H. M., Feyer, A. M., & Paulin, S. M. (2005). Muscoloskeletal discomfort and work-related stress in New Zealand dentists. *Work & Stress, 19*(4), 351–359.

Quick, J. C., & Tetrick, L. E. (2003). *Handbook of occupational health psychology.* Washington, DC: American Psychological Association.

Rabito, F. A., Perry, S., Salinas, O., Hembling, J. M., Schmidt, N., Parsons, P. J., & Kissinger, P. (2011). A longitudinal assessment of occupation, respiratory symptoms, and blood lead levels among Latino day laborers in a non-agricultural setting. *American Journal of Industrial Medicine, 54*(5), 366-374.

Salganik, M. J., & Heckathorn, D. D. 2004. Sampling and estimation in hidden populations using respondent driven sampling. *Sociological Methodology, 34,* 193–239.

Sauter, S. L., Hurrell, J. J., Fox, H. R., Tetrick, L. E., & Barling, J. (1999). Occupational health psychology: An emerging discipline. *Industrial Health, 37,* 199–211.

Sim, J., Lacey, R. J., & Lewis, M. (2006). The impact of workplace risk factors on the occurrence of neck and upper limb pain: A general population study. *BMC Public Health, 6,* 234–244.

Taris, T. W., de Lange, A. H., & Kompier, M. A. J. (2010). Research methods in occupational health psychology. In S. Leka & J. Houdmont (Eds.), *Occupational health psychology* (pp. 269–298). Malden, MA: Wiley-Blackwell.

Thomsen, S. 2000. An examination of non-response in a work environment questionnaire mailed to psychiatric health care personnel. *Journal of Occupational Health Psychology, 5*(1), 204–210.

Wadsworth, E., Dhillon, K., Shaw, C., Bhui, K., Stansfeld, S., & Smith, A. (2007). Racial discrimination, ethnicity and work stress. *Occupational Medicine, 57*(1), 18–24.

15

Quantitative Self-Report Methods in Occupational Health Psychology Research

Erin M. Eatough and Paul E. Spector

Self-report is the dominant method of data collection for research in occupational health psychology (OHP), as it is for many domains of psychology. With self-reports, the subjects of a study provide data about themselves and their experiences. Such reports can be qualitative (e.g., relating details of a stressful incident at work, Keenan & Newton, 1985) or they can be quantitative in which subjects make ratings of items along a continuum to reflect their standing on one or more theoretical constructs. The quantification of variables by subjects themselves, which can be done cheaply and easily, allows for the use of a large variety of inferential statistics to assess simple and complex relationships. Since most studies are concerned with establishing relationships among variables, the quantitative self-report study has become the method of choice for many, if not most, OHP researchers.

Despite the many advantages of using self-reports, there are some serious limitations that one must consider when interpreting results. Many of these limitations are not unique to self-reports, but at times can be even more severe with alternative methods, such as observer reports. Furthermore, there has been a tendency to distort and oversimplify self-report issues by attributing them to common method variance or monomethod bias (e.g., Podsakoff, MacKenzie, Lee, & Podsakoff, 2003). Finally, criticisms of self-report studies sometimes mistakenly attribute limitations that reside in the cross-sectional research design to the measurement method (Spector, 1994). Inability to draw confident causal conclusions, for example, is generally more an issue of design than measurement.

In this chapter we will discuss issues in the use of self-reports, paying particular attention to five questions. First, why are self-reports preferred over possible alternatives and are there times when self-reports are the best method to use? Second, what concerns have been raised about the use of self-report measures? Third, to what extent do self-reports converge with other measures of the same constructs in OHP research? Fourth, how do bias and confounding affect self-reports and how do they distort observed relationships among variables? Fifth, how can we best use self-reports and what sorts of inferences can be made with them? Our discussion will be limited to quantitative methods in which subjects make ratings on one or more items per theoretical construct because this is the dominant method in OHP research. Although much of our discussion will also hold for qualitative methods, there are additional issues with qualitative methods that space does not allow us to include (see Schonfeld and Mazzola's chapter on qualitative methods in this volume).

Why Are Self-Reports Used?

The self-report in quantitative survey research is preferred by many researchers over other methods for a variety of reasons. Cost is perhaps the most practical reason, given that many researchers have few resources to conduct their investigations. All that is required to conduct a self-report study is the distribution and collection of questionnaires that contain self-report items and scales of interest. Large numbers of subjects can be assessed at the same time, so the labor involved per subject can be relatively low compared to other methods. The use of web-based survey resources (e.g., My Survey Lab, Survey Gizmo, Survey Monkey, and Zoomerang), can reduce the cost of printing and postage, and eliminate the need for data entry, making the self-report survey even more attractive.

Beyond the practicalities of conducting research with limited resources, self-reports can be the most accurate means of assessment in many cases. Many of the variables we assess in OHP research are internal states that are best known to the subject himself or herself. Attitudes, cognitions, emotions, and physical discomfort, for example, are difficult to assess with methods that are independent of the individual in question. Thus, we are likely to use self-reports to assess perceptions of working conditions (e.g., role ambiguity or workload), psychological strains (e.g., anger or job dissatisfaction), or physical discomfort (e.g., musculoskeletal pain or stomach upset). It is assumed that these internal states can be assessed directly from subjects who are able and willing to disclose.

On the other hand some internal states might be apparent to individuals close to the subject. For example, coworkers might be aware of how an employee feels about the job, and they might observe displays of emotion. Thus it is possible to get reports by others, and these reports can share variance with the subject's own. For example, Spector, Dwyer, and Jex (1988) found that a supervisor's report of subordinates' job satisfaction correlated .55 with the subordinates' own job satisfaction reports. Furthermore, the pattern of correlations with other variables was similar between the two sources. This suggests that for some internal states, it might be possible to rely on other sources. It should be kept in mind, however, that in some cases subjects might consistently distort their true internal states in both expressions to others and survey responses. An employee, for example, might hide true feelings about work in interacting with colleagues at work as well as in responding to a survey. Although the use of other sources can help eliminate some sources of bias, as we will discuss later, it is not a panacea for limitations with self-reports.

Another alternative to self-reports is the use of implicit measures in which an individual might inadvertently reveal something about himself or herself without intending to do so (R. E. Johnson & Saboe, 2010). Such measures are considered indirect and can assess things that an individual is either unaware of or unwilling to reveal. The development of such measures, however, is difficult and relies on a number of assumptions about what the indirect measure should be assumed to assess. At the current time there are only a few developed measures that have evidence for construct validity that might be used. Indirect measures might show promise, but are not a substitute for direct self-report assessment of most constructs at the current time.

When interpreting self-report measures, it should be kept clearly in mind that they represent the internal states that respondents are willing to reveal. This is a particular issue for measures of an individual's perceptions of the environment, such as job stressors like role ambiguity or workload. Often self-reports are used when the researcher's clear intent is to draw inferences about objective features of the environment. This might occur in a study of the connection between the work environment and emotional strain. However, self-reports are primarily measures of individual perceptions that can be quite subjective and idiosyncratic to the individual. Certainly those idiosyncratic perceptions are vital as they serve a central role in tests of hypotheses and theories that are based on Lazarus's (1991) dominant transactional framework of stress (Perrewé & Zellars, 1999). However, one must be cautious in generalizing results from perceptions to the objective environment.

Concerns in the Use of Self-Reports

Although questions have been raised concerning the use of any kind of self-report, most criticisms have focused on two issues: the use of self-reports as measures of the work environment, and the sole use of self-reports in a study, often referred to as a single-source design. In such a design one includes measures of all variables in a survey so that all data come from the same source (job incumbents) measured with the same method (self-report survey). In this section we will discuss issues with self-reports as measures of the environment. We will discuss the issue of single-source designs in a later section on bias and confounding.

Arguments raised concerning the use of self-reports of the environment are predominately based on issues of validity and relevance. Of particular concern is the possibility that self-reports of the work environment might be contaminated by other internal states, such as emotions and moods, as well as an individual's disposition. Several authors have noted that emotional experience in particular complicates the validity of self-report measures of occupational stressors (e.g., Fox, Spector, Goh, & Bruursema, 2007; Frese & Zapf, 1999; Schaubroeck, 1999). Furthermore, as noted by Schaubroeck (1999), reliance on self-reports to measure both the dependent and independent variables may blur the boundaries between stressors and resulting strains. Use of objective measures can allow for better distinction between the dependent and independent variables (Kasl, 1998). Such a distinction is imperative to draw appropriate conclusions about the relationship between the work environment and potential outcomes of that environment, as well as to identify precisely where intervention is warranted.

Similar to this point, other concerns about practical relevancy have formed another common theme in criticisms of self-report methods. Some critics have firmly argued that if our goal is to increase knowledge about how to improve work environments for the sake of health, measuring the actual, or objective, exposure to stressors offers the most potential for success (e.g., Frese & Zapf, 1999; Schaubroeck, 1999).

Measuring perceptions of stressors only provides information about internal states, which are not typically the focus of intervention. Using objective measures can provide a clearer linkage to the actual job conditions and can therefore provide more information on what needs to be changed (Kasl, 1998). Furthermore, as Schaubroeck (1999) argued, significant health outcomes rarely occur from isolated perceptions of or brief emotional states in response to stressors, which are the constructs measured in common self-report tools. Thus, there should be more focus on chronic exposures that elicit particular emotions for prolonged periods of time.

However, data from secondary sources (e.g., coworkers) or from objective sources (e.g., records), have also been criticized for respective flaws. As noted by Fox et al. (2007) secondary source reports may be based on more limited information than self-reports. For example, Fox et al. (2007) noted that publicly recognized stressors such as interpersonal conflict or incivility are more likely to be perceived by outsiders than more privately experienced stressors such as role ambiguity. Additionally, objective sources of data are often costly and difficult to obtain (Kasl, 1998). Despite such limitations, secondary and objective data are necessary to provide evidence that might be consistent with self-reports and thus provide more confidence in conclusions, or be inconsistent and suggest that one should not accept the self-report results at face value.

Convergence of Self-Reports with Other Measures

Although not typical, some studies supplement self-report measures with other kinds of measures. Whereas most multisource studies merely have nonsubjects (e.g., coworkers or supervisors) complete parallel instruments about the subjects, generally using identical scales with an instruction to make ratings about the target employee, some studies use different instruments or methods to supplement the self-report. For example, Kirmeyer and Dougherty (1988) included in their study both a self-report and a count of hourly tasks completed as measures of workload. In cases such as this one, it is assumed that non-self-report measures are more objective than self-reports and would therefore be free of the subject's subjective bias (Frese & Zapf, 1988). Often such measures are considered somewhat of a standard against which to compare the self-report. In other words, a correlation between the two presumably independent measures is likely to reflect something about the objective nature of the construct. This would hold more for measures of the environment than internal states because such states have no measurable objective reality external to the person. The issues with such states are the person's ability (e.g., awareness) and willingness to disclose, as noted earlier.

In order to investigate how well subject self-reports converge with alternative methods and sources, we conducted a small meta-analysis of studies that reported correlations between self-reports and an alternative source of data on the same variables. We examined articles returned through an electronic database search using relevant keywords (i.e., occupational stressors and multisource). We were able to locate 12 studies that contained convergence correlations for one or more variables. There were a variety of alternative sources used including supervisors, coworkers, and subordinates who provided parallel data using the same instruments as the focal employees, and the use of observers or job analysis methods that used different instruments from the focal employee. Most studies had heterogeneous samples (i.e., had multiple occupations or organizations), and were conducted in the United States.

To be included in the meta-analysis, at least two studies had to be available for each variable. For inclusion, studies were required to compare employee self-reports with at least one alternate data source and if more than one independent source was reported, it was treated as a separate sample. Studies had to include analyses at the individual employee level as opposed to higher levels of aggregation. Environmental variables which met these criteria were workload, interpersonal conflict, constraints (organizational, job, and interpersonal), incivility, and job control/autonomy. We were also able to include counterproductive work behavior (CWB), which reflects a variable that can be considered a behavioral strain that is observable to others.

Table 15.1 shows the results of our meta-analysis. The table shows number of samples, total sample size across all samples, the mean correlation between self-reports and alternate sources, the standard deviation of the correlations, and the range of the correlations. Given the small number of studies, we did not conduct artifact adjustments.

As shown in the table, each of these variables showed a moderate level of convergence between self-reports and other sources/methods, with mean correlations ranging from .30 to .47. Workload showed the highest degree of convergence, whereas job control showed the lowest. Furthermore, there was considerable variance in convergence

Table 15.1 Meta-Analysis of Studies Reporting Convergent Validities for Measures of Occupational Stressors

Variable	Mean *r*	Standard Deviation	Range	Confidence Interval	*k*	*N*
Workload	.47	.11	.35 to .56	.37 to .57	3	363
Interpersonal Conflict	.43	.13	.30 to .66	.36 to .50	6	839
Constraints	.39	.10	.26 to .49	.32 to .46	6	886
Incivility	.39	.14	.29 to .49	.28 to .50	2	302
CWB	.31	.10	.21 to .47	.25 to .37	8	1150
Job Control	.30	.16	.15 to .53	.25 to .35	9	1423
Additional variables with only one study						
Job Demands[1]	.27				1	252
Role Ambiguity[2]	.08				1	135
Work–family Conflict Time-based[3]	.23				1	156
Work–family Conflict Strain-based[3]	.18				1	156
Work–family Conflict Behavior-based[3]	.00				1	156
Working Hours[2]	.83				1	135

Note: CWB = counterproductive work behavior; *k* = number of effect sizes; *N* = total subjects; Mean *r* mean correlation coefficient

1. Rau, Morling, & Rösler (2010)
2. Spector, Dwyer, & Jex (1988)
3. Carlson, Witt, Zivnuska, Kacmar, & Grzywacz (2008)

correlations, with the greatest being for job control that ranged from .15 to .53. Some of the variability for control is due to inclusion of two samples using the Factual Autonomy Scale (Spector & Fox, 2003) that was developed specifically to reduce subjectivity. Convergence for this scale tended to be higher than for other scales, such as the more commonly used autonomy subscale of the Job Diagnostic Survey (Hackman & Oldham, 1975).

The bottom of Table 15.1 summarizes the convergence correlations for variables for which we could find only a single study. As can be seen, convergence ranged from .83 for working hours to 0 for behavior-based work–family conflict. The extent of convergence for all of these variables might well reflect the extent to which they are primarily concrete and factual (e.g., working hours) or abstract and subjective (e.g., role ambiguity). Variables that are factual reflect things that are directly observable and verifiable. One could, for example, video-record a workplace and compute the number of hours each employee is present. Finding a means to observe and verify a variable such as role ambiguity would be far more challenging, as there is unlikely to be a way to directly observe it from recordings of employees. Rather one must infer the level of ambiguity from employee or other reports because ambiguity largely reflects an individual's subjective interpretation of the work environment. Thus one would not expect nonfactual environmental variables that individuals are likely to perceive differently to yield as high convergence across sources as factual variables.

In addition to these field studies, we summarized results of three laboratory studies relating stressful physical conditions to self-reports of those conditions (E. C. Johnson et al., 2010; Petruzzello, Gapin, Snook, & Smith, 2009; Riebe et al., 1997). These studies were concerned with evaluating the efficacy of self-reports of heat exposure, relevant to a number of occupations where heat stress is a concern, including athletes and firefighters. In these studies, correlations between objective measures of laboratory induced stressors (heat, dehydration, lifting tasks, and treadmill tasks) were .56 on average. However, this type of controlled quantification of stressors is difficult to obtain for many of the psychosocial stressors OHP researchers would like to study.

Another issue beyond convergent validity of self-report measures is whether or not results are consistent between self-reports and other measures when correlated with other variables. Eight studies related stressors and CWB to well-being (e.g., anxiety, frustration, and job dissatisfaction), and most reported similar, but smaller, correlations between the alternative-source ratings and well-being. Across the eight studies, 67 comparisons of effect sizes between self-report and alternative-sources stressor and CWB ratings with employee well-being were conducted (see Table 15.2). Of these, only 15 (22%) reported a stronger correlation between the alternative-source and well-being. For example, Penney and Spector (2005) found peer-reports of interpersonal conflict had a larger correlation with job satisfaction (r =. 36) than did self-reports (r = .27). Even though, in the majority of cases, the single source studies that relied entirely on self-report yielded higher correlations, the magnitude of differences varied by variable. Constraints had the largest mean effect size difference between self- and alternative reports at .23, with reports of CWB having, on average, only a .07 effect size difference with well-being.

Table 15.2 Meta-Analysis of Studies Reporting Two Sources of Measurement of Occupational Stressors and Negative Employee Well-Being

Variable	Mean *r*	Standard Deviation	Range	Confidence Interval	*k*	*N*	Average effect size difference
Workload							
Self	.28	.17	.05 to .56	.21 to .35		799	
Alternative	.11	.13	-.08 to .28	.04 to .17	6	799	.17
Interpersonal Conflict							
Self	.33	.12	.08 to .53	.28 to .38		1676	
Alternative	.23	.09	.09 to .36	.14 to .25	11	1532	.10
Constraints							
Self	.42	.11	.21 to .58	.36 to .48		1134	
Alternative	.19	.05	.13 to .26	.14 to .25	8	1134	.23
CWB							
Self	.27	.12	.09 to .41	.22 to .31		1946	
Alternative	.20	.07	.03 to .27	.02 to .34	12	1646	.07
Job Control							
Self	-.21	.12	-.51 to -.04	-.24 to -.18		3828	
Alternative	-.09	.11	-.27 to .04	-.12 to -.05	24	3644	.12
Role Ambiguity							
Self	.29	.18	.07 to .51	.21 to .37		592	
Alternative	.13	.09	.04 to .23	.05 to .21	4	592	.16
Incivility							
Self	.47	.01	.46 to .48	.38 to .56		446	
Alternative	.28	.14	.18 to .38	.17 to .39	2	302	.19

Note: CWB = counterproductive work behavior; *k* = number of effect sizes; *N* = total subjects; Mean *r* = mean correlation coefficient. Well-being outcomes include anxiety, frustration, job dissatisfaction, negative emotion, and physical symptoms.

These results suggest that convergence is very much a function of the specific variable in question. Some variables show considerably higher convergence than others perhaps because they reflect qualities that are either more concrete and therefore less subject to interpretation, or they are more observable (e.g., public behavior versus internal emotional states). Likewise differences in correlations between stressors and strains assessed with the same versus different methods varied according to the variables in question. However, for the most part the patterns of relationships were in the same direction, and only for job control was the all self-report significant and the multisource nonsignificant.

Bias and Confounding of Self-Report Measures

There is widespread suspicion about the use of self-report measures to determine relationships among constructs, generally expressed as concerns about common method variance or CMV (also referred to as common source bias or monomethod bias). The idea of method variance is generally attributed to Campbell and Fiske (1959) who suggested that a certain amount of variance in a set of observations is due to the measurement method itself. If true, this would suggest that measures of multiple constructs assessed with the same method would share a common source of variance attributed to that method, referred to as CMV. This shared source of variance would be expected to inflate observed correlations among measures. Although Campbell and Fiske were talking about methods in general, for some reason, concern about method variance in OHP (as well as other areas of psychology that rely on survey methods) has focused almost entirely on self-report measures used in cross-sectional survey studies. In a survey of 225 editorial board members from three top-tier journals (*Journal of Applied Psychology*, *Journal of Organizational Behavior*, and *Journal of Management*), Pace (2010) found that the majority feel that common method variance is of more concern with self-reports than other methods.

Opinions about CMV vary from those suggesting it is a myth (Spector, 1987) or urban legend (Spector, 2006) to those who feel it is a serious problem (Podsakoff, MacKenzie, Lee, & Podsakoff, 2003). Pace's (2010) survey of editorial board members suggests that there is far from consensus about the extent to which CMV is a problem for self-report studies. For example, her sample was nearly evenly split about whether or not CMV makes it difficult to draw conclusions from self-report studies. About a third of her sample said concerns about CMV would cause them to recommend rejection of a manuscript.

Undoubtedly much of the reason for diverging opinions, even among experienced researchers, is that CMV is a complex issue that has been confounded with a number of other issues concerned with the use of self-reports. Furthermore, the nature of what is meant by method variance, and even what constitutes a method is not always clear. One interpretation of method variance, based on the Campbell and Fiske (1959) perspective, is that the method itself affects observations in a systematic way regardless of the construct being assessed. This might be expected to create a constant method effect that leads to a more or less uniform inflation of observed correlations across measures (Richardson, Simmering, & Sturman, 2009 called this the noncongeneric perspective). Another perspective is that method variance occurs, but that the magnitude of method variance depends on the specific measure, so that some are more subject to method variance than others (Richardson et al., 2009 called this the congeneric perspective). According to this view it is the combination of method plus construct that determines whether there is method variance (Spector & Brannick, 1995).

The noncongeneric perspective is more easily tested than the congeneric. Spector (1987, 2006) has argued that there is little credible evidence that this sort of constant inflation effect actually occurs, and provides evidence to the contrary. For example, if there is a constant inflation, we would expect to find a floor effect in that correlations among all variables within a questionnaire would show some minimum and nonzero correlation. Spector (2006) and Spector and Brannick (2009) demonstrated that this is not the case as many correlations among variables within self-reports surveys are nonsignificant and close to zero. If, on the other hand, method variance is congeneric and a function of both method and the specific construct being assessed, clear evidence that the method itself is a source of correlation inflation is not easy to provide. Expected patterns of relationships, such as stronger correlations for variables measured with the same rather than different methods have a number of feasible alternative explanations other than method variance. For example, it might be that the self-reports are more accurate measures of intended constructs. It has been argued by Dalal (2005), for example, that supervisor ratings of CWB are more subject to halo and other rating errors than self-reports. The complexities involved have made it difficult for researchers who wish to determine if method variance might represent a significant problem for their investigation, because there are no simple procedures that can unequivocally demonstrate if method variance is a problem in a particular circumstance.

Another complicating factor is that the assumption that method variance would necessarily inflate relationships has not been supported by the literature. In a simulation study, Williams and Brown (1994) found that in many cases the existence of method variance would attenuate rather than inflate observed relationships. More recently Lance,

Dawson, Birkelbach, and Hoffman (2010) conducted analyses suggesting that the possible inflating effects of method variance were likely counterbalanced by an approximately equal amount of attenuation due to unreliability of measures. They argued that in many cases observed correlations among self-reports were likely accurate. Beyond simple bivariable linear correlations, simulation studies have shown that tests of moderated regression are not inflated (Evans, 1985) and may well be attenuated (Siemsen, Roth, & Oliveira, 2010) by CMV. Furthermore, Siemsen et al. found in their simulation that CMV did not inflate tests of quadratic terms in multiple regressions and that the potential effects of CMV on linear relationships are reduced as multiple variables are entered into a regression equation.

A number of statistical procedures have been recommended for detecting and adjusting for CMV (see Podsakoff et al., 2003 for a detailed and thorough overview). Many of these methods seem to implicitly assume noncongeneric CMV in that commonalities among a set of measures, or relationships of a set of measures with a marker variable, are treated as if it reflects the level of CMV. Richardson et al. (2009) conducted a simulation study to see whether the use of several of these methods would lead to more accurate conclusions under conditions of no method variance, congeneric method variance, or noncongeneric method variance. They found that these CMV techniques often lead to less accurate results than unadjusted zero-order correlations among measures. These sorts of findings make it difficult to recommend using adjustments to deal with the problem of CMV in self-reports (Conway & Lance, 2010).

How Best to Use Self-Reports

The problem with self-report measures is not that they are used, but rather in how they are used. Far too often researchers call upon this method to address questions that cannot be easily addressed with the self-report alone. These reports reflect an individual's personal and often subjective experience, and can be quite useful when the assessment of such is the researcher's intent (Perrewé & Zellars, 1999). Studies that are designed to assess the impact of the work environment and exposure to potentially health-damaging conditions cannot generally rely on such methods alone. Furthermore, results from self-report studies frequently are misinterpreted, assuming that perceptions reflect objective reality. Self-reports can be a valuable means of studying a variety of OHP topics when used appropriately. We make four suggestions about how to make the most of self-reports in OHP research.

Keep in Mind What Self-Reports Represent

One of the biggest issues in the use of self-reports, that has led to considerable criticism, is the misinterpretation of self-report variables as if they reflect objective reality. Far too often authors seem to equate constructs with their operationalization, as if there is a one-to-one correspondence between them. Thus the correlation between two scales is assumed to reflect the association, and sometimes even a causal connection, between two underlying constructs. For example, in self-report occupational stress studies, one might find a significant positive correlation between say a measure of role ambiguity and a measure of workplace anxiety. A precise conclusion is that individuals who perceive (or report) role ambiguity are likely to experience (or report) feeling anxious. In the absence of additional evidence, it would be mere speculation to conclude that exposure to role ambiguity as an objective environmental condition is associated with anxiety, and even more dangerous to assume a causal direction. It is quite likely that individuals who feel anxious will perceive the environment to be threatening, and perceive high levels of role ambiguity. This is not to say that such individuals do not experience ambiguity, as they very well might. It is just that the self-reports are telling us more about the individual than the environment.

Self-reports should be used when the clear purpose is to determine if variables that reflect subjective experiences are related. Establishing such connections is far from trivial, as an understanding of people's responses to the environment is very much dependent on determining how perceptions and potential reactions might relate (Perrewé & Zellars, 1999). On the other hand, determining these connections only tells us part of the story of how people respond to environments. It is also necessary to establish the connection between exposure to objective features of the environment (Frese & Zapf, 1999; Schaubroeck, 1999), and perceptions as well as potential outcomes.

Do Not Use Self-Reports Alone

Although not every study has to go beyond the investigation of people's internal states, ultimately the field of OHP is interested in determining the connection of the work environment with health, safety, and well-being. To get a complete picture, research studies need to use a variety of assessment methods and sources rather than relying only on self-reports. The use of such methods on both the presumed input (environment) and output (health, safety, and well-being) sides will enable us to draw conclusions that go beyond connecting perceptions to outcomes. Often the biggest advantage is that they can control for biases that may affect the self-report. Mood, for example, might have a biasing effect on all self-report variables of interest in a given study. Data taken from records, such as absence, are unlikely to have been affected by an employee's mood at the time of the survey. Thus a correlation between, say a stressor and absence would not be due to shared biases between the stressor and absence, allowing us to conclude that the stressor and absence are in fact related.

It should be kept in mind, however, that alternative methods are not a panacea, as they have their own limitations that often make them less accurate measures of constructs than self-reports (Frese & Zapf, 1988). Furthermore, when we find a relationship between a self-report and another measure, we are not in a position to draw a causal conclusion. In our stressor-absence example, it is possible that mood causes both perceptions of the stressor (people in bad moods are hypersensitive to stressors and report high levels), and absence (people who are frequently in bad moods avoid work). In order to draw more definitive causal conclusions, we need to rely on different research designs.

Incorporate Time into Research Designs

Self-report studies are often used when researchers are interested in the effects of environmental exposure on health, safety, or well-being. In order to determine effects we need to go beyond cross-sectional studies and incorporate the element of time into designs. This does not mean simply assessing variables of interest at random intervals, say 6 months or a year apart. Rather, one must make observations of the variables of interest at meaningful points in time as the process of interest unfolds to show changes from before to after exposure (Zapf, Dormann, &

Frese, 1996). Incorporating time in this way can be challenging because it can be difficult to determine the point at which exposures occurred. For example, how might one assess employees prior to and after role ambiguity occurs?

In order to incorporate time most effectively, one must be able to identify a sample of employees and assess them before and after an event occurred. Thus one might assess strains at Time 1 prior to announced layoffs and at Time 2 after layoffs have occurred. Changes in variables over time might be attributable to the layoffs, although it is certainly possible that something else that occurred at the same time was the real cause of change. A more conclusive design would allow for a control group of some sort. This might be possible if one can identify a sample of employees, with only a portion of them experiencing the event. Then individuals who were exposed could be compared to those who were not. Although this naturally occurring quasi-experimental design is not as conclusive as a true experiment, it does provide stronger evidence that exposure had an effect than does a cross-sectional design, or a longitudinal design using arbitrary time points.

An example of the naturally occurring quasi-experimental design is a workplace smoking cessation study by Manning, Osland, and Osland (1989). Manning et al. assessed a sample of employees at two points in time, each time asking if the individual smoked. Four groups of employees were identified, smokers who smoked at both times, nonsmokers who did not smoke at either time, quitters who smoked at Time 1 but not Time 2, and starters who smoked at Time 2 but not Time 1. The main group of interest was the quitters, who showed changes in some variables, such as mood, over the course of the study. The other groups served as control groups that ruled out the likelihood that changes displayed by the quitters might have been due to other workplace factors that presumably would have been experienced by all four groups. Of course, this design cannot rule out the possibility that there was something unique about the quitters that led to the changes, or that the changes were the cause of the quitting. For example, perhaps all the Time 1 smokers whose doctors subsequently advised them that their smoking was causing health problems experienced bad moods and other changes that led to their quitting. Nevertheless, this design is an improvement over a more typical cross-sectional design.

Rule Out Feasible Alternatives

Although not unique to self-report studies, a challenge to drawing causal (and other) inferences from studies is the possibility that feasible alternative explanations accounted for results (Shadish, Cook, & Campbell, 2002). In order to rule out that alternative variables accounted for observed results, one must devise a control strategy that can either assess them (thus allowing for statistical control) or incorporate an element in methodology that can eliminate them either through measurement or design.

The assessment strategy is in some ways the easiest. One first identifies the potential variable or variables of concern and includes measures of them in the study. The alternative explanation can be ruled out by comparing results with and without the variables included in analyses, showing little difference in results. For example, one might be concerned that mood is distorting relationships among measures of interest. It would not be difficult to include a mood measure in a self-report study, and then compare zero-order correlations with partial correlations with the mood variable controlled. If the zero-order and partial correlations are close in magnitude, one might conclude that mood is not a feasible alternative explanation for the observed correlation between constructs of interest. This assumes that an appropriate and sound measure of mood was chosen, as the study only rules out what this particular mood measure assesses.

A second approach is to use a measurement technique that can control for the alternative. This might mean using a measure that was developed to control for a potential biasing factor, such as forced-choice scales to control social desirability. It also might mean using a source that might be unaffected by the biasing factor. As noted earlier, a coworker or supervisor might be unaffected by an employee's mood at the time of a survey. Of course, this is an assumption that might or might not be correct. Employees in bad moods might make those moods apparent to their colleagues who are influenced when completing surveys.

The third approach is to use a design that might control for potential alternatives. Separating the assessment of variables over time can reduce the impact of occasion factors, such as mood. This strategy is typically only useful for alternatives that are short lived, such as daily mood fluctuation. A stronger approach is to use an experimental or quasi-experimental design that might control for alternatives.

Concluding Thoughts

Self-report measures have played a prominent role in the development of the OHP field, and are not likely to be abandoned. Almost certainly researchers will continue to submit papers that are based largely on self-reports, and reviewers will continue to complain, but the literature will still be dominated by this method. Rather than abandoning self-reports, the field needs to make better use of them. This can be done by being more cautious in our research reports and in our reviews of the literature about drawing conclusions from self-report studies, and by designing better studies.

The misinterpretation and overinterpretation of cross-sectional self-report studies is widespread, and undoubtedly it has led to uneasiness in the field about self-reports. Sometimes this occurs because the self-report methodology is not really appropriate for the hypotheses being tested. One example is the routine testing of mediator hypotheses, which are statements of causal connections among triplets of variables. Such hypotheses cannot be adequately tested with cross-sectional designs, self-report or otherwise (Stone-Romero & Rosopa, 2008). Misinterpretations also occur in reviews of the literature in which authors claim support for causal hypotheses from studies that were cross-sectional. Often authors of those studies might not have drawn the causal conclusion the reviewer is claiming they did, but merely reported correlations among the variables of interest. Perhaps because the underlying theoretical frameworks of the papers suggested a causal connection, for example, that stressors cause strains, authors feel it is appropriate to claim support for such conclusions from studies that merely showed significant relationships with cross-sectional designs. Clearly, authors need to be more precise in how they interpret results of studies from the literature, and journal reviewers need to be more insistent that conclusions match results in both introduction and discussion sections of papers.

Finally, we need to be more cautious in how self-reports are used in studies. This means that we must design better studies that include both self-report measures of internal states and non-self-report measures of the environment and other variables. Researchers need to fully consider the nature of their variables and whether a self-report can really assess what they wish to assess. Making better use of self-reports won't be easy, but it will be necessary to move the field forward.

References

*Studies included in the meta-analysis

*Bruk-Lee, V., & Spector, P. E. (2006). The social stressors–counterproductive work behaviors link: Are conflicts with supervisors and coworkers the same? *Journal of Occupational Health Psychology, 11*, 145–156.

Campbell, D. T., & Fiske, D. W. (1959). Convergent and discriminant validation by the multitrait–multimethod matrix. *Psychological Bulletin, 56*, 81–105.

*Carlson, D. S., Witt, L. A., Zivnuska, S., Kacmar, K. M., & Grzywacz, J. G. (2008). Supervisor appraisal as the link between family–work balance and contextual performance. *Journal of Business and Psychology, 23*, 37–49.

Conway, J. M., & Lance, C. E. (2010). What reviewers should expect from authors regarding common method bias in organizational research. *Journal of Business and Psychology, 25*, 325–334.

Dalal, R. S. (2005). A meta-analysis of the relationship between organizational citizenship behavior and counterproductive work behavior. *Journal of Applied Psychology, 90*, 1241–1255.

Evans, M. G. (1985). A Monte Carlo study of the effects of correlated method variance in moderated multiple regression analysis. *Organizational Behavior and Human Decision Processes, 36*, 305–323.

*Fox, S., Spector, P. E., Goh, A., & Bruursema, K. (2007). Does your coworker know what you're doing? Convergence of self- and peer-reports of counterproductive work behavior. *International Journal of Stress Management, 14*, 41–60.

Frese, M., & Zapf, D. (1988). Methodological issues in the study of work stress: Objective vs subjective measurement of work stress and the question of longitudinal studies. In C. L. Cooper & R. Payne (Eds.), *Causes, coping and consequences of stress at work* (pp. 375–411). Oxford, England: Wiley.

Frese, M., & Zapf, D. (1999). On the importance of the objective environment in stress and attribution theory. Counterpoint to Perrewe and Zellars. *Journal of Organizational Behavior, 20*, 761–765.

*Goh, A. (2006). *An attributional analysis of counterproductive work behavior (CWB) in response to occupational stress* (Unpublished doctoral dissertation). University of South Florida, Tampa.

Hackman, J., & Oldham, G. R. (1975). Development of the Job Diagnostic Survey. *Journal of Applied Psychology, 60*, 159–170.

Johnson, E. C., Ganio, M. S., Lee, E. C., Lopez, R. M., McDermott, B. P., Casa, D. J., … Armstrong, L. A. (2010). Perceptual responses while wearing an American football uniform in the heat. *Journal of Athletic Training, 45*, 107–116.

Johnson, R. E., & Saboe, K. N. (2010). Measuring implicit traits in organizational research: Development of an indirect measure of employee implicit self-concept. *Organizational Research Methods, 45*, 107-116. doi:10.1177/1094428110363617

Kasl, S. V. (1998). Measuring job stressors and studying the health impact of the work environment: An epidemiolgic commentary. [Comment/Reply]. *Journal of Occupational Health Psychology, 3*, 390–401.

Keenan, A., & Newton, T. (1985). Stressful events, stressors and psychological strains in young professional engineers. *Journal of Occupational Behaviour, 6*, 151–156.

*Kirmeyer, S. L., & Dougherty, T. W. (1988). Workload, tension, and coping: Moderating effects of supervisor support. *Personnel Psychology, 41*, 125–139.

Lance, C. E., Dawson, B., Birkelbach, D., & Hoffman, B. J. (2010). Method effects, measurement error, and substantive conclusions. *Organizational Research Methods, 13*, 435–455.

Lazarus, R. S. (1991). Psychological stress in the workplace. *Journal of Social Behavior & Personality, 6*, 1–13.

Manning, M. R., Osland, J. S., & Osland, A. (1989). Work-related consequences of smoking cessation. *Academy of Management Journal, 32*, 606–621.

Pace, V. L. (2010). Method variance from the perspectives of reviewers: Poorly understood problem or overemphasized complaint? *Organizational Research Methods, 13*, 421–434.

*Penney, L. M., & Spector, P. E. (2005). Job stress, incivility, and counterproductive work behavior (CWB): The moderating role of negative affectivity. *Journal of Organizational Behavior, 26*, 777–796.

Perrewé, P. L., & Zellars, K. L. (1999). An examination of attributions and emotions in the transactional approach to the organizational stress process. *Journal of Organizational Behavior, 20*, 739–752.

Petruzzello, S. J., Gapin, J. I., Snook, E., & Smith, D. L. (2009). Perceptual and physiological heat strain: Examination in firefighters in laboratory- and field-based studies. *Ergonomics, 52*, 747–754.

Podsakoff, P. M., MacKenzie, S. B., Lee, J.-Y., & Podsakoff, N. P. (2003). Common method biases in behavioral research: A critical review of the literature and recommended remedies. *Journal of Applied Psychology, 88*, 879–903.

*Rau, R., Morling, K., & Rösler, U. (2010). Is there a relationship between major depression and both objectively assessed and perceived demands and control? *Work & Stress, 24*, 88–106.

Richardson, H. A., Simmering, M. J., & Sturman, M. C. (2009). A tale of three perspectives: Examining post hoc statistical techniques for detection and correction of common method variance. *Organizational Research Methods, 12*, 762–800.

Riebe, D., Maresh, C. M., Armstrong, L. E., Kenefick, R. W., Castelliani, J. W., Echegaray, M., … Camaione, D. N. (1997). Effects of oral and intravenous rehydration on ratings of perceived exertion and thirst. *Medicine & Science in Sports & Exercise, 29*, 117–124.

Schaubroeck, J. (1999). Should the subjective be the objective? On studying mental processes, coping behavior, and actual exposures in organizational stress research. [Comment/Reply]. *Journal of Organizational Behavior, 20*, 753–760.

Shadish, W. R., Cook, T. D., & Campbell, D. T. (2002). *Experimental and quasi-experimental designs for generalized causal inference.* Boston, MA: Houghton Mifflin.

Siemsen, E., Roth, A., & Oliveira, P. (2010). Common method bias in regression models with linar, quadractic, and interaction effects. *Organizational Research Methods, 13*, 456–476.

Spector, P. E. (1987). Method variance as an artifact in self-reported affect and perceptions at work: Myth or significant problem? *Journal of Applied Psychology, 72*, 438–443.

Spector, P. E. (1994). Using self-report questionnaires in OB research: A comment on the use of a controversial method. *Journal of Organizational Behavior, 15*, 385–392.

Spector, P. E. (2006). Method variance in organizational research: Truth or urban legend? *Organizational Research Methods, 9*, 221–232.

Spector, P. E., & Brannick, M. T. (1995). The nature and effects of method variance in organizational research. In C. L. Cooper & I. T. Robertson (Eds.), *International review of industrial and organizational psychology: 1995* (pp. 249–274). West Sussex, England: Wiley.

Spector, P. E., & Brannick, M. T. (2009). Common method variance or measurement bias? The problem and possible solutions. In D. A. Buchanan & A. Bryman (Eds.), *The Sage handbook of organizational research methods* (pp. 346–362). Thousand Oaks, CA: Sage.

*Spector, P. E., Dwyer, D. J., & Jex, S. M. (1988). Relation of job stressors to affective, health, and performance outcomes: A comparison of multiple data sources. *Journal of Applied Psychology, 73*, 11–19.

*Spector, P. E., & Fox, S. (2003). Reducing subjectivity in the assessment of the job environment: Development of the Factual Autonomy Scale (FAS). *Journal of Organizational Behavior, 24*, 417–432.

*Spector, P. E., Fox, S., & Van Katwyk, P. T. (1999). The role of negative affectivity in employee reactions to job characteristics: Bias effect or substantive effect? *Journal of Occupational and Organizational Psychology, 72*, 205–218.

*Spector, P. E., & Jex, S. M. (1991). Relations of job characteristics from multiple data sources with employee affect, absence, turnover intentions, and health. *Journal of Applied Psychology, 76*, 46–53.

*Stewart, S. M., Bing, M. N., Davison, H., Woehr, D. J., & McIntyre, M. D. (2009). In the eyes of the beholder: A non-self-report measure of workplace deviance. *Journal of Applied Psychology, 94*, 207–215.

Stone-Romero, E. F., & Rosopa, P. J. (2008). The relative validity of inferences about mediation as a function of research design characteristics. *Organizational Research Methods, 11*, 326–352.

Williams, L. J., & Brown, B. K. (1994). Method variance in organizational behavior and human resources research: Effects on correlations, path coefficients, and hypothesis testing. *Organizational Behavior and Human Decision Processes, 57*, 185–209.

Zapf, D., Dormann, C., & Frese, M. (1996). Longitudinal studies in organizational stress research: A review of the literature with reference to methodological issues. *Journal of Occupational Health Psychology, 1*, 145–169.

16

Strengths and Limitations of Qualitative Approaches to Research in Occupational Health Psychology[1]

Irvin Sam Schonfeld and Joseph J. Mazzola

Qualitative methods have a certain therapeutic value for researchers who contribute to occupational health psychology (OHP). These methods help researchers understand the lived experiences of people confronting problems at work, particularly problems that could adversely affect workers' health. Because OHP researchers sometimes get so intensely caught up in research design and data handling, they can lose sight of the purpose behind their research, which is, ultimately, to improve the health of workers. One can observe the value of qualitative observational methods in Peter Chen's (2007) short autobiographical piece. He wrote about a colleague who had just been laid off. Chen listened to the "colleague give voice to his frantic emotions and disbelief" while the two of them walked through the company parking lot on the day the colleague lost his job. Within 30 minutes, the colleague was experiencing a stomachache. Chen wrote, "I felt ashamed and guilty that evening because I just realized that I have been ignoring the true meaning behind the stress data that I have enjoyed analyzing and publishing!" (p. 1).

Most research in OHP involves the use of quantitative methods. OHP researchers are well trained in scale construction, survey development, and regression analysis, as reflected in the fact that the preponderance

1. Author note: We thank Bob Sinclair for his helpful comments during the writing of the chapter. We also thank Pearl Knopf for her comments during the final edits.

of research published in outlets such as the *Journal of Occupational Health Psychology* and *Work & Stress* use these methods. A literature search of the two journals, utilizing keywords related to qualitative research, revealed that of the 272 papers published between 2005 and 2009 only 15 reported qualitative elements. We also note that while we were able to locate a few qualitative studies on job stress, qualitative research in other areas of OHP (e.g., safety) has been extremely rare. Given the potential for rigorous qualitative research and the relatively small number of qualitative studies in the published literature, one purpose of this chapter is to inform the OHP community why qualitative methods are a valuable resource in both research and practice. We also suggest ways to utilize these methods.

Qualitative research encompasses two main categories of methods. The first includes methods in which workers report, either in writing or orally, on their work lives. These methods could include questionnaires containing open-ended questions (e.g., Abouserie, 1996; Schonfeld & Santiago, 1994), interviews (e.g., Arter, 2008; Kinman & Jones, 2005), and focus groups (e.g., Holmes & MacInnes, 2003; Kidd, Scharf, & Veazie, 1996), which are essentially group interviews. The second method category involves either of two types of observation. In one type, the investigator is positioned in a workplace, as unobtrusively as possible, to observe and record activities and conversations of workers (e.g., Ginsberg, Schwartz, Olson, & Bennett, 1987; Kainan, 1994). In the other type, the investigator works at the targeted job in order to observe the work role "from the inside" as well as the roles of coworkers (e.g., Molapo, 2001; Palmer, 1983). These kinds of observational methods have been particularly underutilized, despite their potential for uncovering an extremely rich vein of data.

In the next two sections, we outline several strengths and limitations associated with qualitative research methods and, in doing so, highlight for OHP researchers the tools needed to determine when qualitative methods are most appropriate and useful. Afterwards, we follow with a section highlighting how qualitative methods have been used in unique ways, how they could be used in OHP going forward, and specific challenges the qualitative researcher may encounter. While this chapter will provide some insight into how to conduct OHP-related qualitative research, it is not meant to review the specific steps in conducting qualitative studies. Instead, we direct the reader to recent books written as guides to qualitative research designs (e.g., Creswell, 2006).

Strengths

In this section, we outline seven strengths that qualitative methods offer OHP researchers and practitioners. These strengths include (a) help with item development for quantitative studies, (b) theory development and hypothesis generation, (c) the discovery of stressors and coping strategies that had previously been overlooked, (d) the development of explanations of difficult-to-interpret quantitative findings, (e) insight into why interventions succeed or fail, (f) dependable descriptions of working conditions, and (g) the accumulation of rich descriptions of workplaces that show the human interactions behind the quantitative findings.

Item Development

Qualitative methods are useful in item development for the purpose of scale construction, as well as the development of other types of instruments to be used in quantitative research. Motowidlo, Packard, and Manning (1986) conducted "group discussions" with hospital nurses and asked the nurses to write "brief descriptions of occasions when they felt stressed on the job" (p. 620). The nurses' writings were content-analyzed, and the categories emerging from the descriptions provided the foundation for the development of a scale to assess nursing stress in a study of job stress, support, affective reactions (e.g., depression), and job performance. Similarly, Dewe (1989), using open-ended interviews, examined sources of work stress in five supervisors and five managers who worked in sales offices. He also investigated the coping responses employed in response to the work stressors. Dewe used the results of the qualitative study to develop coping scales for a study of more than 200 sales supervisors and administrators.

In a health-related study, several adults with a variety of healthy and abnormal sleep habits were asked to describe what "good" and "poor" sleep was to them (Yi, Shin, & Shin, 2006). The responses helped in the creation of items for the Sleep Quality Scale. Similar qualitative methodological approaches could be used on a variety of other health and safety topics.

Schonfeld and Feinman (2012) employed qualitative methods to facilitate a different kind of quantitative study. The first author developed a critical incident (CI; Flanagan, 1954) interview, and tailored it to teachers, the targeted occupational group. CIs are "stressful transactions" in terms of antecedents, context, responses provoked, and consequences (O'Driscoll & Cooper, 1994). The interview elicited teachers' descriptions of stressful work-related incidents and the ways in which they attempted to manage each stressful situation. The qualitative data obtained from the CI interview study were content-analyzed. The stressor and coping categories derived from this CI study served as the foundation for a Web-based diary study of stressors facing more than 250 teachers.

When utilizing qualitative methods with the idea of grouping or content-analyzing (Krippendorff, 1980) responses by higher-order themes, there are two main choices at the researcher's disposal. First, many researchers use raters (typically 2 to 4 people) to recognize and sort the responses into categories. When using this approach, it is important to let the responses guide what categories/themes emerge and avoid imposing preconceived notions. In the Schonfeld and Feinman (2012) study, two readers independently read verbatim transcripts of the interviews and categorized stressors and coping behaviors. Agreement between the readers was assessed with the coefficient *kappa* (Cohen, 1960). This technique allows investigators to assess the reliability of the emergent categories.

Alternatively, there are computer programs currently available that content-analyze data by searching for overarching themes. This type of analysis is relatively new to qualitative research and provides researchers with an alternative that can be less labor intensive than the multirater coding described above. Two commonly used programs, Nudist and Atlas/TI, were reviewed by Barry (1998). She found that both programs could expedite content analyses and comprehensively capture the theoretical ideas that emerge from the data.

Theory Development and Hypothesis Generation

Qualitative methods have long been associated with Glaser and Strauss's (1967) grounded theory. Glaser and Strauss emphasized the idea that researchers need to allow theoretically interesting categories and hypotheses to emerge from qualitative data, while approaching the data without preconceived ideas regarding what should emerge. Qualitative research can pave the way to a new theory or hypothesis (that may later be tested with quantitative methods) or can help to further elaborate an existing theory. Schonfeld and Farrell (2010) advanced the view that certain uncontrolled, qualitative observations have played an important role in the history of science. For example, before Jenner discovered a vaccine for smallpox, ordinary people observed that inoculating individuals with small amounts of discharge from the pustules of infected individuals provided immunity from the disease (Hopkins, 1983). These early observations contributed to progress toward a theory of contagion, and helped challenge rival humoral theories of the disease (Miller, 1957).

As part of a longitudinal study (Schonfeld, 2001), novice teachers were given an opportunity to write, with no constraints, about their work experiences. Their descriptions were transcribed and read, and a set of thematic categories was allowed to emerge. The descriptions were reliably assigned to one (and sometimes more than one) of four categories: (a) interpersonal tensions among professionals and lack of support from colleagues and supervisors, (b) happiness with one's job, (c) violence and other safety problems, and (d) classroom management problems (Schonfeld & Santiago, 1994). The thematic categories of happiness with one's job and the presence of support were closely linked. Teachers who were happy with their job reported that their satisfaction was built on the rock of supportive colleagues and supervisors. By contrast, many new teachers were distressed when supervisors neglected their supervisory role either by *not* helping teachers or by being unfair and disrespectful (e.g., "The person who puts stress in my work is my supervisor. She used to walk into my classroom at any time during the first 3 weeks of school to observe me or to give me things.").

Schonfeld and Farrell (2010) augmented these qualitative data by examining others' qualitative research. Qualitative data from the United States (e.g., Blase, 1986; Farber, 1991; Smith & Smith, 2006; Steggerda, 2003) and Canada (Younghusband, 2008) suggest that many teachers are affected by high levels of disrespect and a dangerous level of violence. Based on all the qualitative data, Schonfeld and Farrell hypothesized that working conditions for many teachers are normatively stressful. They advanced the view that individuals entering the teaching profession with reasonably commonplace ideas about workplace courtesy, respect, and supervision are likely to be overtaken by the physically and psychologically draining working conditions found in many schools.

Discovery

Qualitative research has a role to play in the discovery of new phenomena, including stressors and coping behaviors. Kidd et al. (1996) observed that "qualitative methods are preferred to quantitative methods when there is little information known about a phenomenon, the applicability of what is known has not been examined, or when there is reason to doubt the accepted knowledge about a given phenomenon" (p. 225). Firth and Morrison (1986) gave medical students wide latitude by asking them to freely describe both good and bad aspects of their jobs, finding that one of the most stressful aspects of the students' medical work was talking to psychiatric patients. Fischer, Kumar, and Hatcher (2007), in their study of stress in psychiatrists, identified risk factors, such as lack of administrative support, commonly associated with burnout. However, they also discovered what amounted to a ramped-up version of the concept of lack of support in the form of "an aggressive administrative environment," an administrative environment that provoked feelings of vulnerability.

Keenan and Newton (1985) discovered that incidents involving time-wasting were serious workplace stressors for engineers, and that role stressors (e.g., role ambiguity), which many researchers had believed to be common occurrences, were reported less frequently. Polanyi and Tompa (2004) found that lack of meaning or ethics in work is a stressor that had been overlooked in previous research. Hutchinson (1987) identified unique coping responses in nurses, such as self-care activities. In cross-cultural research, Narayanan, Spector, and Menon (1999) found that lack of structure was a major stressor in their Indian sample, a stressor most Western research has not investigated.

Qualitative research can also be a source of discovery in research on safety in the work environment. Kidd et al. (1996) studied safety in farmers, an understudied group in OHP, and found that accident risk was not related to a lack of knowledge about hazards, suggesting that increasing farmers' knowledge about safe work practices would not affect risk. Kidd et al. discovered that farmers in their sample prioritized economic factors above safety concerns in decision making, suggesting that safety interventions need to highlight the economic consequences of accidents when making business decisions.

It is clear that qualitative research paradigms can be applied in underdeveloped areas of research on occupational health and safety and in understudied populations. In contrast, when qualitative methods are employed in a well-explored area, it is likely that theoretical insights will connect with existing theories. Thus, while new discoveries may be made in established research areas, qualitative research has additional value in less developed areas. As new research avenues present themselves, researchers can utilize open-ended research paradigms to make discoveries that should lead to complementary quantitative research.

Interpreting Findings

The fourth strength of qualitative research is that it can help OHP researchers develop explanations of difficult-to-interpret findings. For example, Büssing and Glaser (1999) followed nurses working on wards that had been redesigned "holistically," enabling the nurses to have greater responsibility for fewer patients. The redesigning of work on the wards was intended to enhance the quality of the nurses' professional lives; one would expect the nurses in the redesigned wards to experience lower levels of stressors and strain. With regard to stressors, the nurses on the holistic wards, compared to control nurses who worked on traditional wards organized along Taylorist principles emphasizing highly segmented and repetitive tasks, experienced a significant reduction in three stressors: time pressure, contradictory task goals, and ergonomic stressors. The nurses on the holistic wards, however, experienced significantly higher levels of emotional exhaustion, an ostensibly anomalous finding. Qualitative data helped to make sense of the findings by revealing that the holistic nursing system intensified the nurses' emotional work and interactional stress. Unlike nurses on the traditional wards whose patient contact was more piecemeal, nurses on the holistic wards had no opportunity to withdraw from difficult patients.

In a very different application of qualitative methods for the purpose of better understanding stressful job conditions, Arter (2008) innovatively extended strain theory, Agnew's (1992) theory of antisocial conduct in youth, to working adults, specifically police officers. Although police are at comparatively higher risk for stress-related disorders, not all police assignments are equally stressful. In an excellent example of what Glazer and Strauss would call theoretical sampling, Arter recruited male police officers who worked undercover (the most stressful condition), formerly worked undercover, and never worked undercover, collecting qualitative data on the experiences of the officers, including episodes of deviant behavior (e.g., promiscuity, failure to enforce certain laws). Consistent with strain theory, Arter found the highest levels of deviant behavior in the officers who currently worked undercover and the lowest levels in those who never worked undercover. Arter also tried to understand why not every officer who experienced significant stressors manifested deviant behavior. He found that adaptive coping behaviors (e.g., exercise, seeing family and friends) were related to reduced deviance and maladaptive coping behaviors (e.g., alcohol consumption) to increased deviance.

Finally, in Vinje and Mittelmark's (2007) study of 11 community nurses, qualitative interviews revealed that while job engagement promoted positive outcomes, it was, surprisingly, related to negative outcomes as well. On the positive end, nurses felt that through their engagement in a job that centered on helping people, their lives had greater meaning and they could live out their core values through their work. However, they also indicated a need to be always on top of things, be highly conscientious, live by high ethical standards, and strive hard for excellence in themselves and others. Coupled with the demanding nature of the nursing profession, these characteristics left nurses extremely susceptible to work overload and burnout.

As these examples suggest, it would be advantageous for researchers to collect qualitative data along with quantitative data in order to help explain or describe unusual or unexpected findings. Qualitative methods can add depth to the researcher's understanding of the experiences of workers and, because of the freedom those methods accord respondents, help overcome the researcher's preconceptions.

Insight into the Success or Failure of Interventions

Bunce (1997) underlined the need to understand factors that contribute to the success or failure of workplace interventions designed to promote the health and well-being of workers. In addition to simply implementing an intervention, the process of implementation, as reflected in variables such as the meaning of the intervention for workers and managers, project fatigue in managers, and the cultural maturity of the organization in question, is also important to success and failure (Nytrø, Saksvik, Mikkelsen, Bohle, & Quinlan, 2000).

Although many researchers who study process have used quantitative methods, Saksvik, Nytrø, Gensen, and Mikkelsen (2002) employed qualitative methods in attempting to understand process factors that contributed to the success or failure of seven workplace health interventions in Norway. Qualitative data provided insight into the competence of participating organizations to carry out an intervention and shed light on concealed and informal behaviors that could undermine implementation. Saksvik et al., for example, reported that the "negative" culture at 3 of the 26 post offices studied was an obstacle to the success of the intervention in those 3 locations, with managers accusing employees of showing too little interest and postal workers accusing managers of not showing sufficient initiative.

Saksvik et al. recommended "combining qualitative and quantitative research techniques" to evaluate interventions (p. 53). Qualitative methods are helpful because it is improbable that the dominant natural-science (i.e., experimental) paradigm can solely bear the burden of explaining ongoing processes, particularly "microprocesses," that mediate the relation between an intervention and putative outcomes within the context of a complex, always-evolving work organization (Griffiths, 1999).

Dependability of Workers' Own Reports on Their Work Roles

Workers' descriptions of their work lives constitute an activity that is central to qualitative OHP research. These descriptions provide a dependable vantage point for understanding the stressors and safety problems that affect working people. Schonfeld and Farrell (2010), in keeping with the view of Kasl (1978), suggested that there is some question about the dependability of workers' descriptions of their own work experiences. Kasl (1978), citing research on fighter pilots (Aitken, 1969), police (Kroes, Margolis, & Hurrell, 1974), and air traffic controllers (Smith, 1973), suggested that workers' self-reports on the stressfulness of a work role may be less dependable than originally believed.

We advance the view that workers' descriptions of their jobs constitute a reliable source of information. Although Kasl wrote that RAF fighter pilots were more likely to identify "housing, wife, finances, and children" as sources of personal worry and emotional stress than the dangerousness of the job, the observation was not put into proper context. Aitken (1969) found that the men in the one RAF squadron that had previously experienced a series of fatalities expressed considerably more stress and worry about flying than did the men in the other squadrons, which did not have nearly the same accident experience, a finding that highlights the realism of the men's perceptions. When police officers were asked what was "bothersome" about their jobs, they mentioned administrative hassles (Kroes et al., 1974). When asked about major stressors, they identified civilian deaths and the risk the job holds for themselves and their colleagues (Kirmeyer & Diamond, 1985).

Kasl also cited research on air traffic controllers (Smith, 1973) who, when asked to identify the most disliked aspects of their jobs, indicated job facets such as administration. The job's heavy responsibility and high mental workload were either mentioned infrequently as a disliked job facet or revealed to be an aspect of the job they liked. Smith advanced the view that perhaps high traffic is not a "noxious" condition. One shortcoming of both Kasl's and Smith's interpretation is that mentioning a job condition that a worker liked or disliked is not the same as a worker's being able to accurately describe the facets of the job. Moreover, high levels of traffic are associated with increased risk of hypertension, a more or less silent condition that would be difficult for an individual to subjectively link to immediately observed working conditions.

In research on stressors in factory work, Hugentobler, Israel, and Schurman (1992) observed convergence in their qualitative (individual interviews, focus groups) and quantitative (survey) findings (e.g., job insecurity). Holmes and MacInnes (2003) in a study of prison workers employed two different qualitative methods in identifying workplace stressors (e.g., inmate self-harm). The authors were concerned that among their focus-group participants, imitation could have affected reporting. The results of the individual interviews, however, dovetailed with the focus-group results. The studies by Hugentobler et al. and Holmes and MacInnes underline the realism of the workers' observations. The studies also suggest that multiple methods can be deployed in such a way that the strengths and weaknesses of individual methods can be balanced and, with the convergence of findings, confidence in research results enhanced (Hugentobler et al.).

Rich Description

A seventh strength of qualitative research is that it can provide rich descriptions of stressful workplace transactions that add depth to quantitative data. In concert with a quantitative study, Parkes (1985) assembled qualitative responses of 206 student nurses, with each nurse asked to identify "a recent stressful episode occurring in the course of her work" (p. 946). While a content analysis uncovered six areas of stressful conditions (e.g., insecurity regarding one's professional skills), Parkes also obtained rich descriptions of stressful experiences. For example, nurses felt intense, but unrealistic self-blame after the death of a patient who had been subject to a "minor error or discourtesy."

In their study, Schonfeld and Farrell (2010) included the words of a new female teacher who described events on her job:

> My greatest problem is gaining and maintaining control of my students. Students are constantly getting out of their seats, calling out to each other and throwing paper in class. I admit I have lost control but I also believe that most students have very little respect for anyone.... I feel almost isolated and on most days I get home emotionally and physically drained.

These words underscore the distress of a teacher who has worked with little success to educate her students. Despite the importance of the sophisticated statistical methods required to analyze occupational stress data (Schonfeld & Rindskopf, 2007), qualitative findings clothe in flesh and blood the stressful transactions occurring at the workplace.

Safety researchers often measure the number of accidents, injuries, and deaths that occur in a workplace, but the meaning behind experiencing or witnessing an accident can get lost. Eklöf and Törner (2005) investigated these incidents in a sample of fishermen, an occupation with a high fatality rate (U.S. Bureau of Labor Statistics, 2008). When asked about managing such incidents, one fisherman said, "While it is happening, you are totally focused on sorting out the situation. Afterwards, you joke harshly about it to keep fear at a distance" (p. 366). The remark reveals the crew's unwillingness to take preventive action despite the repetitious nature of their accident experience. While the overall results showed that preventive measures can potentially reduce accident risk, the crew's rich descriptions of incidents added value to the research because the descriptions helped the investigators better understand the workers' experience.

Limitations

Despite the numerous strengths of qualitative methods, they are not without limitations, and here we enumerate five. These include (a) the problem of participant reactivity, (b) the potential to overidentify with study participants, (c) the impracticality of the Glaser-Strauss idea that hypotheses arise from data unsullied by prior expectations, (d) inadequacy with regard to drawing cause–effect conclusions from qualitative data alone, and (e) the Baconian character of the qualitative research enterprise.

Reactivity

The first limitation is the problem of reactivity in the individuals who are observed. People who are observed sometimes change in response to the presence of an observer (Shai, 2002). Reactivity is a concern when a researcher attempts to gain the trust of the participant in order to get an accurate, complete, and rich set of responses, but reactivity has rarely been addressed in qualitative OHP research. Cohen (1989) briefly mentioned the potential for demand characteristics to affect the responses of the executive nurses in her qualitative study of stress and coping.

Since qualitative responses in interviews and questionnaires are often personal and detailed, it is the responsibility of the researcher to make the participant feel at ease. The researcher has to maintain a respectful and friendly relation but at the same time remain objective. The participant must not feel that he or she is being judged, and the confidential nature of the responses must be respected. The idea of maintaining respectful relations can be extended to research based on participant observation. Molapo (2001), in her study of work stress in Black South African gold miners, also addressed the problem of reactivity. Because she went underground regularly in the participant–observation component of her study, after a time her "presence did not really matter" and "everybody treated [her] as part of the crew" (p. 99).

Overidentifying with Study Participants

The second limitation concerns the potential for the researcher to overidentify with study participants, which could affect the investigator's interpretation of qualitative findings. The first author was once a mathematics teacher and was concerned about the potential for his overidentifying with the teachers whom he studied, fearing he might observe more villainy in students riding a teacher than is warranted and more competence in a teacher experiencing classroom management problems than is justified.

Both participant reactivity and researcher overidentification can bias qualitative research results. There are several ways to avoid or check such biases, including the use of structured or semistructured interviews (e.g., Kinman & Jones, 2005), the deployment of multiple independent raters when coding results (e.g., Narayanan et al., 1999), and assessing interrater agreement. The burden is on the qualitative researcher to demonstrate to readers that the results and interpretations have a basis in reality despite the interpretative nature of qualitative research (see Schonfeld & Farrell, 2010). While such evidence could take many forms, some recommendations would include: (a) using probability sampling to ensure the representativeness of the sample, (b) creating sound surveys, interviews, and observations that have a basis in previous research and are planned as carefully as instruments used in quantitative research, and (c) training interviewers/observers/raters in similar techniques to minimize bias.

Theoretically Important Categories Emerging Naturally from Qualitative Data

Glaser and Strauss advanced the idea that qualitative researchers should let theoretically important categories and hypotheses emerge "naturally" from data, unguided by preconceptions. The idea is chimerical. The comparative psychologist David Katz (1937) wrote that "a hungry animal divides the environment into edible and inedible things. An animal in flight sees a road to escape and hiding places. Generally speaking, objects change … according to the needs of the animal" (p. 143). Karl Popper (1957/1963), parrying the point made by Katz, wrote that "objects can be classified, and can become similar or dissimilar, only in this way—by being related to needs and interests. This rule applies not only to animals but also to scientists" (p. 47). Observation is always selective.

One of us faced such a limitation in his own research. As a supplement to a quantitative longitudinal study of new teachers, Schonfeld and Santiago (1994) attempted to keep an open mind and let theoretically important categories emerge from qualitative teacher data. The first author, however, was also aware of several theories of stress. In fact, we cannot imagine that anyone collecting qualitative data on job stress is *not* aware of theories of stress. While trying to keep open minds, and let the categories emerge from the data, Schonfeld and Santiago inevitably coded categories that were consistent with categories that were already visible in the existing OHP literature. For example, one category involved support from coworkers and administrators; another included violence and its threat. Here is just one example of a teacher mentioning both of these factors:

> My supervisor was not helpful. She was daily informed of an insubordinate assistant teacher in my classroom. I was attacked by this person who is almost 100 lbs [heavier] than me and 10 inches taller than I am. The school is not standing behind me even though [administrators] told me this person is being put on probation due to insubordinate behavior in the classroom. (p. 119)

These categories have long been known to OHP researchers and are evident elsewhere in the qualitative stress literature bearing on teachers (Schonfeld & Farrell, 2010). Researchers should be well-versed in the area they are studying. However, it is important that new concepts or themes still be allowed to emerge. If they are not, and preconceived notions rigidly guide the categories, qualitative methods will be of limited value. To be sure, qualitative researchers must straddle a delicate line between awareness of the literature and imposing preconceptions on data.

Testing Causal Hypotheses

A study's capacity to help an investigator draw causal inferences rests more with the nature of a study's design than with the question of whether the data a study generates are quantitative or not. Although the temptation is often present, qualitative research designs are largely inadequate in testing causal hypotheses, especially when uncontrolled, raw qualitative responses are used exclusively. A cautionary example from the history of psychology illustrates the pitfalls of drawing such causal conclusions. Fromm-Reichmann (1948) used clinical case material from her work with a young man diagnosed with schizophrenia. Fromm-Reichmann, who bragged that psychoanalysts used their technique "with the utmost sensitive care and caution" (p. 265), noted that "the schizophrenic is painfully distrustful and resentful of other people, due to the severe warp and early rejection he encountered in important people in his infancy and childhood, as a rule, mainly in a schizophrenogenic mother" (p. 265). Evidence from better controlled quantitative research has not supported the idea that a child's schizophrenia results from poor mothering (Tandon, Keshavan, & Nasrallah, 2008). Fromm-Reichmann did not entertain the hypothesis that the tension she observed in the mother–child relationship was the result, not the cause, of the son's schizophrenia.

With regard to OHP research, one of us (Schonfeld & Ruan, 1991) interviewed a biology teacher who had a great love for her subject. She obtained a job in the Bronx where she taught students who were defiant and verbally abusive, sometimes hurling sexually explicit epithets at each other and the teacher. There was fighting in her class. She tried to teach the subject she loved, but faced great opposition. At the end of the school day she was spent. She became clinically depressed and sought treatment from a psychiatrist. The events suggest that she suffered a reactive depression that developed in response to the great difficulties occurring in her classroom. Although it is tempting to use qualitative data to draw conclusions about causation, it is important to proceed with caution. Additional exploration of her life history indicated that she had suffered her first depressive episode several years before she became a teacher. She suffered a death in her family and a change of domicile the year before she became a New York City teacher, both stressful life events. In light of these additional findings, it would be difficult to conclude that she suffered a depression in reaction to her exposure to difficult classes. It is equally plausible that preexisting psychological distress compromised her effectiveness in managing the class.

Although the temptation for qualitative researchers to draw a cause-effect conclusion exists, such research should not replace appropriate quantitative methods of verification. Qualitative research is ill suited for hypothesis testing.[2]

Baconian Character

The fifth and final limitation is that the Glaser-Strauss idea of collecting qualitative observations is too Baconian in orientation. Glaser and Strauss attempted to address the question of whom and how many to

---

2. There is an exception to the idea that qualitative research is ill suited for hypothesis testing; we refer to studies that employ mixed methodologies that combine qualitative and quantitative methods (Mazzola, Schonfeld, & Spector, 2011). See the section on "Future Directions."

sample in their discussion of theoretical sampling. The endpoint of this sampling, called *theoretical saturation*, is the point at which "no additional data are being found" and the investigator is ready to develop an understanding of the properties of the groups under study. Glaser and Strauss also elaborated the concept of the *depth* of theoretical sampling, which pertains to the amount of data to be collected within a theoretically important group.

The qualitative investigator does not have available power analyses and other statistical means to gauge when he or she has a sufficiently large and categorically diverse sample, which leads to an energetic pursuit of data that has no clearly definable stopping point. The result is an accumulation of facts (see Bacon, 1620/1960). Bertrand Russell (1945) noted that the Baconian idea that an "orderly arrangement of data would make the right hypothesis obvious" is seldom in evidence (p. 544). Russell maintained that without some provisional hypothesis to help guide the collection of facts, the sheer accumulation of facts is, in Russell's word, *baffling*.

With this limitation in mind, we suggest that OHP investigators take sensible precautions to ensure that the qualitative data collected provide a framework for improving investigators' opportunities to develop hypotheses. Although we earlier suggested that the idea of theoretical saturation is elusive, we believe that the idea of theoretical sampling is helpful for getting a fuller picture of what is happening at work. Glaser and Strauss (1967), in their extensive discussion of theoretical sampling, described the importance of "comparing different types of groups within different larger groups" for the purpose of discovering categories relevant to theory development (p. 52). Arter (2008), for example, helped us better understand stress in police officers by selecting officers in three different conditions that can reasonably be suspected to vary in their stressfulness. Another way to ensure that qualitative methods provide fertile ground for developing hypotheses is to have investigators inquire into both (a) satisfying and stressful conditions or (b) safe and unsafe conditions. In this way, work stress and work safety investigators can assess for disconfirming conditions.

Future Directions

Since OHP is a relatively new discipline, there are numerous topics that now or in the near future will fall under the OHP umbrella. As OHP researchers investigate new relationships that bear on work-related safety, stress, and health, they should proceed with appropriate qualitative methods to survey the new territory, generate items for scales, and produce hypotheses that help to inform the design and implementation of quantitative research.

Additionally, qualitative and quantitative methods can be used in conjunction with one another. Researchers too often align themselves with one methodological camp or the other. However, the coordinate application of qualitative and quantitative methods has potential for ferreting out new knowledge. Qualitative and quantitative methods can be combined to inform research in health education (Steckler, McLeroy, Goodman, Bird, & McCormick, 1992). We suggest that OHP can benefit from the coordination of methods in a number of ways: (a) qualitative methods can be used to inform quantitative research (i.e., item generation and discoverability); (b) qualitative methods can help interpret quantitative findings (e.g., Büssing & Glaser, 1999; Vinje & Mittelmark, 2007); and (c) researchers who engage in qualitative research and who have an understanding of relevant quantitative findings, have guidance with regard to the research issues to pursue (e.g., novel extension of strain theory; Arter, 2008).

While qualitative methods have adherents among OHP investigators, there is a trend in OHP research that should be acknowledged. Some researchers have been uniting qualitative and quantitative methods within one study, particularly in stress research (Mazzola, Schonfeld, & Spector, 2011). Such research is especially valuable because the strengths of one method can help balance weaknesses of the other. The two methods can be used in a coordinated fashion to explain or describe a phenomenon (e.g., Mazzola, Jackson, Shockley, & Spector, 2011). In their study, Mazzola and colleagues employed a hybrid methodology that included quantitative and qualitative data on stressors. Liu, Spector, and Shi (2008) also measured stressors using qualitative and quantitative methods in an investigation of cross-national stressor differences.

Another advantage of employing a hybrid methodology is that the investigator can link stressors identified with the help of qualitative methods (but not found on standard scales) to important health and morale outcomes measured quantitatively. For example, Elfering et al. (2005), in a 7-day diary study involving employees at a counseling agency, applied qualitative methods to ascertain the daily incidence of episodically occurring job stressors. Situational well-being (measured quantitatively) in the aftermath of a daily stressor was inversely related to the intensity of chronic stressors (measured quantitatively on the first day). Studies such as the one completed by Elfering et al. are valuable because of the way in which the investigators orchestrated qualitative and quantitative methods for the purpose of making inferences about job stress. We hope to see more such innovative studies in the future because we believe that design refinements that marry qualitative and quantitative methods will further advance the field of occupational health psychology.

Final Observations

At the beginning of this chapter we mentioned the therapeutic value of qualitative methods for OHP researchers. Qualitative findings show the psychological distress that physicians and nurses experience when they lose a patient, expose the suffering of a teacher who was attacked by a student, and describe the somatic symptoms experienced by the man who just lost his job. Qualitative methods help focus our vision on the goal of improving the lives of people who work.

The value of qualitative research also follows from what the philosopher of science Hans Reichenbach (1951) called the "context of discovery." Qualitative research can help OHP researchers engaged in the preliminary work of ascertaining workplace stressors or safety behaviors in order to develop items that can potentially populate structured surveys and interviews. Qualitative research can also help the developers of interventions designed to improve the health of workers discern process variables that may affect the chances of success. Such work constitutes efforts at discovery. Qualitative findings provide a basis for a researcher's intuitions regarding theory development and hypothesis formation.

Like all research methods, qualitative methods have limitations. With an understanding of these limitations (and how to minimize/balance them), OHP researchers can benefit from such methods. It is important to understand that qualitative findings do not establish generalizable cause–effect relations. However, qualitative methods can help a researcher develop a theory of causality and derive hypotheses related to the theory and, thus, motivate quantitative research designed to test the hypotheses. Thus, the challenge for the OHP researcher is to be mindful of what qualitative methods can and cannot do, and exploit their strengths for the benefit of the research enterprise.

References

Abouserie, R. (1996). Stress, coping strategies and job satisfaction in university academic staff. *Educational Psychology, 16*, 49–56.

Agnew, R. (1992). Foundation for a general strain theory of crime and delinquency. *Criminology, 30*, 47–87.

Aitken, R. C. B. (1969). Prevalence of worry in normal aircrew. *British Journal of Medical Psychology, 42*, 283–286.

Arter, M. L. (2008). Stress and deviance in policing. *Deviant Behavior, 29*, 43–69.

Bacon, F. (1960). The new organon. In F. H. Anderson (Ed.), *The new organon and related writings* (J. Spedding, R. L. Ellis, & D. D. Heath, Trans.; pp. 33–268). New York: The Liberal Arts Press. (Original work published 1620)

Barry, C. A. (1998). Choosing qualitative data analysis software: Atlas/ti and Nudist compared. *Sociological Research Online, 3*. Retrieved from http://www.socresonline.org.uk/socresonline/3/3/4.html

Blase, J. J. (1986). A qualitative analysis of sources of teacher stress: Consequences for performance. *American Educational Research Journal, 23*, 13–40.

Bunce, D. (1997). What factors are associated with the outcome of individual-focused worksite stress management interventions? *Journal of Occupational and Organizational Psychology, 70*, 1–17.

Büssing, A., & Glaser, J. (1999). Work stressors in nursing in the course of redesign: Implications for burnout and interactional stress. *European Journal of Work and Organizational Psychology, 8*, 401–426.

Chen, P. (2007). Personal reflection: The meaning of occupational health psychology. *Newsletter of the Society of Occupational Health Psychology, 1*, 1.

Cohen, J. (1960). A coefficient of agreement for nominal scales. *Educational and Psychological Measurement, 20*, 37–46.

Cohen, J. H. (1989). *Nurse executives' psychological well-being: The relationship among stress, social support, coping, and optimism* (Unpublished doctoral dissertation). University of California, San Francisco.

Creswell, J. W. (2006). *Qualitative inquiry and research design: Choosing among five approaches*. Thousand Oaks, CA: Sage.

Dewe, P. J. (1989). Examining the nature of work stress: Individual evaluations of stressful experiences and coping. *Human Relations, 42*, 993–1013.

Eklöf, M., & Törner, M. (2005). Participatory analysis of accidents and incidents as a tool for increasing safety behaviour in fishermen: A pilot intervention study. *Work & Stress, 19*, 360–369.

Elfering, A., Grebner, S., Semmer, N. K., Kaiser-Freiburghaus, D., Lauper-Del Ponte, S., & Witschi, I. (2005). Chronic job stressors and job control: Effects on event-related coping success and well-being. *Journal of Occupational and Organizational Psychology, 78*, 237–252.

Farber, B. A. (1991). *Crisis in education: Stress and burnout in the American teacher*. San Francisco, CA: Jossey-Bass.

Firth, J., & Morrison, L. (1986). What stresses health professionals? A coding system for their answers. *British Journal of Clinical Psychology, 25*, 309–310.

Fischer, J., Kumar, S., & Hatcher, S. (2007). What makes psychiatry such a stressful profession? A qualitative study. *Australasian Psychiatry, 15*, 417–421.

Flanagan, J. C. (1954). The critical incident technique. *Psychological Bulletin, 51*, 327–358.

Fromm-Reichmann, F. (1948). Notes on the development of treatment of schizophrenics by psychoanalytic psychotherapy. *Psychiatry: Journal for the Study of Interpersonal Processes, 11*, 263–273.

Ginsberg, R., Schwartz, J., Olson, G., & Bennett, A. (1987). Working conditions in urban schools. *The Urban Review, 19*, 3–23.

Glaser, B., & Strauss, A. (1967). *The discovery of grounded theory: Strategies for qualitative research*. Chicago, IL: Aldine.

Griffiths, A. (1999). Organizational interventions: Facing the limits of the natural science paradigm. *Scandinavian Journal of Work, Environment & Health, 25*, 589–596.

Holmes, S., & MacInnes, D. (2003). Contributors to stress among prison service staff. *British Journal of Forensic Practice, 5*, 16–24.

Hopkins, D. R. (1983). *Princes and peasants: Smallpox in history*. Chicago, IL: University of Chicago Press.

Hugentobler, M. K., Israel, B. A., & Schurman, S. J. (1992). An action research approach to workplace health: Integrating methods. *Health Education Quarterly, 19*, 55–76.

Hutchinson, S. (1987). Self-care and job stress. *Image: Journal of Nursing Scholarship, 19*, 192–196.

Kainan, A. (1994). Staffroom grumblings as expressed teachers' vocation. *Teaching and Teachers Education, 10*, 281–290.

Kasl, S. V. (1978). Epidemiological contributions to the study of work stress. In C. L. Cooper & R. L. Payne (Eds.), *Stress at work* (pp. 3–38). Chichester, England: Wiley.

Katz, D. (1937). *Animals and men*. London: Longmans, Green.

Keenan, A., & Newton, T. J. (1985). Stressful events, stressors and psychological strains in young professional engineers. *Journal of Occupational Behaviour, 6*, 151–156.

Kidd, P., Scharf, T., & Veazie, M. (1996) Linking stress and injury in the farming environment: A secondary analysis. *Health Education Quarterly, 23*, 224–237.

Kinman, G., & Jones, F. (2005). Lay representations of workplace stress: What do people really mean when they say they are stressed? *Work & Stress, 19*, 101–120.

Kirmeyer, S. L., & Diamond, A. (1985). Coping by police officers: A study of role stress and Type A and Type B behavior patterns. *Journal of Occupational Behaviour, 6*, 183–195.

Krippendorff, K. (1980). *Content analysis: An introduction to its methodology*. Beverly Hills, CA: Sage.

Kroes, W. H., Margolis, B. L., & Hurrell, J. J. (1974). Job stress in policemen. *Journal of Police Science and Administration, 2*, 145–155.

Liu, C., Spector, P., & Shi, L. (2008). Use of both qualitative and quantitative approaches to study job stress in different gender and occupational groups. *Journal of Occupational Health Psychology, 13*, 357–370.

Mazzola, J. J., Jackson, E. M., Shockley, K. M., & Spector, P. E. (2011). Examining stress in graduate assistants: Combining open- and closed-ended survey methods. *Journal of Mixed Methods Research, 5*, 198-211.

Mazzola, J. J., Schonfeld, I. S, & Spector, P. E. (2011). What qualitative research has taught us about occupational stress: A review. *Stress & Health: Journal of the International Society for the Investigation of Stress, 27*, 93-110.

Miller, G. (1957). *The adoption of inoculation for smallpox in England and France.* Philadelphia, PA: University of Pennsylvania Press.

Molapo, M. P. (2001). *A biosocial study of high blood pressure among underground mineworkers in a South African gold mine* (Unpublished doctoral dissertation). Emory University, Atlanta, GA.

Motowidlo, S. J., Packard, J. S., & Manning, M. R. (1986). Occupational stress: Its causes and consequences for job performance. *Journal of Applied Psychology, 71*, 618–629.

Narayanan, L., Menon, S., & Spector, P. (1999). A cross-cultural comparison of job stressors and reactions among employees holding comparable job in two countries. *International Journal of Stress Management, 6*, 197–212.

Nytrø, K., Saksvik, P. O., Mikkelsen, A., Bohle, P., & Quinlan, M. (2000). An appraisal of key factors in the implementation of occupational stress interventions. *Work & Stress, 14*, 213–225.

O'Driscoll, M. P., & Cooper, C. L. (1994). Coping with work-related stress: A critique of existing measures and proposal for an alternative methodology. *Journal of Occupational and Organizational Psychology, 67*, 343–354.

Palmer, C. E. (1983). A note about paramedics' strategies for dealing with death and dying. *Journal of Occupational Psychology, 56*, 83–86.

Parkes, K. R. (1985). Stressful episodes reported by first-year student nurses: A descriptive account. *Social Science Medicine, 20*, 945–953.

Polanyi, M., & Tompa, E. (2004). Rethinking work-health models for the new global economy: A qualitative analysis of emerging dimensions of work. *Work: Journal of Prevention, Assessment and Rehabilitation, 23*, 3–18.

Popper, K. (1963). Science: Conjectures and refutations. In K. R. Popper (Ed.), *Conjectures and refutations: The growth of scientific knowledge.* New York: Basic Books. (Original work published in 1957)

Reichenbach, H. (1951). *The rise of the scientific philosophy.* Berkeley: University of California Press.

Russell, B. (1945). *A history of western philosophy.* New York: Simon & Schuster.

Saksvik, P. O., Nytrø, K., Dahl-Jorgensen, C., & Mikkelsen, A. (2002). A process evaluation of individual and organizational stress and health interventions. *Work & Stress, 16*, 37–57.

Schonfeld, I. S. (2001). One-year longitudinal study of the effects of working conditions social support, coping, and locus of control on depressive symptoms and job satisfaction in women. *Genetic Psychology Monographs, 127*, 133–168.

Schonfeld, I. S., & Farrell, E. (2010). Qualitative methods can enrich quantitative research on occupational stress: An example from one occupational group. In D. C. Ganster & P. L. Perrewé (Eds.), *Research in occupational stress and wellbeing: Vol. 8. New developments in theoretical and conceptual approaches to job stress* (pp. 137–197). Bingley, England: Emerald.

Schonfeld, I. S., & Feinman, S. J. (2012). Difficulties of alternatively certified teachers. *Education and Urban Society, 44,* 215-246.

Schonfeld, I. S., & Rindskopf, D. (2007). Hierarchical linear modeling in organizational research: Longitudinal data outside the context of growth modeling. *Organizational Research Methods, 18,* 417–429.

Schonfeld, I. S., & Ruan, D. (1991). Occupational stress and preemployment measures: The case of teachers. [Special Issue on Occupational Stress, P. Perrewé, Ed.]. *Journal of Social Behavior and Personality, 6,* 95–114.

Schonfeld, I. S., & Santiago, E. A. (1994). Working conditions and psychological distress in first-year women teachers: Qualitative findings. In L. C. Blackman (Ed.), *What works? Synthesizing effective biomedical and psychosocial strategies for healthy families in the 21st century* (pp. 114–121). Indianapolis: University of Indiana School of Social Work.

Shai, D. (2002). Working women/cloistered men: A family development approach to marriage arrangements among ultra-Orthodox Jews. *Journal of Comparative Family Studies, 33,* 97–115.

Smith, D. L., & Smith, B. J. (2006). Perceptions of violence: The views of teachers who left urban schools. *The High School Journal, 89,* 34–42.

Smith, R. C. (1973). Comparison of the job attitudes of personnel in three air traffic control specialties. *Aerospace Medicine, 44,* 918-927.

Steckler, A., McLeroy, K. R., Goodman, R. M., Bird, S. T., & McCormick, L. (1992). Toward integrating qualitative and quantitative methods: An introduction. *Health Education Quarterly, 19,* 1–8.

Steggerda, D. M. (2003). *If I tell them, will they listen? Voices of former teachers* (Unpublished doctoral dissertation). Drake University, Des Moines, IA.

Tandon, R., Keshavan, M. S., & Nasrallah, H. A. (2008). Schizophrenia, "Just the Facts." What we know in 2008. 2. Epidemiology and etiology. *Schizophrenia Research, 102,* 1–18.

U.S. Bureau of Labor Statistics. (2008). *Injury, illnesses, and fatalities: Census of fatal occupational injuries (CFOI)—Current and revised data.* Washington, DC: U.S. Department of Labor.

Vinje, H. F., & Mittelmark, M. B. (2007). Job engagement's paradoxical role in nurse burnout. *Nursing and Health Sciences, 9,* 107–111.

Yi, H., Shin, K., & Shin, C. (2006). Development of the Sleep Quality Scale. *Journal of Sleep Research, 15,* 309–316.

Younghusband, L. J. (2008, March). *Violence in the classroom: The reality of a teacher's workplace.* Paper presented at the Work, Stress, and Health 2008 Conference, Washington, DC.

17

Use of Archival Data in Occupational Health Psychology Research[1]

Gwenith G. Fisher and Janet L. Barnes-Farrell

The term *archival data* refers to data that already exist, such as observations, texts, or other information that predate a planned research project rather than data that are expressly collected for the primary purposes of a particular research project (Fisher & Shultz, 2006; Wang, Barnes-Farrell, & Fisher, 2009). They are often used for secondary data analysis, which constitutes a specific analysis performed after the primary data collection has been completed. Secondary data analyses may represent a reexamination of data for their originally intended purposes, or they may be conducted to address questions that are unrelated to the purposes for which the data were originally collected. As a result, archival data can be used as the focal source of data in a research investigation (e.g., survey data collected and made available as part of a public use data set), or they can be used to supplement another primary data source (e.g., HR or health care administrative records linked to primary survey data; Shultz, Hoffman, & Reiter-Palmon, 2005).

The availability of archival data and the prevalence of scholarly research that makes use of archival data have grown substantially as advances in information technology have afforded the opportunity

---

1. Portions of this chapter have been previously presented at the APA/NIOSH Work, Stress, and Health Conferences, March 2–4, 2006, Orlando, FL and November 2009 in San Juan, Puerto Rico. We sincerely appreciate the helpful comments and suggestions of Jim Grosch, Ken Shultz, Bob Sinclair, and Mike Frone in the writing of this chapter.

for large data sets to be stored and accessed efficiently via the Internet by data users. Some social science fields and other disciplines, such as public health (i.e., epidemiology), have a long history of creating data archives and using archival data for scholarly research. Other fields, such as psychology, have historically had a predilection for relying mainly on primary datasets as the basis for published research studies. Nonetheless, a search in the PsycINFO database of the phrase *archival data* yielded 1,933 search results. In the field of occupational health psychology, although research based on primary data continues to be the norm, archival data are the basis for a significant portion of published scholarly work in mainstream professional journals. Therefore, the authors conducted a review of empirical studies published in the *Journal of Occupational Health Psychology* during the past decade; 14% of these articles utilized archival data as the sole or primary data source in the study. Similarly, Casper, Eby, Bordeaux, Lockwood, and Lambert (2007) indicated that archival data were used in 16% of extant industrial/organizational psychology and organizational behavior work–family research studies.

One of the factors contributing to the increasing use of archival data is the difficulty and expense of collecting certain kinds of data, such as longitudinal data. In these situations, researchers, organizations, and government agencies often see the benefit of joining forces to create archival data sets that can be used to address a variety of research and policy questions. The value of such investments is best realized when the data can be used to meet the needs of many researchers who may have differing but overlapping interests. We suspect that the movement toward investigation of research questions that are interdisciplinary in nature, coupled with increased availability of archival data sets that make such questions amenable to analysis, will provide additional incentives for researchers in fields such as occupational health psychology to give careful consideration to the use of archival data as a vehicle for conducting empirical research.

In this chapter we will describe several types of archival data and how they have been used in occupational health research, discuss unique opportunities and challenges associated with the use of archival data, and present a list of archival data resources that can be used for occupational health research. Throughout the chapter, we will offer examples of published research that have used archival data to address questions about occupational health issues.

Types of Archival Data

Singleton and Straits (2005) presented five categories of archival data. These include social science data archives, public documents and official records, private documents, mass media, and physical, nonverbal materials. These categories provide a useful taxonomy of the basic kinds of archival data that might be considered by a researcher, although some of these categories will be more pertinent than others to OHP researchers.

Social Science Data Archives

Many social science data sets consist of data collected for research purposes and made publicly available to the research community. A number of government agencies have made large investments in the collection of public use data to facilitate research and inform public policy. Increasingly, researchers whose work is funded by these agencies, such as the National Institute of Health (NIH) are required to publicly share data that they have collected in order to maximize the use of a particular data set. Furthermore, some archival data sets have been developed for the express purpose of providing an information-rich resource that can be "harvested" to address a multitude of questions. The development and maintenance of these data sets is expensive and labor intensive. As such, they typically have several characteristics in common: (a) they are intentionally designed to include measures that will be of broad interest and lasting value; (b) they often rely on sophisticated population sampling techniques; (c) they undergo rigorous evaluation of study content and data quality; and (d) they provide support for distributing the data to the research community. A number of these purpose-built archival data sets include information that can be used to explore questions of particular interest to occupational health psychology researchers.

Large-Scale Surveys. In the United States, some public use of data sets is the result of large surveys sponsored by the U.S. Department of Health and Human Services/Centers for Disease Control, the National Institutes of Health (NIH), and the National Science Foundation (NSF), among others. Examples of such data sets include the National Health Interview Survey (NHIS), the National Health and Nutrition Examination Study (NHANES), the Health and Retirement Study

(HRS), the Panel Study of Income Dynamics (PSID), the Wisconsin Longitudinal Study (WLS), Americans' Changing Lives (ACL), the General Social Survey (GSS), the study of Midlife Development in the U.S. (MIDUS), and the National Study of the Changing Workforce (NSCW). Other countries, particularly those in Europe, carry out similar large-scale surveys. In addition, a number of multinational collaborations, such as Eurobarometer and the International Social Survey Program (ISSP) include information relevant to occupational attitudes, health, and behavior. Below we describe a few of these data archives to provide a sense of the nature of these projects and the kinds of information they collect. Some of these studies (e.g., NHIS and NHANES) are considered public health surveillance data sets, where the goal is a "systematic collection, analysis, and interpretation of health data for purposes of improving health and safety" of a given population or group (U.S. Centers for Disease Control and Prevention, 2011). More information about public health surveillance data sets is described by Halperin and Baker (1992).

The Health and Retirement Study (HRS) is a cooperative agreement between the U.S. National Institute on Aging and the Institute for Social Research at the University of Michigan. It constitutes an ongoing nationally representative longitudinal biennial panel study of individuals born in 1959 or earlier (i.e., age 51 or older) that was designed to provide the research community with a large, interdisciplinary data set for investigating the demographic, health, social, and economic implications of aging in the U.S. population. The sample is based on a multistage area-clustered national probability sample with oversamples of Blacks/African Americans, Latinos, and Florida residents. In addition, the sample is updated every 6 years to add a new cohort of respondents ages 51 to 56. To date more than 30,000 individuals have been interviewed as part of the HRS since the first wave in 1992. A growing number of OHP-related studies have been conducted using the HRS. For example, Wang and colleagues identified four categories of antecedents of bridge employment, including individual characteristics, job-related psychological variables, family-related variables, and retirement planning, and found that engaging in bridge employment is associated with better physical and mental health outcomes compared to full retirement (Wang, Zhan, Liu, & Shultz, 2008; Zhan, Wang, Liu, & Shultz, 2009). Stachowski, Fisher, Grosch, Hesson, and Tetrick (2007) used latent growth curve modeling to examine whether the cognitive complexity of one's job was related to the level and rate of change of cognitive ability based on data from the HRS. Barnes-Farrell, Dove-Steinkamp, Golay, Johnson, and McGonagle (2008) studied the role of work/family variables in predicting retirement intentions among a sample of older workers.

The Midlife Development in the U.S. (MIDUS) study is a survey first conducted by the MacArthur Midlife Research Network (1994–1995) among a national sample of more than 7,000 Americans age 25 to 74. A follow-up to the original MIDUS was sponsored in 2002 by the National Institute on Aging. Grzywacz and colleagues have contributed a great deal to the work–family literature using data from MIDUS. For example, Grzywacz (2000) was among the first researchers to illustrate how the work–family interface is not all negative or based solely on conflict. In addition, Grzywacz and Marks (2000) examined various aspects of family relationship quality, work characteristics, work–family spillover, and problem drinking. Their results indicated that problem drinking was associated with higher levels of marital discord and work-related pressure, and that work–family spillover was also related to problem drinking although the direction of the correlation varied by whether work spilled over to family or family spilled over to work.

The Wisconsin Longitudinal Study (WLS) is a publicly available longitudinal study of 10,317 men and women randomly sampled among those who graduated from a Wisconsin high school in 1957 (i.e., born around 1939). Raymo and Sweeney (2006) used data from the WLS to study work–family conflict. In particular, they found that individuals who experienced higher levels of work demands interfering with family demands and family demands interfering with work demands were significantly more likely to want to retire in the next 10 years compared to those who experienced lower levels of work-to-family and family-to-work conflict. In a related study, Coursolle, Sweeney, Raymo, and Ho (2010) used data from two waves of the WLS (i.e., 1993 and 2004) to study the relationship between retirement and emotional well-being and determine how they are related to prior experiences of work/family conflict. Among individuals who previously reported high levels of work–family conflict, retirement is related to better emotional well-being among men, but not necessarily among women.

The U. S. National Study of the Changing Workforce is a nationally representative survey study of working adults in the United States that has been conducted by the Families and Work Institute approximately every 5 years since 1992. It is based on core issues about the quality of working life from the Quality of Employment Survey that was carried out by the U.S. Department of Labor in 1977. The survey gathers information about physical and mental well-being, working conditions, work attitudes, integrating work and family responsibilities, and related issues. Data from various waves of the survey have been used to examine a variety of questions concerning occupational health and work stress. For example, Halpern (2005) used data from the NCSW to demonstrate the value of flexible work policies to employee well-being and reduced organizational costs, including absenteeism, tardiness, and missed deadlines. In another study, Behson (2005) used data from the 2002 wave of the NCSW to provide convincing evidence that informal sources of work–family support are more influential than formal sources of support in mitigating work–family conflict and work stress.

The Eurobarometer is a series of public opinion surveys that have been conducted since 1973 on behalf of the European Commission. Shultz, Wang, Crimmins, and Fisher (2010) recently used data from the Eurobarometer study that assessed working conditions in 15 Western European countries to determine whether there are differences between older and younger workers with regard to the job demands–control model of work-related stress. Shultz et al. (2010) found important age differences in the job demands–control model such that different sources of control may serve to buffer different types of job demands for older vs. younger workers.

Public Social Science Data Archives. In addition to data made publicly available from particular large-scale surveys, another source of archival data includes archives of many different types of studies. The Inter-University Consortium for Political and Social Research (ICPSR) is perhaps the largest and best known of these archives. One example of a data set available from ICPSR is the Americans' Changing Lives (ACL) study, which is a longitudinal study with four waves of data collected between 1986 and 2002 to gather psychological, sociological, mental, and physical health data among individuals in middle and later adulthood. Herzog, House, and Morgan (1991) used the first two waves of the ACL data set to study the role of work and retirement on health and well-being among older adults. They examined patterns of labor force participation and found that individuals whose work status was consistent with their preferences had higher levels of physical health and psychological well-being compared to those whose work status was constrained by other factors. Similarly, Shultz and Wang (2007) used the first three waves of the ACL data set to examine the influence of specific physical health conditions on retirement decisions. They found that different acute and chronic health conditions were associated with retirement, continued work in the same job, or continued work but in a different job.

Another excellent source of archival data for OHP research is the Roper Center for Public Opinion Research at the University of Connecticut. The Roper Center houses the largest collection of public opinion data in the world. For example, the General Social Survey (GSS) data set, collected by the National Opinion Research Corporation, is available here. Firebaugh and Harley (1995) used data from the GSS to study U.S. trends in job satisfaction by various demographic characteristics, including race, gender, and occupation. The Roper Center also archives a number of relevant social science data sets collected from private organizations, such as panel studies of working adults' concerns conducted at intervals by AARP (see Table 17.1).

One important note regarding public data sets is that although the data are publicly available, there may be a cost involved in obtaining access to the data. For example, data from ICPSR are free for individuals whose institutions are members of the Inter-University Consortium, but a cost to obtain the data would likely be incurred for others. The National Study of the Changing Workforce, conducted by the Families and Work Institute, is another archival data set that is publicly available data, but for a fee.

Private Data Sets. Private data sets refer to data already collected by another researcher for another study, data collected by another agency for evaluative or research purposes, and a researcher's own data that he or she gathered for a prior study. Private data sets differ from the aforementioned public use data sets in that they are not made publicly available. However, it may be possible for a researcher to obtain and use private data sets for secondary analysis research. A starting point to obtaining access to private data would be to contact an agency or lead researcher who has made mention of the data set in a publication or technical report. Because the availability of such data sets is dependent on the willingness of the data owner to share a private resource, communicating the value of the research question to be addressed and developing a trusting relationship with the data owner is likely to play a key role in gaining access to private data sets.

One example of a private data set is data on workers' compensation claims maintained by many private health insurance companies. Such data sets are rich with demographic, economic, and health data, including a description of the nature of the injury/illness and any recurrence that may have occurred. Most insurance companies are understandably concerned about the confidentiality of these data sets, but may be open to sharing them once a data use agreement has been established. Another example of private data set use is illustrated by Shultz and his colleagues' research using the U.S. Navy career development data set (Spiegel & Shultz, 2003; Shultz, Taylor, & Morrison, 2003; Taylor, Shultz, Morrison, Spiegel, & Green, 2007). These three studies examined naval officers' work-related attitudes, occupational attachment, skills, and preretirement planning in relation to retirement satisfaction and adjustment.

Public Documents, Data Sets, or Official Records

Many sources of public documents and records are available on individuals, including recorded births and deaths (e.g., the National Death Index), and U.S. Census Bureau data (which provides individual-level data, as well as employer information via the U.S. Census Bureau Business Register database). The U. S. Bureau of Labor Statistics conducts a number of surveys (e.g., Census of Fatal Occupational Injuries and the Current Population Survey) which might be of particular interest to OHP researchers (see Table 17.1).

Schubauer-Berigan, Couch, and colleagues conducted a series of studies to evaluate the effects of occupational exposure to beryllium. Schubauer-Berigan et al. (2011) conducted a cohort mortality study among beryllium processing plant workers. In particular, these researchers used employment history records from two companies operating beryllium processing plants to code complete work history data and link the work history records to job exposure matrices, Social Security Administration records, and the National Death Index to obtain information pertaining to employee mortality and cause of death.

The Occupational Information Network (O*NET; n.d.) is a publicly available data set developed by the U.S. Department of Labor that serves as the "primary source of occupational information" in the United States (O*NET Resource Center). This contains several hundred variables with detailed information about specific occupations, including detailed characteristics of work and workers in size domains, including worker characteristics, worker requirements, experience requirements, occupational requirements, workforce characteristics, and occupation-specific information. Development of the O*NET began in the late 1990s and has replaced the Dictionary of Occupational Titles (DOT), which was last revised in 1991.

During the last decade, the O*NET has been used in a growing number of occupational health psychology studies (e.g., Alterman et al., 2008; Cifuentes et al., 2007; Cifuentes, Boyer, Lombardi, & Punnett, 2010; Ford & Tetrick, 2011; Liu, Spector, & Jex, 2005). For example, Alterman et al. (2008) demonstrated how the O*NET database could be used to identify job dimensions as proxy measures of psychosocial and environmental factors at work, and then related these dimensions to health outcomes using data from two other archival data sets containing health measures. In addition, Ford and Tetrick (2011) used data from the O*NET to measure occupational hazards, and related those occupational hazards to psychological empowerment and organizational identification in predicting occupational safety performance. Cifuentes et al. (2007) compared O*NET ratings of working conditions to self-reported survey measures of psychosocial working conditions among health care workers. Their results demonstrated a good level of agreement between these sources, indicating that the O*NET database can serve as a useful source of job level psychosocial exposure within the context of the demand/control and effort/reward models.

Private Documents or Records

Unlike social science data archives and other public documents or private research data sets, private records were not necessarily collected for research purposes. This source of archival data involved collecting data about individuals for one's own sake, although these records may be collected or maintained by organizations. Examples of such data include human resources and other company or organization records, medical records, school records, financial statements, and credit history. Some records which are collected by government agencies may also be publicly available. Other sources are truly private records which are subject to specific privacy laws and regulations. As a result, such private records are likely to require special permission for obtaining access to the records.

Company records of workplace injuries and illnesses constitute an example of private documents or records. The Occupational Safety and Health Act was passed in 1970 to govern occupational safety and health in the U.S. federal government and the private sector. Employers covered by this act are required to maintain occupational injury and illness records (Seligman, Sieber, Pedersen, Sundin, & Frazier, 1988). However, usefulness of company records relies on record-keeping compliance. Seligman et al. (1988) reviewed data collected by the National Occupational Exposure Survey (NOES) to assess compliance with the U.S. Occupational Safety and Health Administration (OSHA) record-keeping requirements. The NOES, conducted by NIOSH in the early 1980s, used a probability sample of 4,490 facilities which covered 523 industries and 410 difference occupations. Seligman et al. (1988) reported that 75% of employers with 11 or more employees maintained OSHA Form 200 (i.e., an annual log and summary of workplace injuries and illnesses). Organization size was the most important factor in predicting compliance with having OSHA Form 200. Specifically, 96% of large organizations (i.e., those with 500 or more employees) had these records, compared with 62% of smaller organizations (i.e., those with 11 to 99 employees). One concern about the use of such records for occupational health research is therefore the issue of possible underreporting.

Mass Media

The number of mass media sources and access to such "data" has rapidly grown with the development of technology. This includes not only newspapers, magazines, television, and movies, but other content available on the Internet, including blogs and videos (e.g., YouTube; Jones, 2010). In the contemporary climate, social media (e.g., blogs, Facebook, Twitter) may be considered another possible source of such data, especially to the extent that posts are made publicly available on the Internet. Mass media have been identified by researchers as a possible source of archival data (e.g., Jones, 2010; Singleton & Straits, 2005). Donnelly (1982) cites mass media as having played a critical role in increasing public awareness of occupational hazards during the 1960s, leading to the establishment of the Occupational Safety and Health Act of 1970.

Research using or examining mass media (e.g., evaluating the role of mass media in communicating health information or affecting various health behaviors) is much more common in public health research than in occupational health psychology. However, this constitutes a class of data that appears to be worthy of further research investigation and may serve as a potentially rich avenue for OHP researchers to explore.

Physical, Nonverbal Materials

Based on the typology of archival data presented by Singleton and Straits (2005), physical, nonverbal materials (i.e. artifacts of human behavior) constitute another source of archival data. We are not aware of any applications in the extant occupational health psychology literature; however, in the social sciences they have been frequently used by anthropologists.

We have just described the various types of archival data and given a few examples of archival data used in occupational health psychology research. In the next two sections we will describe the opportunities afforded by as well as some challenges inherent in the use of archival data.

Unique Opportunities Afforded by Using Archival Data

Available Data

The first and perhaps most obvious advantage to using archival data is that the data already exist. Obtaining readily available data may save a significant amount of time and expense. This can be quite nontrivial when conducting a research project, especially with limited time and budget constraints (Shultz et al., 2005). In many cases, archival data may be obtained at no cost to the data user. As we will see in the section on "Challenges," using data collected by others may come with costs and consequences, but thoughtful opportunistic use of existing data can be an efficient way to move a research program forward quickly. It can also be a way to pilot ideas before investing additional time and resources in the design and implementation of a separate primary data collection effort. Table 17.1 provides a list of some publicly available archival data resources which may be of interest to occupational health researchers.

Data with Specific Research Designs or Methodology

In addition to having data readily available, the use of an archival data set may facilitate one's research by incorporating a particular research design or methodology that is particularly challenging to implement. For example, some archival data sets may include longitudinal data, with repeated or multiple observations among participants. A study

Table 17.1 Selected Archival Data Resources

Resource	Description	Website
U.S. Department of Labor		
Occupational Information Network (O*NET)	Standardized information about occupation characteristics and work conditions	http://dol.gov http://online.onetcenter.org/
Bureau of Labor Statistics		
Current Population Survey	Monthly household survey; provides labor force characteristics of U.S. civilian population	http://www.bls.gov/data/
Census of Fatal Occupational Injuries	Demographic and industry information on occupational injuries and illnesses	http://www.census.gov/cps/
National Longitudinal Studies	Longitudinal surveys of men and women that include work history and other life events	http://www.bls.gov/iif/oshcfoi1.htm
National Longitudinal Study of Youth 1979	Longitudinal study of men and women who were 14–22 in 1979, including work and life events	http://www.bls.gov/nls/ http://www.bls.gov/nls/nlsy79.htm
U.S. Census Bureau		http://www.census.gov/
American Community Survey	Information for U.S. households about many topics including work and health	http://www.census.gov/acs/www/
Data Ferret	Online application that provides access to multiple years of many publicly available datasets	http://dataferrett.census.gov/
U.S. Centers for Disease Control and Prevention		http://www.cdc.gov
Behavioral Risk Factor Surveillance system (BRFSS)	Telephone health survey system that tracks preventive health practices and risk behaviors of adults linked to chronic diseases and injuries	http://www.cdc.gov/brfss/
National Center for Health Statistics (NCHS)	Links to wide variety of health-related data	http://www.cdc.gov/nchs/
National Health and Nutrition Examination Survey (NHANES)	Program of studies on health and nutritional status of adults and children	http://www.cdc.gov/nchs/nhanes.htm

(continued)

Table 17.1 Continued

Resource	Description	Website
National Health Interview Survey (NHIS)	Household interview data on a broad range of health topics	http://www.cdc.gov/nchs/nhis.htm
National Institute for Occupational Safety and Health	Workplace surveillance data by industry sector and gateway to other relevant survey databases	http://www.cdc.gov/niosh/data/
Institute for Social Research (ISR)—University of Michigan		http://www.isr.umich.edu/home/
Americans' Changing Lives (ACL) study	National longitudinal panel survey to study middle and later life among Black and White Americans on a variety of social science issues	http://www.isr.umich.edu/acl/
Inter-university Consortium for Political and Social Research (ICPSR)	Data archive. Gateway to search and access multiple datasets that available for download or online analysis	http://www.icpsr.umich.edu/icpsrweb/ICPSR/
ICPSR Thematic Collections	Access to the National Comorbidity Survey and other studies that are thematically relevant to occupational health issues	http://www.icpsr.umich.edu/icpsrweb/ICPSR/partners/archives.jsp
Health and Retirement Study	Longitudinal study of individuals and couples over the age of 50, including work, retirement, health and well-being issues	http://hrsonline.isr.umich.edu/
Michigan Census Data Research Center	Secure access to confidential unpublished U.S. Census Bureau demographic and health data for qualified researchers with approved research proposals	http://www.isr.umich.edu/src/mcrdc/
National Archive on Computerized Data on Aging (NACDA)	Access to a library of electronic data on aging for secondary analysis	http://www.icpsr.umich.edu/icpsrweb/NACDA/
Panel Study of Income Dynamics	Longitudinal household study, including employment, health and other topics	http://psidonline.isr.umich.edu/
Population Studies Center	Datasets on health, aging, race and gender, labor force participation, occupations, transitions to and from work and other related topics.	http://www.psc.isr.umich.edu/dis/data/

Resource	Description	Website
Quality of Employment Survey	A 1977 study of the working conditions in the U.S. among more than 1,500 workers	http://www.icpsr.umich.edu/icpsrweb/ICPSR/studies/7689
International Social Survey Program (ISSP)—Work Organizations II	Data archive. Archives and integrates data and documentation for merged cross-national data sets from the International Social Survey Programme. Surveys gather information on a variety of recurring and special topics including work and health	http://www.issp.org/index.php
Roper Center for Public Opinion Research - University of Connecticut	Data archive. Search for and access national and international survey data sets on a wide variety of topics; locate summary data from individual questions of interest pulled from a compendium of surveys	http://www.ropercenter.uconn.edu/
Harvard—MIT Data Center	Gateway to access other data archives, including ICPSR and Roper Center as well as major public collections	http://www.hmdc.harvard.edu/
American Association of Retired Persons (AARP)	Links to surveys on aging, work and retirement and relevant research databases. Contact information for gaining access to AARP-sponsored survey data	http://www.aarp.org/research/
Employment and Disability Institute—Cornell University	Searchable database of datasets relevant to disability and rehabilitation	http://www.disabilitystatistics.org
Eurobarometer Survey	Access to annual reports and interactive trend reports for household interviews of adults in European nations on a variety of topics including working conditions, health and well-being; link to data archives for secondary analysis	http://ec.europa.eu/public_opinion/archives_en.htm
McGraw-Hill Research Resources for the Social Sciences—Social Science Data Archives	List of social science data archives with links to websites	http://www.socsciresearch.com/r6.html

(*continued*)

Table 17.1 Continued

Resource	Description	Website
National Opinion Research Center	Access to multiple publicly available data archives for projects carried out by NORC	http://www.norc.org/
General Social Survey	Standard 'core' of demographic, behavioral, and attitudinal questions plus special topics of interest including work and health; cross-national module that is part of ISSP	http://www.norc.org/GSS+Website/
National Study of the Changing Workforce	Public use data files from multiple waves of a survey work and family issues for of self- and organization-employed workers	http://familiesandwork.org/site/work/main.html
Social Science Information System—University of Amsterdam	List and gateway to national and international data archives and related websites of interest to social science researchers, listed by country	http://www.sociosite.net/databases.php
Whitehall Studies	Longitudinal study of stress and health in a cohort of working men and women in the U.K.	http://www.ucl.ac.uk/whitehallII/
Wisconsin Longitudinal Study	Longitudinal study of more than 10,000 men and women who graduated from high school in Wisconsin during 1957. Includes information on work experiences, health, well-being, and many other topics.	http://www.ssc.wisc.edu/wlsresearch/

by Tucker et al. (2009) provides an excellent illustration of how OHP research has used archival data in this fashion. They used an archival data set containing six waves of longitudinal data to examine work demands, control, and counterproductive work behavior among U.S. soldiers. Tucker et al. (2009) did not find any main effects of work overload and work stress, and found contradictory results regarding the interaction between work overload and control on soldier indiscipline.

Similarly, many archival data sets have the advantage of large sample sizes, and may be larger than what a researcher could feasibly collect on his or her own. Barling, Kelloway, and Iverson (2003) used data from a large, randomly drawn sample of Australian employees to study the relationship between high quality work, job satisfaction, and occupational injuries. The 1995 Australian Workplace Industrial Relations Survey (AWIRS95) randomly sampled employees from a stratified sample of more than 2,000 workplaces with a minimum of 20 employees. Barling et al.'s (2003) results indicated that high quality jobs had a direct and indirect effect on occupational injuries.

In addition, some data sets may include what some may consider more "objective" data, such as administrative data, biomarkers, or other forms of data that may be difficult to obtain as part of a primary data collection. For example, the O*NET database, described earlier, is considered to be an objective source of occupational information (Cifuentes et al., 2007; Liu et al., 2005). Another example is the Health and Retirement Study (HRS), which now includes biomarkers, physical performance measures, as well as linkages to CMS (Medicare) claims files and the National Death Index. Up to 10 waves of data are presently available (i.e., 1992 through 2010). As we mentioned previously, a growing number of studies in occupational health psychology have used data from the HRS (e.g., Barnes-Farrell et al., 2008; Stachowski et al., 2007; Wang et al., 2008; Zhan et al., 2009). Recently, Mezuk, Kershaw, Hudson, Lim, and Ratliff (2011) examined psychosocial predictors (job strain and workplace discrimination) on hypertension, and used resting seated blood pressure measurements that were added to the HRS in 2006.

Population/Subpopulation Representative

Depending on the population of interest in one's research study, using an archival data set may help a researcher obtain a data set that is representative of the population of interest. For example, for those interested in examining older adults in the U.S., the Health and Retirement Study is an excellent data resource. The HRS sample frame is a national probability sample that is representative of adults in the U.S. age 51 or older. Some archival data sets are quite large, and may be useful for studying specific sub-populations, such as a particular gender, race/ethnic group, or occupational sector. Using a data set that is representative of a population can be particularly useful for evaluating the prevalence of particular phenomena. In addition, analyzing data from a population-representative sample can serve to increase the generalizability of the results. As an illustration, Wang et al.'s (2008) research on the antecedents of bridge employment using data from the HRS can be generalized to the U.S. population of older workers based on the use of the HRS data and the national probability sample frame employed by the HRS.

Access to Special Populations

An archival data set may provide a researcher with data from a special population that may be especially challenging to obtain in a primary data collection. For example, some archival data sets have been collected with a specific purpose or population in mind (e.g., survivors of prostate cancer who are currently employed). Also, some data sets are quite large and have purposefully incorporated systematic oversampling of some low-frequency population sectors. A subset of the data set could be used to study a particular population of interest (e.g., individuals of a particular age, country or region, race/ethnic group, health status, or employment sector) or to make comparisons between important subpopulations. For example, Shultz et al. (2010) used the cross-national Eurobarometer survey data to compare older versus younger workers with regard to the job demands–control model. Tucker et al.'s (2009) study examined U.S. soldiers, including soldiers in units in garrison, on peacekeeping deployments, and in training rotations.

Broader Scope of Variables

One distinct advantage of using an archival data set is that such data sets may include a broad scope of variables with data on a variety of topics. Many archival data sets include data across a range of disciplines, which can facilitate interdisciplinary or cross-disciplinary research. The HRS is a terrific exemplar of this characteristic. It is an interdisciplinary data set that was developed by economists, sociologists, psychologists, epidemiologists, and medical doctors to broadly measure issues affecting the aging U.S. population. This has afforded researchers the opportunity to address boundary-spanning questions that would otherwise be difficult to study. As an example, the HRS has been able to facilitate OHP research examining occupational complexity related to cognitive decline (Stachowski et al., 2007); antecedents of bridge employment (Wang et al., 2008); as well as the study of work–family interference and enhancement (Barnes-Farrell et al., 2008; Fisher, 2006).

Some archival data sets may include data that are challenging to obtain or sensitive (e.g., health records), yet quite valuable for research. It may prove quite difficult for an individual researcher to obtain access to this kind of information, but existing sources may contain these data. There are many examples of data sets containing this type of information. For example, Kivimäki et al. (2005) conducted a prospective study of more than 5,000 employees to examine optimism and pessimism as predictors of change in health in specific circumstances. They used data from the 10-Town Study, coordinated by the Finnish Institute of Occupational Health and used employer-maintained records of sickness absence days as the measure of health. The HRS data set, which has been mentioned several times, contains a number of variables pertaining to individuals' health status, and can also be linked to Medicare and Social Security records.

Objective Rather Than Subjective Data

There are many sources of objective data that have been used in the study of occupational health psychology, including sickness absence data, performance measures, accidents, and death records (Kompier, 2005; Schubauer-Berigan et al., 2011). When used in conjunction with self-report data, objective data may serve to minimize or alleviate concerns of common method bias (Podsakoff, MacKenzie, Lee, & Podsakoff, 2003). As such, these records are not susceptible to some of the measurement errors associated with more subjective measures, such as self-report bias (Spector, 1994). However, even objective sources of data are subject to a variety of errors. This issue is discussed later in this chapter with regard to challenges in OHP research.

Suitable for Student Research Projects

Archival data sets may provide an excellent source of data for an undergraduate thesis or for graduate students working on a thesis or dissertation (Shultz et al., 2005). As many thesis advisors recognize, one of the practical challenges of carrying student research projects to completion is the time that must be devoted to designing and implementing a data collection effort, particularly when the research question is one that is not amenable to laboratory study with student participants. Many excellent data sets already exist, and may be quite suitable for answering students' particular research questions. The use of an archival data set may help students successfully complete their work much more quickly and easily than if they need to collect the data themselves. This may be particularly advantageous if the primary goal of the research project is to gain experience in developing and testing research hypotheses, rather than the goal of developing research design and primary data collection skills. Some disciplines, such as psychology or survey methodology, may value students developing the skills inherent in designing research and collecting data. However, in other social science disciplines (e.g., economics), this seems to be a less central focus, and secondary data analysis based on archival data is valued as an important approach to conducting research in its own right. In addition, one might argue that students' own data collection lead to using small convenience samples which may result in misleading conclusions based on low power (high Type II errors), or research that has limited chances of publication, particularly in a high-quality journal. Perhaps the development of skills in identifying, retrieving, merging and managing large data sets may be as important, or more important, than the "traditional" skills emphasized by designing and collecting one's own data.

To date, 240 dissertations and theses have been written using data from the HRS (HRS 2011), so it is clear that appropriate archival data sets can be useful as the basis of student projects. With this in mind, we encourage research advisors for OHP projects to give serious consideration to whether there are existing data sets that can open doors for their students to empirically test their research ideas in a rigorous fashion.

Challenges to Using Archival Data

Although archival data sets offer many advantages to the OHP research community, there are many challenges associated with the use of such data sets. These issues should be considered prior to a decision to use archival data in general or when considering the use of a particular data set (Zaitzow & Fields, 2006).

Ethical Issues

The use of archival data may pose additional ethical issues compared to a primary research data set in which the data were collected directly by the researchers analyzing the data. First, researchers analyzing data from a secondary data source should try to determine whether data were gathered in a manner consistent with ethical guidelines in their field. Second, researchers granting or obtaining use of data for secondary analysis need to ensure that the purpose of the secondary analysis does not deviate from what participants may have been told when the data were collected. For example, respondents may have been told that the data were being collected for Purpose X, but a researcher may want to use the stored data for Purpose Y. If Purpose Y is different from what respondents were initially told or Purpose Y exposes participants to an increased risk of which they were not previously informed and for which they did not give consent, then the use of the data for Purpose Y may be considered inappropriate or unethical. This issue may also arise if the ethical principles and practices of those proposing to use the data for secondary analysis are based on different ethical standards compared to those who originally collected the data. When researchers repurpose their own data that was originally collected for other purposes, they take on the additional responsibility of reporting to journal editors about any prior use of the data and clarifying any overlap between that work and the purposes for which the data will be used in a secondary analysis.

In addition, some data sets require specific permission from the original data gatherers/owners in order to access the data, or in order to obtain clearance for the study by the Institutional Review Board. In some cases, additional procedures may need to be taken to ensure anonymity with regard to response protocols and/or confidentiality of the data being analyzed. For example, the HRS data set can be linked to Social Security and Medicare records. However, these additional administrative data sources are considered restricted data that are only obtainable under certain circumstances; extra care must be used to be sure these data are not linked to other data (e.g., geographical information), which increases the risk that individual respondents may be identified.

Although the secondary analysis of an archival data set may be considered "exempt from continuing review" by the secondary researcher's Institutional Review Board, the ethical responsibility for the use and IRB approval of the archival data set still belongs to the researchers who are considering use of the data set. We encourage researchers to become familiar with their institution's policy regarding secondary data analysis and to follow proper Institutional Review Board protocols. It may seem easy to overlook this important issue when not directly collecting data from human participants.

Large, Complex Databases

Many archival data sets consist of large, complex databases that may be difficult to navigate. Because someone else designed and prepared the data set, it will take extra time to become familiar with its structure and format (Shultz et al., 2005). These databases may not be user-friendly in terms of acquiring and setting up the data, navigating the documentation, or both. Although some data sets may be easy to use immediately upon download, others require a more involved set-up process. For example, some data are distributed in an ASCII format, and statistical program statements must be run to convert the ASCII-formatted data to a usable data file. Another issue with large databases, especially those based on population surveys, is that the use of complex sample survey designs may result in the need to use sample weights, and pose particularly complex sampling and analytical challenges when combining multiple data sets which are based on different complex sample survey designs.

The ease with which one can readily assemble and make use of an archival data set will depend on the extent and quality of the documentation provided with the data set. There is great variability across data sets regarding the amount and quality of information available that serves as documentation. Ideally the data set itself should be accompanied by a codebook which provides detail about each variable in the data set (e.g., question test and explanations of code values), as well as a description of the methodology used in collecting the data, sources of measures, and information about measurement properties.

Because of such complexities, the use of large archival data sets typically involves more work on the front end of a research project in order to organize the data set in a particularly useful manner compared to using a data set that one collected and organized oneself. Archival data use typically requires skills in data management, such as merging and subsetting large data files, constructing variables, and recoding values assigned to variables. Sometimes the data management tasks require a considerable amount of time to assemble and organize a data set for analysis before the data are ready to be used for analysis. We particularly encourage potential archival data users to budget sufficient time for data management tasks prior to commencing the actual data analysis needed to answer the specific research questions of interest. Likewise, in the interest of broadly facilitating effective secondary analysis, we recommend that OHP researchers who are collecting primary data become more systematic about the way they document the methodology and features of their own data sets.

Complex Sample Survey Designs and Need for Sample Weights

Many desirable archival data sets (e.g., those including, but not necessarily limited to, large-scale population-based surveys) have samples derived from complex probability sampling. Such sampling techniques can be useful for generalizing to an entire population of interest, but can vastly increase the complexity of data analyses. Training in occupational health psychology does not typically involve learning about survey sampling, nor the issues associated with analyzing data from studies with a complex sample survey design, including probability sampling, clustering, and stratification. However, misuse of data collected using complex sampling techniques can result in inaccurate results and misleading conclusions. Therefore, we encourage individuals using archival data sets with complex sample designs which require the use of sample weights (and possibly additional adjustments for variance estimation due to clustering and stratification) to become knowledgeable about these issues.

There are many different kinds of sample designs. Understanding the sample design is important because it has implications for data analysis. A probability sample refers to the concept that elements of a sample frame are selected using chance methods. Elements in the sample frame have a nonzero probability of being selected. A probability sample is quite useful when making statistical inferences from a sample about a population, because the probability of selection from the population is known. A simple random sample is when all individuals in the sample frame have an equal probability of being selected. A stratified sample is when the population is classified into subpopulations based on supplementary information (e.g., geographical location or individuals' demographic characteristics, such as age, race, or gender), and then the sample is selected separately for each stratum. Proportionate stratification takes place when units (e.g., individuals or households) are selected from each stratum in proportion to the stratum's frequency in the population. Clustered sampling refers to a sampling technique in which frame elements are selected jointly rather than selecting sample frame elements individually (Groves et al., 2009). Geographical area clustering is probably the most common type of sample clustering. Clustered geographical sampling is frequently used to reduce survey costs because the cost involved in conducting a simple random sample across a large geographical area is usually prohibitive. For example, when conducting a survey in a wide geographical area (e.g., across the nation), it is often significantly more cost-effective to cluster interviewing in particular geographical areas rather than to select a simple random sample across the wide geographical area.

Sample weights are used to adjust sample design characteristics (e.g., sampling probability, clustering, and stratification) so that statistical computation of point estimates and variances in point estimates accurately represent the population. Typically sample weights are derived by taking the inverse probability with which each individual was selected into the sample. Additional adjustments can be made to sample weights to account for unit nonresponse (Gelman & Carlin, 2002).

Sample weights are critical for the analysis of complex sample survey data found in many large, population-based public use data sets. We will use sample weights in the HRS as an example. The HRS is intentionally geographically stratified and clustered in order to obtain a nationally representative sample while minimizing the data collection cost. In addition, the HRS sample includes oversamples of Blacks, Hispanics, and Florida residents. Failure to use sample weights in the analysis would result in these oversampled individuals' responses being overemphasized relative to non-Blacks, non-Hispanics, and non-Florida residents. In other words, sample weights adjust the proportion to which these responses count based on the probability with which these individuals represent others in the population as a whole. Failure to include the clustering and stratification variables in an analysis would result in underestimated variance estimates (e.g., estimated standard errors that are lower than what they should be with the sample clustering and stratification taken into account).

With regard to data analysis, some statistical software packages are much more capable of handling complex sample survey design variables (sample weights, stratification, and cluster variables) than others. Stata, SAS, and Sudaan are generally able to incorporate sample weights and sample design variables in the analysis. Recently SPSS has developed a complex samples module that is currently available as an add-on to the base SPSS program, but it is still rather primitive. For additional information about sampling and data analysis issues, see Lehtonen and Pahkinen (2004), Korn and Graubard (1999), Lohr (2010), Sterba

(2009a), and Trzesniewski, Donnellan, and Lucas (2011). In addition, Sterba (2009b) provides an online appendix regarding statistical software with some examples.

Lack of Desirable Measurement Properties

Imagine this scenario: You obtained an existing, free data set that includes valuable, population-representative data on your topic of interest. You're all set to complete your dream study, right? Upon looking further at the data set, you may find that measures of interest are missing or less than optimal for measuring a certain construct. Because you did not design the study and develop the measures yourself, the measures may not be the same ones you would have chosen to measure the same construct. However, it is sometimes the case that the way in which studies are able to cover a broader scope of variables and obtain information on a variety of topics is that each construct measured in the study is measured with only a few, or in some cases, a single item. Nonetheless, once the data are collected and part of an archival data set, the user is only able to use the data that are available. Careful consideration of the way key variables are operationalized in the data set will sometimes lead to the conclusion that the data set is not suitable for the researcher's purposes. In other cases, the researcher may decide that the data set affords important insights into the research question, albeit at the cost of reduced measurement precision for one or more of the variables that are part of the intended study.

Researchers may need to be creative to develop a measure of constructs they are interested in assessing. This will require a clear understanding of the construct they wish to assess and thorough understanding of information available in the archival data set. A careful mapping of research variables onto the extant variables can sometimes provide a proxy measure for the construct of interest. Of course it must be recognized that researcher-constructed variables are approximations with unknown psychometric quality and may be subject to criticism on that basis. However, although a data set may not present measures with ideal measurement properties, it may still prove to be quite useful for research and such methodological limitations may not constitute a deal-breaker. For example, in using the Eurobarometer data set to examine the demand control model of work stress among older workers, Shultz et al. (2010) were constrained by having only single-item measures of work stress. Although the authors were well aware of the limitations of having single-item measures, the strengths offered by having data from 15 countries and other characteristics of the study were able to outweigh this particular measurement limitation. Similarly, Barling et al. (2003) encountered single-item measures of occupational injuries when using data from the Australia Workplace Industrial Relations Survey database. In their case, other strengths of the research design appeared to outweigh this methodological limitation.

Because measurement rigor can play a key role in the publication fate of a research study, our recommendation is that researchers who are considering the use of archival data begin by thinking carefully about what they *need* to measure, followed by thinking creatively about alternative ways to accomplish that goal within the confines of the available data set. Then they should be prepared to educate editors and reviewers as to the suitability of their choices, and the strength of the research contribution afforded by this approach. They may also want to consider bolstering the strength of their arguments by using multiple operationalizations for the key construct or conducting an additional small-scale validation study to support the use of a nonvalidated measure.

Lack of a Theoretical Framework

Occasionally OHP researchers conducting their own study may design a study with a specific theoretical framework in mind. However, some archival data sets may lack a specific theoretical framework which could be useful for OHP research. For example, public health data sets (e.g., NHIS) often have a main purpose of obtaining generalizable health data about the population without regard to a specific theoretical framework that may be more amenable to hypothesis testing.

Missing Data

Related to the aforementioned issue of a lack of desirable measurement properties, some data sets, particularly longitudinal data sets, may also contain missing values. Researchers using any data set should spend time in understanding the pattern of missing data, and apply the most appropriate statistical methods for treating the missing data. It is particularly important for initial exploration of the data set to include a thorough study of distributions of all variables of interest and missing data patterns that might influence the usability of the data for the researcher's intended purposes.

Objective Rather Than Subjective Measures

Lerner and Lee (2006) described methods for asking individuals about the impact of their health on work, and indicated that employer archival data offer an alternative to such self-report methods. For example, employer administrative data may include information about absenteeism attributable to health reasons, performance measures, and workplace accidents. Objective measures such as physiological data, health care records, and administrative data (e.g., human resources records or accident data) may not suffer from self-report bias. However, it is important to note that they are not necessarily error-free, and in many cases may be inferior to certain kinds of subjective data. These data are only as good as the methods used to collect and record the data. For example, data entry errors (e.g., a clerk entering an incorrect ICD-9 code in health care records) or computer programming errors may be present in administrative records. In addition, in some cases, it may be difficult to obtain estimates of reliability for objective sources of data. We suggest that researchers carefully consider the measurement properties of any measures employed in a particular research study and carefully consider the strengths and limitations of each source with regard to answering the research question.

"Old" Data

One additional shortcoming of using some archival data sets is that the data may have been collected a while ago. As a result, journal editors or reviewers may be concerned that the data are too old to be relevant or generalizable to current phenomena. The extent to which this may be a problem will depend on the topic of investigation and the perceptions of reviewers. We encourage researchers who are considering the use of older data sets to identify possible generalizability concerns early in the research process as well as reasons why older data may still be appropriate for investigating a particular occupational health issue.

Lack of Suitability for Student Research Projects

Although ability to access data that are suitable for investigating student research questions has already been mentioned as an advantage of archival data sets, the fact that such data sets do not provide students with data collection experience can likewise mean that they may be viewed as unsuitable or unacceptable for student research projects. There appear to be large differences in disciplines regarding the extent to which students are permitted or encouraged to use archival data. For example, some graduate programs (e.g., some in psychology or survey research methodology) permit students to carry out research based on archival data sets, but they prohibit or discourage the use of archival data sets for conducting research for degree projects, such as a doctoral dissertation. The primary reason is that students in these fields must demonstrate the ability to design and collect their own data, and obtain the experience in doing so. On the other hand, some social science disciplines (e.g., sociology, economics) often encourage the use of archival data because research questions may necessitate having population-level data for analysis that would be difficult or cost-prohibitive for an individual to collect on his or her own. In some cases, existing archival data sets have been developed for others in the field to analyze, and one of the key competencies a student is expected to develop is the skill set to appropriately and effectively use such data sets to conduct research. Prior to embarking on a thesis or dissertation project, students considering the use of archival data should clearly establish whether it would be appropriate in relation to formal department policies as well as the preferences of their advisor/committee. We also recommend that faculty who are members of departments in which use of archival data may be controversial establish clear guidelines about this issue.

Focusing on the Research of Interest

The last challenge we would like to address regarding the use of archival data is the problem of "too much." Having a rich data set containing variables across a broad array of topics can be almost intoxicating. If the data set is approached in an unfocused way, the endless array of possibilities may make it difficult for a researcher to narrow the scope of a study to a reasonable analysis. In our experience, archival data use is most successful when a researcher has a particular question in mind and then obtains data from an appropriate existing source to answer that question. There are some cases when valuable research is conducted *after* identifying a particular archival data set. However, researchers need to exercise caution and not turn a valuable research study into a fishing expedition or let the available data drive the research agenda. Likewise, the fact that a research question *can* be addressed with a particular data set does not imply that the research should be conducted. As always, the possibility of meaningful research contribution should drive this decision; one must first determine whether a particular research question has already been addressed with the archive.

Archival Data Resources for Occupational Health Researchers

Many archival data sets exist that assess variables of potential interest to occupational health researchers, and the number of such data sets continues to grow. Many of these resources are provided by agencies in the U.S. government (e.g., the U.S. Department of Labor, the National Institutes of Health) specifically with the purpose of making data available to researchers for secondary analysis. In some cases, funding for research studies may be contingent upon primary investigators agreeing to share data resources with others in the research community for the collective good. Therefore researchers will share their data sets with others through various mechanisms. Some limitations of archival data sets can be addressed (at least in part) by linking or merging different archival data sets that were created for different purposes (e.g., Alterman et al., 2008, who linked health and occupational data together from O*NET and NHIS). In other words, combining multiple archival data sets can allow a researcher to examine relationships that may not be possible if only a single archival data set is used. Multiple archival data sets can also be used to compare data across different countries (Crimmins, Kim, & Solé-Auró, 2011).

The initial challenge for most researchers is identifying and locating these data sets. The list of archival data sets that we have provided in Table 17.1 will not be comprehensive by the time this chapter is published. However, it provides an excellent starting point in a search for archival data sources that include variables studied by occupational health psychologists. Some of these resources are large public archives, such as the ICPSR at the University of Michigan's Institute for Social Research or the Roper Center for Public Opinion Research at the

University of Connecticut, that house many different kinds of publicly available data sets. The search capabilities installed at the websites for these archives can be particularly useful tools for researchers who are interested in locating newly archived data sets or data sets that focus on particular issues or populations. Published work that has been conducted with many of these data sets has been mentioned throughout the chapter. We encourage OHP investigators who are considering the use of archival data in their research to read some of this work as inspiration, and then take a look at the data sets from which they were drawn to get a sense of the unexplored possibilities that are available to be tapped by interested and motivated researchers.

Although a full discussion of some of the methodological and statistical issues involved in analyzing archival data is beyond the scope of this chapter, we recommend a few resources which may be useful for researchers interested in using archival data. For example, Firebaugh (1997) wrote about issues related to using data from repeated cross-sectional surveys. He gives examples from his own work using data from the General Social Survey (GSS) regarding how one can successfully analyze data from repeated surveys to look at "social change" over time. Finkel (1995) published a book describing causal analysis with panel data (e.g., repeated observations among the same sample of individuals). Finally, Zaitzow and Fields (2006) published a chapter for graduate students and research assistants about the use of archival data.

Concluding Comments

Archival data offer an excellent, and perhaps underutilized, opportunity for research in the area of work, stress, and health. Here are a few parting suggestions to OHP researchers who are new to the use of archival data in their research: (a) Identify and focus your research question before you begin; (b) determine the appropriateness of the data set for answering your research question; (c) investigate key aspects of the data, including the sampling, operationalization of key constructs, levels of measurement, unit of analysis, and variables and values; (d) budget plenty of time for data acquisition, data file formatting, and data file management activities; (e) investigate IRB issues and requirements early on; and (f) network with others who have used the same data source.

There are many issues for a researcher to consider prior to diving into an archival data set, from evaluating the relevance of the data set for the research of interest to considering the ease of use and measurement properties of the data available. Archival data offer an additional option for obtaining data to answer important research questions pertaining to occupational health. Archival data have many strengths and limitations associated with their use, and therefore it behooves researchers to always choose the most appropriate research methods for answering the research questions of interest.

References

Alterman, T., Grosch, J., Chen, X., Chrislip, D., Petersen, M., Krieg, E., Jr., ... Muntaner, C. 2008. Examining associations between job characteristics and health: Linking data from the occupational information network (O*NET) to two US National Health Surveys. *Journal of Occupational and Environmental Medicine, 50*, 1401–1413.

Barling, J., Kelloway, E. K., & Iverson, R. D. (2003). High-quality work, job satisfaction, and occupational injuries. *Journal of Applied Psychology, 88*, 276–283.

Barnes-Farrell, J. L., Dove-Steinkamp, M., Golay, L., Johnson, N. C., & McGonagle, A. (2008, November 12–14). How does the nature of the work–family interface influence planned retirement age of men and women? In G. G. Fisher (Chair), *The graying of the American workforce: Implications for occupational health psychology.* Symposium presented at the European Academy of Occupational Health Psychology, 2008, Valencia, Spain.

Behson, S. (2005). The relative contribution of formal and informal organizational work–family support. *Journal of Vocational Behavior, 66*(3), 487–500.

Casper, W. J., Eby, L. T., Bordeaux, C., Lockwood, A., & Lambert, D. (2007). A review of research methods in IO/OB work–family research. *Journal of Applied Psychology, 92*, 28–43.

Cifuentes, M., Boyer, J., Gore, R., d'Errico, A., Tessler, J., Scollin, P., ... PHASE in Healthcare Research Team. (2007). Inter-method agreement between O*NET and survey measures of psychosocial exposure among healthcare industry employees. *American Journal of Industrial Medicine, 50*, 545–553.

Cifuentes, M., Boyer, J., Lombardi, D. A., & Punnett, L. (2010). Use of O*NET as a job exposure matrix: A literature review. *American Journal of Industrial Medicine, 53*, 898–914.

Coursolle, K. M., Sweeney, M. M., Raymo, J. M., & Ho, J. H. (2010). The association between retirement and emotional well-being: Does prior work–family conflict matter? *The Journals of Gerontology: Series B Psychological Sciences, 65B*, 609–620.

Crimmins, E. M., Kim, J. K., & Solé-Auró, A. (2011). Gender differences in health: Results from SHARE, ELSA and HRS. *European Journal of Public Health, 21*, 81–91.

Donnelly, P. G. (1982). The origins of the Occupational Safety and Health Act of 1970. *Social Problems, 30*, 13–25.

Finkel, S. E. (1995). *Causal analysis with panel data*. Thousand Oaks, CA: Sage.

Firebaugh, G. (1997). *Analyzing repeated surveys* (Quantitative Applications in the Social Sciences, series no. 07-115). Thousand Oaks, CA: Sage.

Firebaugh, G., & Harley, B. (1995). Trends in job satisfaction in the United States by race, gender and type of occupation. In R. L. Simpson & I. H. Simpson (Eds.), *Research in the sociology of work: Vol. 5. The meaning of work* (pp. 87–104). Greenwich, CT: JAI.

Fisher, G. G. (2006, November). Positive and negative work/life spillover among older workers in the United States. In C. A. Bulger (Chair), *Furthering our understanding of aging and health: Results from the U.S. regarding quality of work and nonwork life*. Symposium presented at the 7th Conference of the European Academy of Occupational Health Psychology, Dublin, Ireland.

Fisher, G. G., & Shultz, K. S. (2006 March 2–4). Methodological issues: Using archival data to conduct occupational stress and health research to study older workers. In G. G. Fisher (Chair), *Using archival data: Research examples studying issues among older workers*. Symposium presented at the Work, Stress, and Health Conference, Orlando, FL.

Ford, M. T., & Tetrick, L. E. (2011). Relations among occupational hazards, attitudes, and safety performance. *Journal of Occupational Health Psychology, 16*, 48–66.

Gelman, A., & Carlin, J. B. (2002). Post-stratification and weighting adjustments. In R. M. Groves, D. A. Dillman, J. L. Eltinge, & R. J. A. Little (Eds.), *Survey nonresponse* (pp. 289–302). New York: Wiley.

Groves, R. M., Fowler, F. J., Couper, M. P., Lepkowski, J. M., Singer, E., & Tourangeau, R. (2009). *Survey methodology* (2nd ed.). New York: Wiley.

Gryzwacz, J. G. (2000). Work–family spillover and health during midlife: Is managing conflict everything? *American Journal of Health Promotion, 14*, 236–243.

Grzywacz, J. G., & Marks, N. (2000). Reconceptualizing the work–family interface: An ecological perspective on the correlates of positive and negative spillover between work and family. *Journal of Occupational Health Psychology, 5*, 111–126.

Halpern, D. (2005). How time-flexible work policies can reduce stress, improve health, and save money. *Stress & Health, 2* (3), 157–168.

Halperin, W., & Baker, E. L. (Eds.). (1992). *Public health surveillance*. New York: Van Nostrand Reinhold.

Health and Retirement Study (HRS). (2011). Retrieved from http://hrsonline.isr.umich.edu/cgi-bin/getbib.pl

Herzog, A. R., House, J. S., & Morgan, J. N. (1991). Relation of work and retirement to health and well-being in older age. *Psychology and Aging, 6*(2), 202-211.

Jones, C. (2010). Archival data: Advantages and disadvantages for research in psychology. *Social and Personality Psychology Compass, 4*, 1008–1017.

Kivimäki, M., Vahtera, J., Elovainio, M., Helenius, H., Singh-Manoux, A., & Pentti, J. (2005). Optimism and pessimism as predictors of change in health after death or onset of severe illness in family. *Health Psychology, 24*, 413–421.

Kompier, M. (2005). Assessing the psychosocial work environment—"Subjective" versus "objective" measurement. *Scandinavian Journal of Work and Environmental Health, 31*, 405–408.

Korn, E. L., & Graubard, B. I. (1999). *Analysis of health surveys*. New York: Wiley.

Lehtonen, R., & Pahkinen, E. (2004). *Practical methods for design and analysis of complex surveys* (2nd ed.). New York: Wiley.

Lerner, D. J. & Lee, J. (2006). Measuring health-related work productivity with self-reports. In R. C. Kessler & P. E. Stang (Eds.), *Health and work productivity: Making the business case for quality health care* (pp. 66–87). Chicago, IL: University of Chicago Press.

Liu, C., Spector, P. E., & Jex, S. M. (2005). The relation of job control with job strains: A comparison of multiple data sources. *Journal of Occupational and Organizational Psychology, 78*, 325–336.

Lohr, S. L. (2010). *Sampling: Design and analysis* (2nd ed.). Boston, MA: Brooks/Cole.

Mezuk, B., Kershaw, K. N., Hudson, D., Lim, K A., & Ratliff, S. (2011). Job strain, workplace discrimination, and hypertension among older workers: The Health and Retirement Study. *Race and Social Problems, 3*, 38–50.

O*NET Resource Center. O*NET Online. (n.d.). http://www.onetcenter.org/overview.html

Podsakoff, P. M., MacKenzie, S. B., Lee, J. Y., & Podsakoff, N. P. (2003). Common method biases in behavioral research: A critical review of the literature and recommended remedies. *Journal of Applied Psychology, 88*, 879–903.

Raymo, J. M., & Sweeney, M. M. (2006). Work–family conflict and retirement preferences. *Journals of Gerontology: Social Sciences, 61B*, S161–S169.

Schubauer-Berigan, M. K., Couch, J. R., Petersen, M. R., Carreón, T., Jin, Y., & Deddens, J. A. (2011). Cohort mortality study of workers at seven beryllium processing plants: Updates and associations with cumulative and maximum exposure. *Occupational and Environmental Medicine, 68*, 345–353.

Seligman, P. J., Sieber, W. K., Pedersen, D. H., Sundin, D. S., & Frazier, T. M. (1988). Compliance with OSHA record-keeping requirements. *American Journal of Public Health, 78*, 1218–1919.

Shultz, K. S., Hoffman, C. C., & Reiter-Palmon, R. (2005). Using archival data for I-O research: Advantages, pitfalls, sources, and examples. *The Industrial-Organizational Psychologist, 42*(3), 31–37.

Shultz, K. S., Taylor, M. A., & Morrison, R. F. (2003). Work related attitudes of naval officers before and after retirement. *International Journal of Aging and Human Development, 57*, 261–276.

Shultz, K. S., & Wang, M. (2007). The influence of specific physical health conditions on retirement decisions. *International Journal of Aging and Human Development, 65*, 149–161.

Shultz, K. Wang, M., Crimmins, E., & Fisher, G. G. (2010). Age differences in the demand-control model of work stress: An examination of data from 15 European countries. *Journal of Applied Gerontology, 29*, 21–47.

Singleton, R. A., & Straits, B. C. (2005). *Approaches to social research* (4th ed.). New York: Oxford University Press.

Spector, P. E. (1994). Using self-report questionnaires in OB research: A comment on the use of a controversial method. *Journal of Organizational Behavior, 15,* 385–392.

Spiegel, P. E., & Shultz, K. S. (2003). The influence of pre-retirement planning and transferability of skills on naval officers' retirement satisfaction and adjustment. *Military Psychology, 15,* 284–306.

Stachowski, A., Fisher, G. G., Grosch, J., Hesson, J., & Tetrick, L. (2007, April). Job complexity: Prevention of cognitive functioning decline? In G. G. Fisher (Chair), *Work and the retirement process: International perspectives on older workers.* Symposium presented at the 22nd annual conference of the Society for Industrial and Organizational Psychology, New York.

Sterba, S. K. (2009a). Alternative model-based and design-based frameworks for inference from samples to populations: From polarization to integration. *Multivariate Behavioral Research, 44,* 711–740.

Sterba, S. K. (2009b). Online appendix to accompany Sterba, S. K. Retrieved from http://www.vanderbilt.edu/peabody/sterba/pubs/Online.Appendix_Sterba.2009.pdf

Taylor, M. A., Shultz, K. S., Morrison, R. F., Spiegel, P. E., & Greene, J. (2007). Occupational attachment and met expectations as predictors of retirement adjustment of naval officers, *Journal of Applied Social Psychology, 37,* 1697–1725.

Trzesniewski, K. H., Donnellan, M. B., & Lucas, R. E. (Eds.). (2011). *Secondary data analysis: An introduction for psychologists.* Washington, DC: APA.

Tucker, J. S., Sinclair, R. R., Mohr, C. D., Adler, A. B., Thomas, J. L., & Salvi, A. D. (2009). Stress and counterproductive work behavior: Multiple relationships between demands, control, and soldier indiscipline over time. *Journal of Occupational Health Psychology, 14,* 257–271.

U. S. Centers for Disease Control and Prevention. (2011). *Workplace Safety and Health Topics: Surveillance.* http://www.cdc.gov/niosh/topics/surveillance

Wang, M., Barnes-Farrell, J., & Fisher, G. G. (2009, November 5–8). Using archival data to understand occupational health psychology issues: Methodological challenges and opportunities. In L. Kath (chair), *Research methodology: Issues in occupational health psychology research.* Symposium presented at the Work Stress, and Health conference, San Juan, PR.

Wang, M., Zhan, Y., Liu, S., & Shultz, K. S. (2008) Antecedents of bridge employment: A longitudinal investigation. *Journal of Applied Psychology, 93,* 818–830.

Zaitzow, B. H., & Fields, C. B. (2006). Archival data sets: Revisiting issues and considerations. In F. T. L. Leong & J. T. Austin (Eds.), *The psychology research handbook: A guide for graduate students and research assistants* (2nd ed., pp. 326–341). Thousand Oaks, CA: Sage.

Zhan, Y., Wang, M., Liu, S., & Shultz, K. S. (2009) Bridge employment and retirees' health: A longitudinal investigation. *Journal of Occupational Health Psychology, 14,* 374–389.

18

An Overview of Multilevel Modeling in Occupational Health Psychology

Lisa M. Kath, Scott C. Roesch, and Mark G. Ehrhart

Multilevel models go by a variety of aliases, including *hierarchical linear models, random coefficient models, nested models*, and *mixed-effects models*. What defines a multilevel model is that it explicitly accounts for data that are nested (i.e., grouped or clustered) in some way. It is fairly common for data collected in occupational health psychology (OHP) research to be nested; for example, employees can be nested within departments, which are then nested within locations. Data like these, residing at different levels of analysis, require specific analytic techniques to model the nested structure. This chapter is intended to be a brief overview of multilevel modeling, including its application to theory development, measurement issues associated with multilevel data, analysis of nested data, and a glimpse of some of the more complicated issues that multilevel researchers may face.

Conceptual Issues

Because OHP research frequently includes people working in groups or as part of a larger organization, it is important to consider the influences of these shared group experiences when developing OHP theories. For example, a group of employees may work for the same supervisor, and that supervisor may be lax about safety issues. Theorizing about these employees' safety-related attitudes or experiences solely at the individual level fails to recognize that these employees who work for this supervisor share some common input about safety from their supervisor. It is important to remember that individual attitudes and behaviors of employees are shaped by both the individual and an individual's social context (group), and that employees in similar situations are likely to be affected in similar ways (for a thorough review of organizations as multilevel systems, see Kozlowski & Klein, 2000).

There are countless OHP research topics that naturally benefit from a multilevel mindset. For example, safety researchers have shown that group-level safety climate can have an influence on how well safety procedures are actually carried out (e.g., Zohar, 2000; Zohar & Luria, 2005). There is an emerging interest in the influence of climate/norms on the incidence of incivility/bullying/workplace aggression (Cortina & Magley, 2003; Pearson, Andersson, & Porath, 2005; Walsh, Magley, Reeves, Davies-Schrils, Marmet, & Gallus, in press). Researchers interested in work–family/work–life balance have examined the effects of family-friendly climates on the experiences of work–family conflict (e.g., O'Neill et al., 2009). Several group influences on individuals' workplace stress, like job type, social support, leadership, or organizational interventions, have been studied as well—to read more about multilevel modeling in organizational stress research, see Bliese and Jex (2002). Finally, some OHP researchers have adopted an experience sampling method (ESM) approach, where the responses nested within individuals can be modeled using multilevel modeling. Examples can be found in research on work–family conflict (e.g., Wang, Liu, Zhan, & Shi, 2010) and on emotional labor (e.g., Totterdell & Holman, 2003). The list could go on, but these are just some of the ways in which OHP research has been influenced by multilevel thinking.

Group Membership Issues and Nesting Terminology

Besides identifying what situational influences may be of interest, it is also important in multilevel theory development to think about the types of nesting that are relevant to the constructs/relationships of interest. For example, an employee may be part of a functional group (e.g., chemists) and a project team (with chemists, biologists, crystallographers, etc.). This employee may also be nested within a department as well as a geographic location, which are then nested within an organization. To further complicate matters, the employee may also change group memberships over time. In developing your multilevel theory, it is important to spend some time thinking about which memberships and levels are most salient for the phenomena you are interested in studying. For an excellent primer on multilevel theory development, please consult Mathieu and Chen (2011).

From here on out, we will focus almost exclusively on the most common multilevel nesting structure, where the individual is the lowest level of analysis (Level 1) and the group (used loosely to describe any work unit, including the organization) is the next highest level (Level 2). Other nesting structures have multiple responses from an individual as Level 1 or have group-level data as Level 1. So even though we use the terms *individual level* and *Level 1* interchangeably, and *group level* and *Level 2* interchangeably, in reality, the designation of Level 1 and Level 2 depends on the study design. The term *linking variable* indicates the variable used to identify the nesting structure (typically the group identification number, but for ESM would be a participant identification number). Please consult Bliese and Ployhart (2002), Hedeker and Gibbons (2006), or Singer and Willett (2003) to read more on using multilevel modeling for repeated measures data (i.e., multiple observations nested within individuals).

Is Multilevel Modeling Necessary?

Before multilevel modeling was prevalent in OHP (and other organizational) research, researchers often conducted one of two types of analyses that ignored the nesting of the data. The first analysis of this type is called "aggregating up," which means that individual-level variables are averaged to reflect group-level variables and everything is analyzed at the group level. Doing so is not a problem when the level of theory is at the group level. For example, you might be interested in studying group climate predicting group outcomes. In this case, aggregating up is appropriate. However, the problem occurs when the level of theory is at the individual level and then the data are "aggregated up" and analyzed at the group level for the sake of analytic convenience. This approach can be problematic for two reasons. First, the variability in responding at the individual level is lost, which is extremely important if the level of theory is at the individual level and there is more individual-level variability than group-level variability. Second, by averaging across individual responses, the sample size equals the number of groups (rather than the number of individuals), thus reducing statistical power (Kim, 2009).

The second approach is when each individual in a group is assigned the value associated with his/her group and everything is analyzed at the individual level (called disaggregating group-level variables or a cross-level operator [CLOP] analysis). This approach leads to a violation of the independence assumption. When individual participants are treated as independent entities it results in smaller standard errors, but this increases the probability of Type I errors (inappropriately rejecting the null hypothesis). There is recent evidence that shows that ignoring nonindependence can increase Type II errors in some instances (Bliese & Hanges, 2004).

Moreover, when adopting one of the two approaches listed above, inferential biases can result by ignoring a nested data structure (Bliese & Hanges, 2004; Bliese & Jex, 2002). Making individual-level inferences about group-level findings (ecological fallacy) and group-level inferences about individual-level findings (atomistic fallacy) can both lead to erroneous conclusions (for more on this issue, please consult Bliese & Hanges, 2004). In closing, we have to admit that it is possible that your conclusions would not change regardless of how you analyze your data. However, if you believe, as we do, in helping organizations make decisions based on sound analysis of quality data, please keep reading.

If your data do not include any nesting whatsoever, multilevel modeling is not even possible. However, it is unlikely that employees will be completely unaffected by shared, local influences. Before moving forward with analyses, a decision must be made about whether those local influences are likely to affect constructs or relationships of interest. One way to examine whether this might be an issue would be to calculate an intraclass correlation coefficient (ICC(1), also referred to as a variance partition coefficient) for any constructs of interest. The intraclass correlation is a measure of the variance between groups relative to the total variance (Bliese, 2000). Stated differently, the ICC(1) answers the question, "How much of the variance in this measure is predicted by group membership?" (McCoach, 2010). If the amount of variance is significant (e.g., if the associated *F* test for the ANOVA below is significant), then this is a good indication that multilevel modeling should be used.

Computing the ICC(1) is quite easy: simply run a one-way ANOVA with the linking (or group) variable as the predictor and the measure of interest as the outcome. The ICC(1) value can be determined by the following formula, where K is the average number of people in the groups:

$$\frac{MS_{between} - MS_{within}}{MS_{between} + (K - 1)\, MS_{within}}$$

If the ICC(1) value is very small *and* the associated *F* test for the ANOVA is nonsignificant, it is reasonably safe to ignore the nesting in the data and proceed with individual-level analyses. Of course, that leads to the question of what is considered "very small," and the best answer to that is probably that it depends on the context being studied. Nevertheless, some guidelines have been suggested by LeBreton and Senter (2008), specifically that a value of .01 be considered a small effect size and .10 a medium effect size for ICC(1). In addition, they also noted that "values as small as .05 may provide prima facie evidence of a group effect" (p. 838). So although there are no hard guidelines on this issue, an ICC(1) of .05 or higher may be a good indication that the nesting should be accounted for in the analyses.

Multilevel Construct Conceptualization and Measurement

Conceptualization and Item Design. As researchers begin to consider studying a variable at the group level, it is important to consider several questions: What does the construct mean at the individual level? And what does the construct mean at the group level? In what ways are these meanings similar or different? These are not simple questions to be glossed over; they require some serious thought. When the meaning of a construct is very similar across levels, this is referred to as isomorphism. For example, a researcher could ask employees about their confidence in their ability to effectively handle safety-related issues, or their safety self-efficacy. Similarly, a researcher could ask individuals about their group's ability to handle safety issues, or their collective safety efficacy perceptions, which could then be averaged to create a measure of group safety efficacy (the group's shared perceptions of its ability to effectively handle safety issues). The meaning of the construct is very similar across levels; the primary difference is the focus on the individual versus the focus on the group as a whole.

However the meanings of constructs across levels can vary greatly. For example, examining the role of gender in predicting incivility perceptions would tell us whether women report different levels of incivility than men. By aggregating gender to the workgroup level, the construct means something very different; it captures the gender diversity of the group, which may also be related to incivility, either directly or as an interaction with individual-level gender (e.g., perhaps each gender reports more incivility when they are a minority in the group).

Although the difference between gender and group gender diversity is fairly straightforward, for other constructs, the difference is not as clear. For instance, what does it mean to take a group mean of individual group members' levels of extraversion? Stated differently, is there such a thing as team personality (e.g., Halfhill, Sundstrom, Lahner, Calderone, & Nielsen, 2005)? This concept of the meaning of constructs at different levels is covered in greater detail by Bliese (2000).

After thinking carefully about the construct of interest and what it might mean at different levels of analysis, the next step is to design items that are consistent with relevant theory and that measure the construct at the appropriate level. There are different ways that an individual-level response might be aggregated to the group level, each of which has implications for writing survey items (Chan, 1998; Chen, Mathieu, & Bliese, 2004). Most multilevel research on OHP topics uses direct consensus or referent shift composition models. In both of these models, the mean of individual responses is taken to represent the group as a whole; the primary distinction is the wording of the items. The *direct consensus* method involves measuring the construct from the perspective of the respondent. For example, a respondent may be asked to what extent she or he agrees with the following statement: "Safety is a primary concern of mine." The *referent shift* method, however, involves measuring the construct from the perspective of the group to which the data will be aggregated. For example, a respondent may be asked to what extent she or he agrees with the following statement: "Safety is a primary concern in my work group." Although the difference may seem minor, there are important theoretical and practical implications. If the underlying theory states that the construct is a shared perception of a group attribute, then the referent shift model aligns best with this conceptualization. Furthermore, research has shown that the referent shift model may lead to better support for aggregation (Klein, Conn, Smith, & Sorra, 2001). What is most important is that the aggregation method aligns with the theoretical definition of the higher-level construct. To read more about other types of composition models, such as additive models (no assumption of agreement) or dispersion models (measure of variability in groups), please consult Chan (1998) and Chen et al. (2004).

Aggregation. Most OHP researchers are well-versed in getting data at the individual level through methods such as surveys, physiological measures, or interviews. However, special consideration should be taken when considering measurement at the group level. In some situations, there are naturally occurring (or true) group-level measures, such as team size (i.e., how many employees are in each group) or team life span (i.e., total time the team as a whole has existed). The hallmark of these measures is their inability to be disaggregated. However, more often, it is necessary to aggregate individual-level measures to create a group-level measure.

When aggregating the mean of individual responses, there are some statistical checks that can be conducted to see if aggregation to the group level is appropriate. If the assumption is, for example, that there exists such a thing as group safety climate, one would expect that group members would generally agree in their evaluations of something like management attitudes toward safety. To test this statistically, multilevel researchers calculate an index of within-group agreement. One of the most common measures of within-group agreement in organizational research is $r_{wg(j)}$ (James, Demaree, & Wolf, 1984). A standard rule of thumb is that the average or median $r_{wg(j)}$ should be at or above .70 as justification for aggregation, although some have argued that the appropriate level of agreement depends on the research question being investigated (LeBreton & Senter, 2008). More information about $r_{wg(j)}$, other measures of agreement, and additional issues that should be considered (e.g., the null distribution to use when calculating $r_{wg(j)}$) can be found in a review article by LeBreton and Senter (2008).

In addition to reporting $r_{wg(j)}$ values or some similar index of within-group agreement, intraclass correlation (ICC) values are also often reported as justification for aggregation of individual-level measures to the group level. Technically, ICC(1) combines information on within-group agreement and interrater reliability, and low values can result even when there is high agreement if between-group variability is low. Some researchers also report an ICC(2) value, which is defined as the reliability of the group means. The ICC(2) provides an index of how stable the group means are and is interpreted similar to other reliability measures (i.e., generally aiming for .70 or higher). As noted by Bliese (1998), even when ICC(1) values are high, it may be difficult to achieve a large ICC(2) value if the number responses per group is small.

In some fields, such as health psychology, reporting the ICC(1) value is common practice for the of justification for aggregation of individual-level measures to represent group-level constructs, whereas reporting an $r_{wg(j)}$ value is not common practice. However, the norm in other fields, such as industrial-organizational psychology, is to report at least an ICC(1) and $r_{wg(j)}$ value (or some other within-group agreement measure). We strongly suggest that you take some time to get a feel for the norms for reporting aggregation statistics in the specific area in which you wish to publish.

Construct Validation. Several authors have written about the issues of construct validation when dealing with multilevel models (Chen et al., 2004), which we will summarize and illustrate with a hypothetical example. To establish the construct validity of group safety climate, a researcher would need to first define the construct and its measurement at the individual and group levels, including consideration of issues such as dimensionality and appropriate aggregation method. Definitions of group safety climate have typically emphasized a "sharedness" of perceptions (e.g., Zohar, 1980, 2000), which would indicate that aggregation would be based on a composition model, typically a mean of individual responses to referent shift items (see "Conceptualization and Item Design" above). The dimensionality of safety climate has been examined and debated at the individual level (e.g., Flin, Mearns, O'Connor, & Bryden, 2000), but we do not know of any attempts to establish dimensionality of the construct at the group level.

Next, the psychometric properties of the construct would need to be assessed *at each level of analysis*. That is, measures of internal consistency (e.g., Cronbach's alpha) should be reported for individual safety climate measures and aggregated group safety climate, the factor structure (e.g., Dyer, Hanges, & Hall, 2005; Zyphur, Kaplan, & Christian, 2008) should be examined at each level, and within-group agreement (e.g., $r_{wg(j)}$) should be reported as an examination of the "sharedness" of these individual perceptions. After that, the variability of the constructs between units would need to be assessed (e.g., ICC), and the constructs placed in their appropriate nomological networks. For safety climate, the nomological network could include both group-level outcomes of safety climate (e.g., department accident rates) and individual-level outcomes (e.g., safety motivation). For more details about how to conduct construct validation for your higher-order construct, please consult Berson, Avolio, and Kahai (2003); Chen, Bliese, and Mathieu (2005); and Chen, Mathieu, and Bliese (2004).

Different Types of Multilevel Models

Multilevel models can take on a variety of forms. In this chapter, we will focus on three general types of models that are the most common: (a) single-level models (direct and interaction effects), (b) cross-level direct effect models, and (c) cross-level moderation models. With *single-level models*, all constructs of interest reside at the same level of analysis. For example, if employees are nested within departments, and you are interested in how their individual safety motivation predicts their individual participation in voluntary safety activities (safety behavior), both safety motivation and safety behavior are at the individual level (see Figure 18.1). If motivation or behavior varies significantly depending on what department the employee is in, these data should be analyzed using multilevel modeling to account for the nested structure of the data. In this situation, multilevel modeling is used only to prevent analyses from violating regression assumptions; it is not used to explicitly test relationships among constructs at different levels of analysis.

An alternative single-level model would exist if employees are nested within departments, and you are interested in how the aggregated climate perceptions of employees (at the department level) predict departmental outcomes (e.g., safety climate predicts department injuries). In this case, both climate and outcomes are at the department level of analysis. Although the level of measurement is different from the level of analysis, making it necessary to provide analyses that support the aggregation of the data to the department level, the analysis itself is a single-level model (the department level). Unless there is a higher level of nesting that needs to be taken into account (for example, if those departments are nested within locations, shifts, or organizations), then multilevel modeling techniques are probably not necessary.

With cross-level models (both direct effects and moderation types), the constructs of interest reside at different levels of analysis. Theoretically, it is more common for higher-level constructs to influence lower-level constructs than vice versa, although the reverse (or even reciprocal causality) is certainly possible (Mathieu & Chen, 2011). It is easiest analytically when the outcome in the cross-level model resides at the lowest level of analysis; only recently are methods emerging that allow researchers to test models with individual-level predictors of group-level outcomes (Croon & van Veldhoven, 2007).

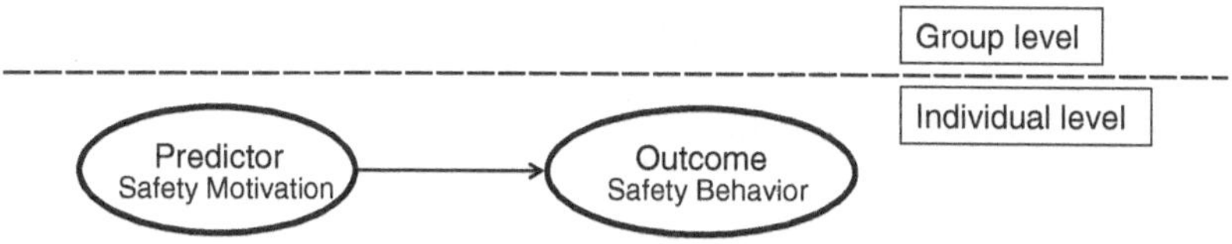

Figure 18.1 Example of a single-level model.

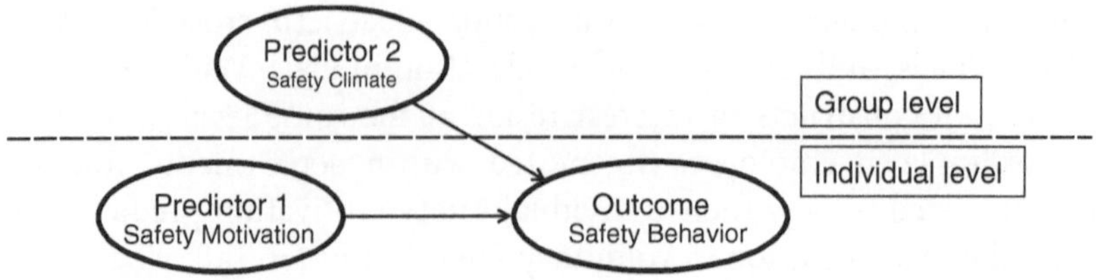

Figure 18.2 Example of a cross-level direct effect model.

In *cross-level direct effects models*, also referred to as intercepts-as-outcomes models (Hofmann, 1997), there is a group-level measure predicting an individual-level outcome (see example in Figure 18.2). There may also be one or more predictors at the individual level, but a cross-level model must have at least one group-level predictor. As with standard multiple regression, there can be as many predictors (at either level) as theory and statistical power allow. In our example in Figure 18.2, the focus is on the effect of group safety climate (Level 2) on individual group members' safety behavior (Level 1). In addition, you could include safety motivation as an individual-level predictor of safety behavior (as shown) or any of a number of control variables (e.g., individual affect or group size; not shown) at Level 1 and/or 2.

With *cross-level moderation models*, also known as slopes-as-outcomes models (Hofmann, 1997), most often there is a group-level variable moderating the relationship between two individual-level variables (see example in Figure 18.3), although there are other alternatives, such as a group-level variable moderating the relationship between a group-level predictor and an individual-level outcome. An example of a cross-level moderation model would be when safety climate moderates the relationship between individual-level safety motivation and individual-level safety behavior. In other words, this type of model allows for the investigation of whether the strength (i.e., slope) of the individual-level relationship varies depending on the group context.

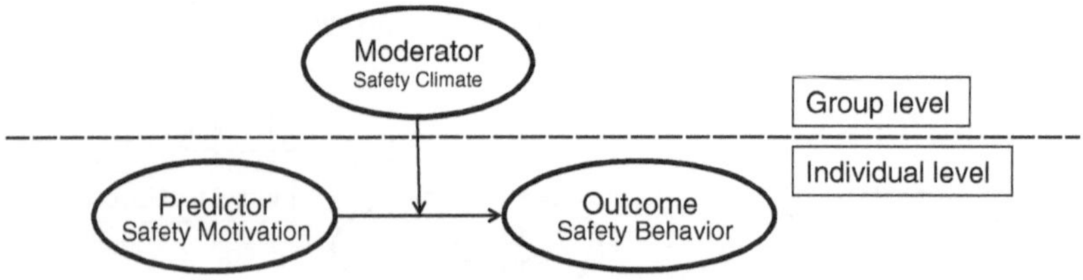

Figure 18.3 Example of a cross-level moderation model.

Data Analysis

Currently, multilevel models are most often conceptualized and analyzed as close cousins of standard multiple regression. That is, these models can have multiple predictors, but only one outcome variable at a time. Advances are being made in multilevel (or multigroup) structural equation modeling/path analysis (Bauer, 2003; Chan, 2010; Cheung & Au, 2005; Kaplan, Kim, & Su-Young, 2009; Kim, 2009; Lüdtke et al., 2008; Mehta & Neale, 2005; Skrondal & Rabe-Hesketh, 2005), but this chapter is only going to focus on multilevel models as more direct extensions of multiple regression. Furthermore, in our explanations of multilevel modeling, we will assume that you have a solid working knowledge of standard multiple regression, including the testing of interaction effects.

Translating the Conceptual Model to Equations

We recommend that you draw out your model before starting data analysis, as in Figures 18.1, 18.2, and 18.3, so you can clearly see what variables are at which levels. This will help you set up your data sets for each level as well as the equations that represent your multilevel model. The idea of equations may be scary for some, but rest assured: it is really not that much more complicated than regression equations, which is something most OHP researchers are very familiar with.

We will start with a standard regression equation: $Y = \beta_0 + \beta_1X_1 + \beta_2X_2 + e$. The Level 1 outcome is Y, the intercept term is β_0, the other β_n values are the regression coefficients, and e is the error or residual. In multilevel modeling, what happens with the baseline (or Level 1) regression equation is that each β gets its own bonus (Level 2) equation, as below:

Level 1: $Y_{ij} = \beta_{0j} + \beta_{1j}(X_1) + \beta_{2j}(X_2) + r_{ij}$

Level 2: $\beta_{0j} = \gamma_{00} + u_{0j}$

$\beta_{1j} = \gamma_{10} + u_{1j}$

$\beta_{2j} = \gamma_{20} + u_{2j}$

The Level 1 equation houses any Level 1 predictors (X_n) in your multilevel model. The Level 2 β_{0j} equation is called the intercept equation, and it houses any Level 2 variables that predict the outcome (Y_{ij}) directly. The β_{1j} equation is the first slope equation, and it houses any Level 2 variables that predict the relationship between X_1 and Y_{ij}. Similarly, the

β_{2j} equation is the second slope equation, and it houses any Level 2 variables that predict the relationship between X_2 and Y_{ij}. Regression coefficients (γs) in multilevel modeling are often referred to as fixed effects. Finally, the r_{ij} and the u_{nj} terms are the residual variance terms, also referred to as random effects. These residual variance terms enable the partitioning variance as coming from within groups or between groups.

As an example, here is a simple, *single-level model* (see Figure 18.1) with safety motivation predicting voluntary safety behavior at the individual level:

Level 1: Safety Behavior = $\beta_{0j} + \beta_{1j}$ (Safety Motivation) + r_{ij}
Level 2: $\beta_{0j} = \gamma_{00} + u_{0j}$
$\beta_{1j} = \gamma_{10} + u_{1j}$

Although it may feel like there should be something in the Level 2 equations to distinguish multilevel modeling from regular regression, please note that multilevel modeling software will automatically take into account the nesting in your data without your having to do anything else beyond indicating which variable in your data sets is the linking (or grouping) variable.

Now let's look at a *cross-level direct effects model,* as depicted in Figure 18.2. If we add in a group-level predictor (Safety Climate) as well, which is predicting the outcome directly, the proper place for that predictor is in the Level 2 intercept (β_{0j}) equation.

Level 1: Safety Behavior = $\beta_{0j} + \beta_{1j}$ (Safety Motivation) + r_{ij}
Level 2: $\beta_{0j} = \gamma_{00} + \gamma_{01}$ (Safety Climate) + u_{0j}
$\beta_{1j} = \gamma_{10} + u_{1j}$

The higher-order predictor (Safety Climate) went into the Level 2 intercept equation, with a gamma (γ_{01}) as a coefficient. Close inspection will reveal that this equation is also a regression equation: the γ_{00} value is an intercept term, and the γ_{01} value is a regression coefficient. What this model tests is the effects of individual-level safety motivation and group-level safety climate on individual-level safety behavior.

If we are interested in the interaction between group-level safety climate and individual-level safety motivation, that brings us to the last model type, the *cross-level moderation model.* Start by thinking about how moderated relationships are tested in standard multiple regression. In the final equation, both the predictor and the moderator are main or first-order effects, and the cross-product term is the interaction effect:

$$Y = \beta_0 + \beta_1 X_1 + \beta_2 X_2 + \beta_3 X_1 X_2 + e$$

Moving to a multilevel model, the first things to add are the main effects. The individual-level predictor (safety motivation) goes in the Level 1 equation, as before. Similarly, the group-level predictor (safety climate) goes in the Level 2 intercept (β_{0j}) equation. Now, with both main effects entered, the interaction term is ready to be entered. Unlike multiple regression, this is not done by directly creating a multiplicative term and adding it to the equation. As depicted in Figure 18.3, the group-level moderator is predicting the relationship between the individual-level predictor and the outcome, which is represented by β_{1j} in the Level 1 equation. So, if the group-level moderator (safety climate) goes in the β_{1j} equation, this model will test whether safety climate predicts the *relationship* between individual-level safety motivation and the outcome (safety behavior). In other words, this model tests whether the predictor-outcome relationship is dependent on the moderator, just like in moderated regression, except across levels:

Level 1: $\text{Safety Behaviors} = \beta_{0j} + \beta_{1j}\,(\text{Safety Motivation}) + r_{ij}$

Level 2: $\beta_{0j} = \gamma_{00} + \gamma_{01}\,(\text{Safety Climate}) + u_{0j}$

$\beta_{1j} = \gamma_{10} + \gamma_{11}\,(\text{Safety Climate}) + u_{1j}$

As we will discuss in more detail in the "Interpreting Output" section below, if the γ_{11} term is significant, there most likely is a significant cross-level interaction. More on testing whether you actually have a true cross-level interaction is included in the section on "Centering" below.

One final note about these equations: inclusion of the u_{nj} terms in the Level 2 slope equations can be considered optional. In general, it is important to understand that as you increase the number of random effects (u_{nj} terms) in a model, the number of parameters that you are estimating increases dramatically. This can cause estimation problems and the model may not converge (i.e., a final solution will not be found). One approach is to estimate random effects only for substantive predictors (i.e., leave the u_{nj} term in only those equations), and treat any covariates in the models as only fixed effects (i.e., do not include the u_{nj} term in those equations).

Preparing the Data

Preparing Data Files for Importation. There are a number of software packages used to analyze multilevel models (see Roberts & McLeod, 2008 for a review). The most commonly used one is HLM (Raudenbush,

Bryk, & Congdon, 2004), and this is the software for which specific advice will be given throughout the chapter. The most straightforward way to prepare for multilevel model analysis is to create two separate data sets: one for Level 1 data and the other for Level 2 data, with files sorted by the linking variable.

Looking at the equations based on the model being tested (see "Translating Conceptual Model to Equations" above), check that the Level 1 and 2 data sets have all the variables in the Level 1 and 2 equations, respectively. If there are Level 1 interaction terms, the predictors will need to be centered (see "Centering" below) and multiplied to create a cross-product for the Level 1 dataset. If you have a cross-level moderation model, please consult the "Centering" section below for additional terms you need to create. Run descriptive statistics on the model variables; this is an easy way to make sure the software read the data into its data set properly.

Centering. Primarily an issue with predictor variables at Level 1, centering is an important and sometimes tricky issue in multilevel modeling. To determine which type of centering is appropriate, it is necessary to think about what research question is being answered by the analyses. Centering decisions can be greatly aided by consulting Enders and Tofighi (2007), but we will summarize some of the major points. Many of the variables studied in OHP research do not have meaningful zero points. For example, it is common to use a Likert-type scale, with a minimum value of 1, to measure constructs such as role overload, work–family conflict, or management attitudes toward safety. Centering (subtracting the mean from each data point) is a way to make the zero point meaningful, which can aid in the interpretation of multilevel models, just as in multiple regression. There are two major types of centering in multilevel models: grand-mean centering and group-mean centering. In grand-mean centering, the overall sample mean is subtracted from each data point, and in group-mean centering (what Enders and Tofighi refer to as centering within cluster), the focal individual's group mean is subtracted from his or her individual data point. With grand-mean centering, the interpretation of a regression coefficient is relative to the entire sample; with group-mean centering, the interpretation of a regression coefficient is relative to the group to which an individual belongs. Beyond interpretation, there are some statistical considerations that occur with centering that we outline briefly below.

For *single-level models*, Enders and Tofighi (2007) recommend group-mean centering of Level 1 predictor(s). The reason for this is that when grand-mean centering is used, the regression coefficient (and associated variance components) for a Level 1 predictor variable is a blend of within-group and between-group variability. We would like to have a "pure" measure of within-group variability, and group-mean centering Level 1 predictors gives us this.

For *cross-level direct effects models*, Enders and Tofighi (2007) recommend grand-mean centering Level 1 predictors when the primary research question involves the Level 2 predictor, with any Level 1 predictors included as covariates. A variable that is grand-mean centered, again, is a blend of within-group and between-group variability, so controlling for that Level 1 variable will remove both the within-group and between-group variability associated with the outcome. What remains is the direct effect of the Level 2 variable (grand-mean centered) on the outcome.

For *cross-level moderation models*, in which the interaction between a Level 1 predictor and a Level 2 predictor is of interest, Enders and Tofighi (2007) recommend using group-mean centering for the Level 1 predictor variable and grand-mean centering for the Level 2 predictor variable—however, this is one of the rare cases where using grand-mean centering of a Level-1 predictor is algebraically equivalent to using group-mean centering. Remember that group-mean centering results in a pure index of within-group variability. Because of this, between-group variability needs to be reintroduced back into the statistical model to ensure that the analyses are testing a true cross-level interaction, instead of confounding the results with between-groups effects (you can read more about this issue in Enders & Tofighi, 2007 or Hofmann & Stetzer, 1998). This is done by creating a variable that represents the mean value of the Level 1 predictor for each group and entering this as a Level 2 predictor variable in the intercept (β_0) equation (see below). More specifically, simply aggregate your Level 1 predictor ($aggX_1$), create a cross-product of the aggregated predictor and moderator variables ($aggX_1{*}Z_1$), and include both Level 2 terms in the Level 2 intercept equation, as below:

Level 1: $Y_{ij} = \beta_{0j} + \beta_{1j} (X_1) + r_{ij}$

Level 2: $\beta_{0j} = \gamma_{00} + \gamma_{01} (aggX_1) + \gamma_{02} (Z_1) + \gamma_{03} (aggX_1{*}Z_1) + u_{0j}$

$\beta_{1j} = \gamma_{10} + \gamma_{11} (Z_1) + u_{1j}$

Now, the γ_{11} coefficient will represent a clean test of the cross-level interaction.

Running the Analyses

This section will describe using HLM software (Raudenbush et al., 2004), but as noted above, there are several software packages that can estimate these types of models. Moreover, please consult one of the multilevel books referenced earlier for information on statistical assumptions (and how to check them) with nested data. After opening the HLM software, under File, choose "Make new MDM file" and choose the input file format appropriate in your situation. Assuming you are estimating a two-level model with continuous predictors/outcomes, choose HLM2. In the following window, give your new HLM data set a name (with .mdm as the suffix) and "Save your mdmt file." Browse to choose your Level 1 and Level 2 files. Once your files have been associated, choose the variables that will go in your HLM file. This is where you will let HLM know which variable is your linking (or grouping) variable by checking the box next to "ID." Next, check your descriptive statistics against those we suggested you run after you prepared your data sets and confirm that the number of Level 1 and Level 2 units are as expected. Once you are sure your data were read properly, you can start building your model according to your equations.

Interpreting Output

As before, we are focusing on output generated by the HLM software program (Raudenbush et al., 2004), but the general principles apply to any output. First, confirm the type of centering and the equations were as intended. HLM output (versions 6 and 7) indicates the equations and the type of centering used under the heading "Summary of the model specified." In HLM 6, the output reports β's with the letter "B" and γ's with the letter "G."

Next, run a null or intercept-only model (i.e., a model without predictors) to get a sense of where the variability of your nested data structure is. Sigma squared (σ^2) is a measure of within-group variance, whereas Tau (τ_{00}) is a measure of between-group variance. You can recalculate your ICC(1) value by using this simple equation: $\tau_{00} / (\tau_{00} + \sigma^2)$. Looking at the output, there is a table of Taus, and τ_{00} is in the upper left corner of the table (i.e., the first value in the line starting with "INTRCPT1, β_0"), just below the σ^2 value.

Then, add predictor variables and check to see if relationships of interest are statistically significant. The table that is labeled, "Final estimation of fixed effects" will have all the gamma coefficients, standard errors, T-ratio (the ratio of the coefficient to standard error), degrees of freedom, and the all-important *p*-value. From the equations you wrote, you should be able to tell which coefficient is associated with your research question of interest.

In the case of the single-level model, it can be more difficult to determine which gamma is associated with your research question, because the Level 1 equation only has betas:

Level 1: $Y_{ij} = \beta_{0j} + \beta_{1j}(X_1) + \beta_{2j}(Z_1) + {}_{3j}(X_1{*}Z_1) + r_{ij}$

Level 2:

$$\beta_{0j} = \gamma_{00} + u_{0j}$$
$$\beta_{1j} = \gamma_{10} + u_{1j}$$
$$\beta_{2j} = \gamma_{20} + u_{2j}$$
$$\beta_{3j} = \gamma_{30} + u_{3j}$$

Obviously, to see if the interaction is significant, you want to know about the β_{3j} coefficient, but your output will be silent on this issue, only giving you a pile of gammas. But if you look at the Level 2 equations, you can see that the gamma term associated with the β_{3j} coefficient is γ_{30}. That is the gamma coefficient that will indicate whether your interaction is significant.

If you have concerns that the assumptions underlying the multilevel statistical model (e.g., a normal distribution of random effects) are incorrect, you may want to consider the "Final estimation of fixed effects," with the secondary label "with robust standard errors." As outlined by Raudenbush and Bryk (2002, pp. 276–280), robust standard errors adjust for violations of these assumptions. However, this is only appropriate when the highest level of the nested data structure (typically, the group level) has a relatively large sample size.

Estimating Variance Explained/Effect Sizes

You may be interested in how much variance is explained by a certain predictor (or set of predictors) or the amount of variance explained by your model as a way to examine the practical significance of your findings (Aguinis et al., 2010). Multilevel modeling programs do not provide an R^2 value directly, but they all provide information that provides an analog to R^2. All you have to do is (a) run your model with and without the predictor(s) you are interested in; (b) make note of your σ^2 and τ_{00} in each model; and then (c) calculate the variance explained at each level like this: at Level 1, $R^2 = (\sigma^2_{\text{without predictor}} - \sigma^2_{\text{with predictor}}) / \sigma^2_{\text{without predictor}}$, and at Level 2, $R^2 = (\tau_{00\text{without predictor}} - \tau_{00\text{with predictor}}) / \tau_{00\text{without predictor}}$. To approximate total R^2, use your ICC(1) value to create weights for the Level 1 and Level 2 R^2 values. That is, multiply your Level 2 R^2 value by ICC(1), and your Level 1 R^2 value by (1 – ICC(1)), and add the two together. In addition to these pseudo-R^2 values (as they are often called) as an indicator of effect size, some promising recent developments (Feingold, 2009) have been made that convert unstandardized regression coefficients into measures of effect size (Cohen's *d*), but because they are so new, they are not yet used widely. More information about estimating variance explained can be found at http://ssc.utexas.edu/software/faqs/hlm#HLM_4 or by consulting Kreft and de Leeuw (1998), Raudenbush and Bryk (2002), Singer (1998), or Snijders and Bosker (1994, 1999).

Graphing Significant Interactions

We recommend graphing significant interactions; as before, the conceptual process is similar to that of multiple regression. Perhaps the most straightforward way to graph interactions is to rerun your model with completely uncentered predictors and use HLM's graphing function. That way the X- and Y- axes are easy to interpret. However, if your results change when you use uncentered predictors, you will probably want to stick with your centered predictors and work a little harder to figure out what your X- and Y-axes mean. In HLM 6 or 7, just click on "Basic Settings" in the main toolbar and make sure that your graph file name is something you will be able to identify/locate. Run your analysis as usual. Then go to File → Graph Equations → Model Graphs to get to the graphing window.

In "X focus," choose the Level 1 or Level 2 variable you would like to have displayed on the X-axis. You can set the range of the X-axis in this section as well. In "Z focus," choose the variable(s) you would like to have as moderators (i.e., separate lines) and identify the appropriate range(s). If you click on "Range/Titles/Color," you can set minimum and maximum values for X and Y. We recommend graphing Y at plus/minus one standard deviation from the mean. Finally, in this subwindow, you can also create titles for each axis as well as for the graph as a whole. In "Other settings," you can change how other predictors in the model are treated. Then click "OK" to generate your graph. You can save this graph as a metafile for use in other documents. As an alternative, you can use a handy website that will graph cross-level two-way interactions when you input in a variety of values (Preacher, Curran, & Bauer, 2006). This website will also provide points to plot, so you can use Excel or your favorite graphing program.

Advanced Multilevel Modeling Issues

In the pages above, we have outlined the fundamental principles and approach used in multilevel modeling. As with most statistical approaches, there are additional issues to consider that are more complex conceptually and statistically. We briefly review and provide references for four additional issues that all multilevel modelers should have at least cursory knowledge of: (a) moving beyond two-level models, (b) outcomes that are *not* continuous, (c) mediation models, and (d) power analysis. Although these topics are advanced, a conceptual and statistical nexus can be drawn to multiple regression in most cases.

Modeling More Than Two Levels

We have relegated ourselves to relatively simple two-level models throughout this chapter, but one could easily imagine OHP applications that have three levels of nested data. For example, a reasonable data structure could be individuals (Level 1) nested within supervisors (Level 2) nested within departments (Level 3). Conceptually and statistically, we have the same considerations that we had with the two-level models. However, power can be greatly diminished when moving beyond two levels (Hedges & Rhoades, 2010). It is recommended to determine, using ICC(1), that significant variability exists at each level of a three-level nested data structure for an outcome of interest before predictive models are developed. If variability is minimal, one can reasonably omit that level of the nested data structure.

Noncontinuous Outcome Variables

Previously, we have assumed that the outcome of interest has been continuous (i.e., measured on an interval or ratio scale), but this need not be the case. Hierarchical generalized linear models (or generalized linear mixed models) refer to multilevel models where the outcome at Level

1 is not continuous (Atkins & Gallop, 2007; Bliese, Chan, & Ployhart, 2007; Hedeker, 2007; Hedeker & Gibbons, 2006; O'Connell, Goldstein, Rogers, & Peng, 2008; Raudenbush & Bryk, 2002, chapter 10). Noncontinuous outcome variables have regression analogs: binary outcomes (e.g., turnover, mortality) require logistic regression, general categorical outcomes (e.g., types of workplace injuries) use multinomial logistic regression, ordinal outcomes (e.g., work schedule/shift preferences) require ordinal regression, and count data (e.g., number of days absent from work, number of workplace injuries) use Poisson regression. The relationship between predictor variables and outcomes of this nature are *nonlinear*. To estimate the multilevel regression models, a link function is specified that transforms the nonlinear relationship between predictor variables and these outcomes to a linear relationship. Although this requires a different statistical estimation method, the basics of the multilevel modeling framework outlined above still generally apply. To choose the appropriate link function and statistical estimation method for your noncontinuous outcome, consult the sources listed above.

Multilevel Mediation

Mediation models can also be tested in a multilevel modeling framework. When all variables reside at Level 1, the model is referred to as a 1-1-1 mediation model (Bauer, Preacher, & Gil, 2006). Models where the predictor is at Level 2 and the mediator and outcome variables are at Level 1 are called 2-1-1 mediation models. For example, one could hypothesize that the relationship between a group-level variable (e.g., safety climate) and the individual-level outcome (e.g., safety participation) is mediated by an individual-level variable (e.g., safety motivation). Similarly, a predictor and mediator could be at Level 2 and the outcome at Level 1; this type of model is referred to as a 2-2-1 mediation model. Analyses of these types of models are outlined in Pituch, Stapleton, and Kang (2006).

Conceptually, establishing mediation is similar whether you have a nested data structure or not. Of primary interest is establishing that the compound paths (the predictor-to-mediator path and the mediator-to-outcome path) that compose the mediated effect are statistically significant (see MacKinnon, Fairchild, & Fritz, 2007, for the various ways that this can be established). The primary difficulty in testing for mediation with multilevel data is the calculation of the standard error terms for the mediated effect, as these error terms differ depending on whether or not random effects (among other things) are estimated (Kenny, Korchmaros, & Bolger, 2003). For a full discussion and examples of multilevel mediation, see Krull and MacKinnon (2001), MacKinnon (2009, chapter 9), and Zhang, Zyphur, and Preacher (2009). It is also possible to combine mediation and moderation in a multilevel framework; see Bauer, Preacher, and Gil (2006) for more details on multilevel moderated mediation models.

Power Analysis

Multilevel models, because of their increased complexity, require larger sample sizes than nonnested models (Maas & Hox, 2005; Mathieu & Chen, 2011; Snijders & Bosker, 1993). Researchers are working on ways to estimate the statistical power of multilevel models, which are complicated because nested data have a sample size at each level. As outlined by Scherbaum and Ferreter (2009), the ICC(1) has a direct influence on the standard error terms for parameters of interest in multilevel models (e.g., a cross-level interaction term), which obviously influences the test statistic and associated *p*-value associated with said parameter. To get a sense of how this is done and why this is important, consider the following formula for the design effect, which is an indication of the degree to which the standard error value for a target effect is underestimated when assuming Level 1 individuals are independent (i.e., when not taking into account the nested structure of the data; Kish, 1987):

$$\text{design effect} = \text{square root } [1 + (\text{ICC}(1))*(n_j - 1)],$$

where ICC(1) is defined as before and n_j is the average number of individuals within each group (cluster). If the ICC(1) value is 0 (no clustering effect), the design effect is 1, and the standard errors are accurate. However, as the ICC(1) value increases, the design effect, which provides an index of how inaccurate the standard errors are as a function of clustering, also increases. Any calculation of power or necessary sample size needs to explicitly incorporate ICC(1) and standard error values, as is done in computer programs such as Optimal Design (Raudenbush, Spybrook, Congdon, Liu, Martinez, 2011), PinT (Snijders, Bosker, & Guldemond, 2007), and ML. PowerTool (Mathieu, Aguinis, Culppeper, & Chen, in press), as well as power figures (Scherbaum & Ferreter, 2009) and tables (Hedges & Rhoades, 2010).

Conclusion

The goal of this chapter was to provide an introduction to multilevel theory and analysis for OHP researchers. Space would not allow for diving into all the details of specific analyses or special cases, but there are plenty of resources available to you when you are ready to take that deeper dive. For now, we hope that this chapter prompted those researchers who have tended to focus on the individual level in their areas of research to expand their thinking and research questions to include additional levels of analysis. In addition, for those researchers who have been interested in multilevel models but overwhelmed by the analyses for testing those models, we hope that we have been able to break it down in an accessible way and, we hope, convince you that it is not only possible, but necessary in many cases. Expanding your research across multiple levels is an exciting step: we hope you enjoy the adventure!

References

Aguinis, H., Werner, S., Abbott, J. L., Angert, C., Joon Hyung, P., & Kohlhausen, D. (2010). Customer-centric science: Reporting significant research results With rigor, relevance, and practical impact in mind. *Organizational Research Methods, 13*(3), 515–539.

Atkins, D. C., & Gallop, R. J. (2007). Rethinking how family researchers model infrequent outcomes: A tutorial on count regression and zero-inflated models. *Journal of Family Psychology, 21*(4), 726–735.

Bauer, D. J. (2003). Estimating multilevel linear models as structural equation models. *Journal of Educational and Behavioral Statistics, 28*(2), 135–167.

Bauer, D. J., Preacher, K. J., & Gil, K. M. (2006). Conceptualizing and testing random indirect effects and moderated mediation in multilevel models: New procedures and recommendations. *Psychological Methods, 11*(2), 142–163.

Berson, Y., Avolio, B. J., & Kahai, S. (2003). Level specification: Using triangulation in a grounded theory approach to construct validation. In F. Dansereau & F. Yammarino (Eds.), *Multi-level issues in organizational behavior and strategy* (Vol. 2, pp. 85–112). Oxford, England: JAI.

Bliese, P. D. (1998). Group size, ICC values, and group-level correlations: A simulation. *Organizational Research Methods, 1*(4), 355–373.

Bliese, P. D. (2000). Within-group agreement, non-independence, and reliability: Implications for data aggregation and analysis. In K. J. Klein & S. W. J. Kozlowski (Eds.), *Multilevel theory, research, and methods in organizations: Foundations, extensions, and new directions* (pp. 349–381). San Francisco, CA: Jossey-Bass/Pfeiffer.

Bliese, P. D., Chan, D., & Ployhart, R. E. (2007). Multilevel methods: Future directions in measurement, longitudinal analyses, and nonnormal outcomes. *Organizational Research Methods, 10*(4), 551–563.

Bliese, P. D., & Hanges, P. J. (2004). Being both too liberal and too conservative: The perils of treating grouped data as though they were independent. *Organizational Research Methods, 7*(4), 400–417.

Bliese, P. D., & Jex, S. M. (2002). Incorporating a mulitilevel perspective into occupational stress research: Theoretical, methodological, and practical implications. *Journal of Occupational Health Psychology, 7*(3), 265–276.

Bliese, P. D., & Ployhart, R. E. (2002). Growth modeling using random coefficient models: Model building, testing, and illustration. *Organizational Research Methods, 5*(4), 362–387.

Chan, D. (1998). Functional relations among constructs in the same content domain at different levels of analysis: A typology of composition models. *Journal of Applied Psychology, 83*(2), 234–246.

Chan, D. (2010). Advances in analytical strategies. InS. Zedeck (Ed.), *APA handbook of industrial and organizational psychology: Vol 1. Building and developing* (Vol. 1, pp. 85–113). Washington, DC: American Psychological Association.

Chen, G., Bliese, P. D., & Mathieu, J. E. (2005). Conceptual framework and statistical procedures for delineating and testing multilevel theories of homology. *Organizational Research Methods, 8*(4), 375–409.

Chen, G., Mathieu, J. E., & Bliese, P. D. (2004). A framework for conducting multilevel construct validation. In F. Dansereau & F. Yammarino (Eds.), *Research in multilevel issues: The many faces of multi-level issues* (pp. 273–303). Oxford, England: Elsevier Science.

Cheung, M. W. L., & Au, K. (2005). Applications of multilevel structural equation modeling to cross-cultural research. *Structural Equation Modeling, 12*(4), 598–619.

Cortina, L. M., & Magley, V. J. (2003). Raising voice, risking retaliation: Events following interpersonal mistreatment in the workplace. *Journal of Occupational Health Psychology, 8*(4), 247.

Croon, M. A., & van Veldhoven, M. J. P. M. (2007). Predicting group-level outcome variables from variables measured at the individual level: A latent variable multilevel model. *Psychological Methods, 12*(1), 45–57.

Dyer, N. G., Hanges, P. J., & Hall, R. J. (2005). Applying multilevel confirmatory factor analysis techniques to the study of leadership. *The Leadership Quarterly, 16*(1), 149–167.

Enders, C. K., & Tofighi, D. (2007). Centering predictor variables in cross-sectional multilevel models: A new look at an old issue. *Psychological Methods, 12*(2), 121–138.

Feingold, A. (2009). Effect sizes for growth-modeling analysis for controlled clinical trials in the same metric as for classical analysis. *Psychological Methods, 14*(1), 43–53.

Flin, R., Mearns, K., O'Connor, P., & Bryden, R. (2000). Measuring safety climate: Identifying the common features. *Safety Science, 34*(1), 177–192.

Halfhill, T., Sundstrom, E., Lahner, J., Calderone, W., & Nielsen, T. M. (2005). Group personality composition and group effectiveness: An integrative review of empirical research. *Small Group Research, 36*(1), 83–105.

Hedeker, D. (2007). Multilevel models for ordinal and nominal variables. In J. De Leeuw & E. Meijer (Eds.), *Handbook of multilevel analysis* (pp. 237-274). New York: Springer.

Hedeker, D., & Gibbons, R. D. (2006). *Longitudinal data analysis*. Hoboken, NJ: Wiley-Interscience.

Hedges, L. V., & Rhoades, C. (2010). *Statistical power analysis in education* research (NCSER 2010-3006). Washington, DC: National Center for Special Education Research. http://ies.ed.gov/ncser/.

Hofmann, D. A. (1997). An overview of the logic and rationale of hierarchical linear models. *Journal of Management, 23*, 723–744.

Hofmann, D. A., & Stetzer, A. (1998). The role of safety climate and communication in accident interpretation: Implications for learning from negative events. *Academy of Management Journal, 41*(6), 644–657.

James, L. R., Demaree, R. G., & Wolf, G. (1984). Estimating within-group interrater reliability with and without response bias. *Journal of Applied Psychology, 69*(1), 85–98.

Kaplan, D., Kim, J.-S., & Su-Young, K. (2009). Multilevel latent variable modeling: Current research and recent developments. *The Sage handbook of quantitative methods in psychology* (pp. 592–612). Los Angeles, CA: Kaplan.

Kenny, D. A., Korchmaros, J. D., & Bolger, N. (2003). Lower level mediation in multilevel models. *Psychological Methods, 8*(2), 115–128.

Kim, J. S. (2009). Multilevel analysis: An overview and some contemporary issues. In A. Millsap &. A. Maydeu-Olivares (Eds.), *The Sage handbook of quantitative methods in psychology* (pp. 337–361). Thousand Oaks, CA: Sage.

Kish, L. (1987). *Statistical design for research*. New York: Wiley.

Klein, K. J., Conn, A. B., Smith, D. B., & Sorra, J. S. (2001). Is everyone in agreement? An exploration of within-group agreement in employee perceptions of the work environment. *Journal of Applied Psychology, 86*(1), 3–16.

Kozlowski, S. W. J., & Klein, K. J. (2000). A multilevel approach to theory and research in organizations. In K. J. Klein & S. W. J. Kozlowski (Eds.), *Multilevel theory, research, and methods in organizations: Foundations, extensions, and new directions* (pp. 3–90). San Francisco, CA: Jossey-Bass.

Kreft, I., & de Leeuw, J. (1998). *Introducing multilevel modeling*. Thousand Oaks, CA: Sage.

Krull, J. L., & MacKinnon, D. P. (2001). Multilevel modeling of individual and group level mediated effects. *Multivariate Behavioral Research, 36*(2), 249–277.

LeBreton, J. M., & Senter, J. L. (2008). Answers to 20 questions about interrater reliability and interrater agreement. *Organizational Research Methods, 11*(4), 815–852.

Lüdtke, O., Marsh, H. W., Robitzsch, A., Trautwein, U., Asparouhov, T., & Muthén, B. (2008). The multilevel latent covariate model: A new, more reliable approach to group-level effects in contextual studies. *Psychological Methods, 13*(3), 203–229.

Maas, C. J. M., & Hox, J. J. (2005). Sufficient sample sizes for multilevel modeling. *Methodology: European Journal of Research Methods for the Behavioral and Social Sciences, 1*(3), 86–92.

MacKinnon, D. P. (2009). *Introduction to statistical mediation analysis.* New York: Erlbaum.

MacKinnon, D. P., Fairchild, A. J., & Fritz, M. S. (2007). Mediation analysis. *Annual Review of Psychology, 58*, 593–614.

Mathieu, J. E., Aguinis, H., Culpepper, S. A., & Chen, G. (in press). Understanding and estimating the power to detect cross-level interaction effects in multilevel modeling. *Journal of Applied Psychology.* doi 10.1037/a0028380

Mathieu, J. E., & Chen, G. (2011). The etiology of the multilevel paradigm in management research. *Journal of Management, 37*(2), 610–641.

McCoach, D. B. (2010). Hierarchical linear moderling. In G. R. Hancock & R. O. Mueller (Eds.), *The reviewer's guide to quantitative methods in the social sciences* (pp. 122–140). New York: Routledge.

Mehta, P. D., & Neale, M. C. (2005). People are variables too: Multilevel structural equations modeling. *Psychological Methods, 10*(3), 259–284.

O'Connell, A. A., Goldstein, J., Rogers, J., & Peng, C. J. (2008). Multilevel logistic models for dichotomous and ordinal data. In A. A. O'Connell & D. B. McCoach (Eds.), *Multilevel modeling of educational data* (pp. 199–243). Charlotte, NC: Information Age.

O'Neill, J. W., Harrison, M. M., Cleveland, J., Almeida, D., Stawski, R., & Crouter, A. C. (2009). Work–family climate, organizational commitment, and turnover: Multilevel contagion effects of leaders. *Journal of Vocational Behavior, 74*(1), 18–29.

Pearson, C. M., Andersson, L. M., & Porath, C. L. (2005). Workplace incivility. In S. Fox & P. E. Spector (Eds.), *Counterproductive work behavior: Investigations of actors and targets.* (pp. 177–200). Washington, DC: American Psychological Association.

Pituch, K. A., Stapleton, L. M., & Kang, J. Y. (2006). A Comparison of single sample and bootstrap methods to assess mediation in cluster randomized trials. *Multivariate Behavioral Research, 41*(3), 367–400.

Preacher, K. J., Curran, P. J., & Bauer, D. J. (2006). Computational tools for probing interaction effects in multiple linear regression, multilevel modeling, and latent curve analysis. *Journal of Educational and Behavioral Statistics, 31*, 437–448.

Raudenbush, S. W., & Bryk, A. S. (2002). *Hierarchical linear models: Applications and data analysis methods* (2nd ed.). Thousand Oaks, CA: Sage.

Raudenbush, S. W., Bryk, A. S., & Congdon, R. (2004). *HLM 6 for Windows.* Lincolnwood, IL: Scientific Software International.

Raudenbush, S. W., Spybrook, J., Congdon, R., Liu, X., Martinez, A. (2011). Optimal Design Software for Multi-level and Longitudinal Research (Version 3.01) [Software]. Available from http://pikachu.harvard.edu/od/ or from www.wtgrantfoundation.org

Roberts, J. K., & McLeod, P. (2008). Software options for multilevel modeling. In A. A. O'Connell & D. B. McCoach (Eds.), *Multilevel modeling of educational data* (pp. 427–467). Charlotte, NC: Information Age.

Scherbaum, C. A., & Ferreter, J. M. (2009). Estimating satistical power and required sample sizes for organizational research using multilevel modeling. *Organizational Research Methods, 12*(2), 347–367.

Singer, J. D. (1998). Using SAS PROC MIXED to fit multilevel models, hierarchical models, and individual growth models. *Journal of Educational and Behavioral Statistics, 23*(4), 323–355.

Singer, J. D., & Willett, J. B. (2003). *Applied longitudinal data analysis: Modeling change and event occurrence.* New York: Oxford University Press.

Skrondal, A., & Rabe-Hesketh, S. (2005). *Generalized latent variable modeling: Multilevel, longitudinal, and structural equation models.* Boca Raton, FL: Chapman & Hall/CRC.

Snijders, T. A., & Bosker, R. J. (1993). Standard errors and sample sizes for two-level research. *Journal of Educational Statistics, 18*(3), 237–259.

Snijders, T. A., & Bosker, R. J. (1994). Modeled variance in two-level models. *Sociological methods and research, 22*(3), 342–363.

Snijders, T. A., & Bosker, R. J. (1999). *Multilevel analysis: An introduction to basic and advanced multilevel modeling.* London: Sage.

Snijders, T. A., Bosker, R. J., & Guldemond, H. (2007). *Power analysis in two-level designs* (PinT) (Version 2.12). Available from http://www.stats.ox.ac.uk/~snijders/multilevel.htm#progPINT

Totterdell, P., & Holman, D. (2003). Emotion regulation in customer service roles: Testing a model of emotional labor. *Journal of Occupational Health Psychology, 8*(1), 55–73.

Walsh, B. M., Magley, V. J., Reeves, D. A., Davies-Schrils, K. A., Marmet, M. D., & Gallus, J. A. (in press). Assessing work group norms for civility: The development of the Civility Norms Questionnaire—Brief. *Journal of Business and Psychology.* doi: 10.1007/s10869-011-9251-4

Wang, M., Liu, S., Zhan, Y., & Shi, J. (2010). Daily work–family conflict and alcohol use: Testing the cross-level moderation effects of peer drinking norms and social support. *Journal of Applied Psychology, 95*(2), 377–386.

Zhang, Z., Zyphur, M. J., & Preacher, K. J. (2009). Testing multilevel mediation using hierarchical linear models. *Organizational Research Methods, 12*(4), 695–719.

Zohar, D. (1980). Safety climate in industrial organizations: Theoretical and applied implications. *Journal of Applied Psychology, 65*(1), 96–102.

Zohar, D. (2000). A group-level model of safety climate: Testing the effect of group climate on microaccidents in manufacturing jobs. *Journal of Applied Psychology, 85*(4), 587–596.

Zohar, D., & Luria, G. (2005). A multilevel model of safety climate: Cross-level relationships between organization and group-level climates. *Journal of Applied Psychology, 90*(4), 616–628.

Zyphur, M. J., Kaplan, S. A., & Christian, M. S. (2008). Assumptions of cross-level measurement and structural invariance in the analysis of multilevel data: Problems and solutions. *Group Dynamics: Theory, Research, and Practice, 12*(2), 127–140.

19

Person-Centered Analysis

Methods, Applications, and Implications for Occupational Health Psychology

Mo Wang, Robert R. Sinclair, Le Zhou, and Lindsay E. Sears

> There are two types of people in the world: Those who think there are two types of people, and those who don't. (Robert Benchley)

As Benchley's Law of Distinction makes clear, people often tend to explain others' behavior by sorting them into groups, sets of people who share some common characteristics. In organizational theory, such groups are often referred to as profiles or configurations that "denote any multidimensional constellation of conceptually distinct categories that commonly occur together" (A. D. Meyer, Tsui, & Hinnings, 1993, p. 1175). As Meyer et al. note, while in theory any particular set of attributes could combine together in an infinite number of combinations, in practice, sets of attributes tend to fall into small numbers of patterns. Thus, "just a fraction of the theoretically conceivable configurations are viable and apt to be observed empirically" (A. D. Meyer et al., 1993, p. 1176).

Mintzberg's (1979) theory of organizational forms provides an example of configural thinking. He argued that organizations tend to take on one of a small number of structural configurations that encompass a wide range of variables, such as the relative size of production and support staff, the degree to which work processes are highly standardized, and the nature of authority relationships. For instance, large organizations with very few layers of management are rare or nonexistent; small organizations are never highly bureaucratic. Although Mintzberg's model deals with configurations of "objective" characteristics of organizations, it is possible to apply the same kind of reasoning to individual characteristics to study configurations or patterns of psychological characteristics that a group of people share in common. Such person-centered research involves empirical efforts to identify a set of groups that best represent the patterns of covariation among the variables of interest and then investigate differences between these groups on other measures, such as safety performance behaviors or health outcomes.

Studies using person-centered methods are rare in occupational health psychology (OHP), but offer a great deal of potential for generating increased understanding of existing bodies of literature as well as generating new theoretical perspectives. With this enormous potential benefit in mind, the goal of this chapter is twofold. First, we conduct a detailed review of contemporary statistical methods for conducting configural research. Then, we use the literature on organizational commitment profiles to illustrate several of the decisions and issues that arise in configural research. We use the terms *configural, profile-based,* and *person-centered* interchangeably, to refer to research that has the general goal of identifying differences between people in their patterns of scores on a set of variables, with each pattern typically referred to as a profile or a configuration (Wang & Hanges, 2011).

Person-Centered Analytical Methods

Understanding person-centered methods begins with an appreciation of how they differ from variable-centered methods. When using a variable-centered, or dimension-centered, approach, researchers are usually interested in describing or explaining the interrelatedness among variables (e.g., the correlation between workload and stress) and inferring the underlying processes or causes that account for these associations. In occupational health research, typical variable-centered statistical methods include ANOVA, regression analysis, factor analysis, and structural equation modeling. Even a casual glance at the OHP literature would show that, similar to organizational research in general (Wang & Hanges, 2011), the vast majority of quantitative OHP research uses variable-centered methods.

Variable-centered methods are useful when the research question is about whether and how variables, observed or latent, relate to each other and generate the observed covariances. However, these methods are limited when the research question is about whether individuals are from qualitatively different subpopulations or how subpopulations differ from each other in terms of the relationships among variables. In other words, variable-centered methods cannot tell us whether the observed covariances among variables result from a mixture of several covariances within different subpopulations or how the interrelatedness differs across subpopulations. Instead, methods taking a person-centered approach or methods integrating variable-centered and person-centered analyses are more appropriate for these endeavors. Person-centered analysis, such as latent class procedures (e.g., latent class analysis [LCA], and latent profile analysis [LPA]), classify individuals into subpopulations based on observed scores on categorical (for LCA) or continuous variables (for LPA; Lubke & Muthén, 2005). Such methods offer several advantages over traditional variable-centered methods. First, person-centered analysis and integrated analysis allow researchers to directly test the assumption of population heterogeneity/homogeneity in the relationships among variables. Occupational health psychology research usually assumes that the populations studied are homogeneous. Based on this assumption, usually a single set of parameters is used to describe the phenomena of interest. Accordingly, researchers typically use variable-centered methods to estimate a single set of parameters for the single population from which the data were sampled. However, the homogeneity assumption underlying these studies does not always hold and efforts to identify and explain population heterogeneity may lead to important theoretical developments. For example, whereas most research has treated part-time workers as a single undifferentiated group, Martin and Sinclair (2007) identified eight distinct groups with very different patterns of job attitudes, turnover behavior, and demographic characteristics. The large amount of heterogeneity among part-time workers suggests that treating them as a single group for research purposes may yield information that is essentially meaningless because it may not apply to any particular worker.

For observed population heterogeneity, subpopulation (usually called "group") membership is quantified by a categorical variable. Research questions are usually about whether the theoretical model differs across these observed subpopulations; for example, whether there are gender differences in general stress level and whether job satisfaction measures are equivalent across cultures. When subpopulation membership is unobserved, a latent categorical variable (also called a latent class variable) can be used to represent the subpopulation membership. In model-based latent variable analysis, the subpopulation membership, or the value of the latent class variable, is not a zero/one estimate. Instead, based on the observed data, the probability of each individual being classified into each latent class is estimated. Moreover, the interrelatedness among variables within each latent class is estimated by model parameters. Therefore, research questions can be asked about whether individuals can be reliably classified into qualitatively different classes and how classes differ from each other; for example, whether two subpopulations can be identified from the response patterns of scores on diagnostic criteria (Muthén, 2006).

Second, latent class procedures or person-centered analysis using latent variables are model-based procedures. In latent class procedures, model parameters (e.g., number of latent classes, within-class covariance structure) are specified a priori. Models are evaluated and compared using consistent criteria (e.g., information criteria, the adjusted likelihood ratio test, and bootstrapped likelihood ratio test) based on the maximum likelihood estimates obtained. Therefore, latent class procedures and person-centered analyses using latent variables are theory-driven methods and are especially suitable for comparing theories from different perspectives. This advantage represents a great improvement over traditional methods such as cluster analysis which relied heavily on subjective judgments of researchers to determine the appropriate numbers of categories.

Finally, from a practical perspective, latent class procedures are more reliable and interpretable in terms of class classification results. These methods estimate the posterior probabilities of individuals belonging to different latent classes. Therefore, the final classification results are more interpretable. Moreover, using estimated model parameters, it is possible to calculate how likely additional observations are from each subpopulation. This is an advantage over nonmodel-based clustering methods which usually generate sample specific classification results that rely heavily on the scores observed from the particular sample. As we will show below in our discussion of commitment profile research, this is a common concern with many prior applications of person-centered methods.

In the following sections, we introduce several different latent class methods and provide empirical examples to illustrate their usefulness in occupational health research.

Latent Profile Analysis and Latent Class Analysis

Sometimes researchers are interested in classifying individuals into unobserved subpopulations based on their observed scores on some continuous or categorical variables. As we will show below, several recent studies in the literature on commitment profiles have adopted this approach. Latent profile analysis (LPA) and latent class analysis (LCA) can achieve this purpose by modeling the relationship between a latent class variable and several continuous/categorical observed outcome variables. Specifically, in both LPA and LCA, there is a latent categorical variable representing the unobserved subpopulation membership. The latent categorical variable has effects (estimated as loadings) on observed continuous/categorical outcome variables. Within subpopulations (i.e., latent classes), the observed outcome variables are assumed to be independent from each other, which is called the assumption of local independence. Therefore, the observed interrelatedness among outcome variables is caused by the mixture of subpopulations whereas the observed within-class interrelatedness is considered as measurement error or residuals.

Growth Mixture Modeling

Conventional latent growth modeling (LGM) has been used in studies examining the growth trajectory of a variety of outcome variables, such as job performance (e.g., Ployhart & Hakel, 1998), newcomer adaptation (e.g., Chan & Schmitt, 2000), and employee well-being (e.g., Gorgievski-Duijvesteijn, Bakker, Schaufeli, & van der Heijden, 2005). Studies using LGM usually assume that there is a single population of theoretical interest; thus, a single set of growth parameters is estimated to model the growth curves for all individuals. Multigroup-LGM is the method to use when researchers are interested in comparing the growth curves of observed subpopulations. When subpopulation membership cannot be observed, growth mixture modeling (GMM) can be used to model unobserved latent classes, the growth trajectories for each class, and the effects of covariates on latent class membership and growth parameters. Specifically, GMM can identify unobserved subpopulations which differ qualitatively in their growth curves. Therefore, it can be used to test models predicting multiple change paths from multiple theoretical perspectives at the same time (e.g., Wang, 2007). Moreover, as with all models under the FMM framework, covariates, categorical or continuous, latent or observed, can be included in GMM and may be treated as antecedents predicting latent class membership and growth parameters (Lubke & Muthén, 2005; Muthén, 2001) or as outcomes of latent class membership (e.g., Muthén, 2004).

Growth mixture models can be specified by modeling two components. The first component models different latent growth curves, each of which corresponds to a specific latent class (i.e., defining the latent class membership). The second component models predictors of the latent class membership. Specifically, a multinomial logistic regression (when there are more than two latent classes) or a simple logistic regression (when there are only two latent classes) model is used to link the predictors to the latent class variable. GMM uses expectation-maximization (EM) algorithm to estimate means and variances-covariances of growth parameters, factor loadings of observed outcomes on growth factors, residuals, and posterior probabilities of individuals belonging to each latent class (Muthén & Shedden, 1999). GMM can handle missing data on outcome variables using full-information maximum-likelihood (FIML) estimation assuming data are missing at random (MAR; Little & Rubin, 2002). When evaluating growth mixture models, it is necessary to consider multiple criteria, including values of information criteria, aLRT test results, entropy value, theoretical interpretability, and practical usefulness (Muthén, 2004; Wang & Bodner, 2007). In addition, GMM generally performs better in classification when covariates are included in modeling (Lubke & Muthén, 2007). Technical details about model specification, model estimation, and model selection of GMM procedures are covered in other places (e.g., Muthén, 2001; Wang & Bodner, 2007).

Wang's (2007) research on retirees' psychological well-being change patterns can be used to illustrate how GMM can be applied in occupational health research. Theories from different perspectives predict different longitudinal change patterns of retirees' well-being. Empirical research also reports inconsistent findings of retirees' well-being change curves in single samples. Therefore, it is possible that subpopulations may exist, each with a different growth pattern of well-being. As such, Wang (2007) used GMM to analyze repeated measures of retirees' well-being to identify latent subpopulations and to examine effects of covariates on subpopulation memberships.

Based on theories, different growth curves (maintaining, U-shape, and improving) were specified for different latent classes. Growth mixture models with different numbers of latent classes were estimated. According to information criteria values, aLRT test results, and entropy values, a three-class mixture growth model fit the data best. These results suggested that there are three subpopulations in retirees whose changes in well-being follow different trajectories. Specifically, approximately 70% of retirees experienced minimum psychological well-being changes (i.e., maintaining pattern); about 25% of retirees experienced negative changes in psychological well-being during the initial transition stage, but showed improvements afterwards (i.e., U-shape pattern); and about 5% of retirees experienced positive changes in psychological well-being (i.e., improving pattern).

Results from GMM with covariates further showed that retirees, who (a) held a bridge job; (b) were more actively engaged in retirement planning; and (c) were married and whose spouses were present and not working, were more likely to be classified into the maintaining pattern class. Retirees, who (a) retired from highly physically demanding jobs; (b) retired from highly stressful jobs; and (c) had low job satisfaction at prior employment, were more likely to be classified into the recovering pattern class. Retirees, who (a) experienced objective health declines during the retirement transition; (b) had unhappy marriages; and (c) retired earlier than they expected, were more likely to be classified into the U-shape pattern class. Overall, these predictive effects suggest that the multiple longitudinal change patterns of retiree's psychology well-being during the retirement transition and adjustment process can be predicted.

Mixture Latent Markov Modeling

Another type of occupational health research question concerns patterns of change or transition between discrete statuses over time (e.g., alcohol use history, employment status change, development of counterproductive work behaviors; Wang & Chan, 2011). In these studies, the outcome variable can be modeled as an observed or latent categorical variable, the repeated observations of which manifest as a Markov chain. When the categorical outcome is assumed to be measured without error; that is, it truly reflects categories on underlying categorical variable, simple Markov chain modeling can be used (Langeheine & van de Pol, 2002). Statistical methods using latent variables to model the longitudinal categorical data are called latent transition analyses or latent Markov modeling, which assume that the observed categories reflect the category of an underlying latent categorical variable with some measurement error (Wiggins, 1973).

Both simple Markov modeling and latent Markov modeling assume that the observed transition pattern is from a single population; thus, a single set of parameters in a single Markov chain can describe all individuals. However, similar to the continuous growth curves we discussed above, population heterogeneity may exist. Mixture Markov modeling or mixture latent Markov modeling is used to model multiple Markov chains for unobserved subpopulations. Mixture latent Markov modeling combines features of latent class analysis and latent Markov modeling (Langeheine & van de Pol, 2002). As with conventional latent Markov models, in mixture latent Markov models, the observed category is a function of both the latent categorical variable and the measurement error. Thus, the total observed transition patterns can be modeled as the reflections of fewer possible latent transition patterns. In other words, minor discrepancies between some similar observed patterns are modeled as fluctuation in observation due to measurement error. As with other mixture modeling methods, mixture latent Markov models can include multiple types of covariates, either as predictors or outcomes of the latent class variable (Wang & Chan, 2011). As a model-based statistical method, mixture latent Markov modeling also has the advantages of other model-based methods discussed above, such as consistent evaluation criteria and model-based classification results.

Mixture latent Markov models can be specified as an extension to conventional latent Markov models. For a mixture latent Markov model with K latent classes, K times of the original number of parameters in conventional latent Markov models are estimated plus K mixture parameters which denote the proportions of individuals in each latent class. All the original parameters in the conventional latent Markov models are estimated as conditional on each latent class. Similar to GMM, when covariates are included, multinomial or simple logistic-regression models can be specified to model the path between covariates and latent class variables. Mixture latent Markov models can also be estimated using EM algorithms (Langeheine & van de Pol, 2002). Missing data on outcome variables can also be handled by using FIML estimation when assuming MAR. For model selection, when the data have no missing values, the Pearson chi-square test can be used to evaluate model fit by comparing the observed (counted from data) and model-based frequencies (calculated based on individuals' latent class membership) of each transition pattern. It is also recommended to consider multiple criteria when evaluating and selecting models, including values of information criteria, aLRT or Lo-Mendell-Rubin LRT results, bootstrap likelihood test results, entropy values, theoretical interpretability, and practical usefulness (Lo, Mendell, & Rubin, 2001; Muthén, 2004; Wang & Bodner, 2007). Same as selecting growth mixture models, classification results are more accurate when covariates are included in modeling mixture latent Markov models (Muthén, 2004; Wang & Bodner, 2007). However, it should be noted that how well likelihood-based tests perform in comparing mixture models and single-class models with more than one latent categorical variable is unknown. Technical details about model specification, model estimation, and model selection of mixture latent Markov modeling procedures can be found in other references (e.g., Langeheine & van de Pol, 2002; Wang & Chan, 2011).

Wang and Chan (2011) demonstrated applying mixture latent Markov modeling to study employees' employment status change patterns. According to previous theories and empirical findings, retirees' post-retirement employment statuses (PESs) might form three qualitatively different patterns corresponding with three subpopulations. The first subpopulation may never be employed again after retirement. The second subpopulation may always be employed after retirement. The third subpopulation (movers) may transit between employment and unemployment but the general tendency is to transit from being employed to unemployed. Moreover, retirees' years of education might predict the PESs change patterns, given that more educated individuals are more likely to be asked for professional advice and expertise after retirement. Therefore, the more years of education that retirees received, the more likely that they will stay in employment than always being unemployed or transit out of jobs.

To test the predictions, data of 994 retirees' observed PESs from four time points and their years of education were analyzed by mixture latent Markov modeling. Based on information criteria and entropy values, the three-class model fit the data best. Estimates of reliability parameters in the three-class model showed that observed categories were reliable in reflecting latent categories for both employed and unemployed responses. Out of the total sample, 43.1% and 42.2% of retirees were classified into never employed and always employed classes respectively. For those who were identified as movers (14.8% out of total sample), the probabilities of transition from being employed to unemployed was higher than the probabilities for the reverse in all three transitions. Moreover, years of education was negatively related to the log odds of being in the never employed class versus movers' class. In support of the hypothesis, this result suggested that for retirees who have more years of education, it is more likely for them to stay employed after retirement than to transition from being employed to unemployed, as compared to retirees with a shorter education history.

Issues and Decisions in Person Centered Research: Commitment Profiles as an Exemplar

The literature on organizational commitment profiles provides useful examples of several of the issues researchers should consider before embarking on person-centered research. Commitment is a well-established construct in organizational science with many studies showing links between commitment and outcomes such as job satisfaction, job performance, and turnover (Cooper-Hakim & Viswesvaran, 2005). J. P. Meyer and Maltin (2010) reviewed the extensive body of research concerning the relationship between organizational commitment and employee health and well-being. They offer ample evidence to suggest that at least some forms of commitment are related to employee well-being (some positively, some negatively) and they review other evidence showing how employees' reactions to workplace stressors may depend on the nature of their organizational commitment.

J. P. Meyer and Herscovitch (2001) offered a general definition of commitment as "the force that binds an individual to a course of action relevant to a particular target" (p. 301). The commitment literature has addressed several potential commitment targets (cf. Cohen, 2003; Vandenberghe, 2009), including the global organization, one's occupation/career, particular stakeholders within or around the organization (e.g., supervisors, teams, unions), and ideas and concepts (e.g., change, innovation).

J. P. Meyer and Allen (Allen & Meyer, 1990; Meyer, 2009; J. P. Meyer & Allen, 1997) describe three specific types of binds, or "mind-sets," including affective, continuance, and normative commitment. *Affective commitment* (AC) represents the shared values, identification with, and emotional attachment to a particular focus. *Continuance commitment* (CC) represents the perceived costs of leaving the target in relation to the perceived benefits of staying, such that CC reflects the extent to which the benefits of staying outweigh the benefits of leaving. Lastly, *normative commitment* (NC) is driven by a felt obligation toward a target, and involves feelings that one should stay because of social norms.

One of the fundamental issues in the commitment literature concerns the interplay between different commitment mind-sets and targets (cf.

J. P. Meyer & Herscovitch, 2001; J. P. Meyer, Stanley, Herscovitch, & Topolnytsky, 2002); often referred to as a person's commitment profile (Meyer, Stanley, & Parfyonova, 2012). The idea that employees might form different *patterns* of work-related attachments can be traced back to some of the earliest literature on commitment. Gouldner (1958) proposed the first commitment profile typology distinguishing between those high in organizational commitment (locals) and those high in occupational commitment (cosmopolitans). Although most research strayed away from this early person-centered thinking, recent methodological and conceptual developments have reinvigorated interest in commitment profiles. This literature highlights some of the promises and pitfalls of person-centered research, as well as some of the critical considerations when conducting person-centered research.

What Variables Should Be Used to Form the Profiles?

The first decision a researcher needs to make in person-centered analyses concerns the choice of input variables used to form the profiles. Most people are familiar with the old saying "garbage in=garbage out," highlighting the idea that when poor quality inputs are used in a process, that process is unlikely to yield high quality outputs. Of course, good measurement is a prerequisite for any study; however, one of the important issues in profile research is that the particular pattern of profiles obtained is highly dependent on the number and nature of the variables included.

As the number of variables increases in a profile model; the number of possible profiles increases exponentially. Whereas a two variable model has 9 possible combinations of high, medium, and low scores on the two variables, a three variable model has 27 possible combinations. And, as the number of profiles grows, it becomes increasingly difficult to make theoretically defensible predictions about expected differences between profiles. Whereas a four profile model involves 6 comparisons of pairs of profiles, a six profile model quickly jumps to 15 comparisons of profile pairs. Two ways researchers can deal with this issue are (a) to develop hypotheses about specific profile pairs and (b) to test hypotheses about how each profile differs from all other profiles as a set (Martin & Sinclair, 2007 provides examples of both of these).

These issues are certainly important in the commitment literature as studies vary greatly in the choices of input variables. Nearly all studies include profiles of AC and most include CC. Some studies include

NC and others do not, with varying degrees of rationale offered for the decision. The global organization is the most common target, although several studies include measures of occupational commitment. Some of this literature focuses on occupational AC and CC (e.g., Sears, 2010), other studies investigate commitment to both the career and the organization (e.g., Becker & Billings, 1992; Carson, Carson, Roe, Birkenmeir, & Phillips, 1999; Tsoumbris & Xenikou, 2010). Another uncommon variation is to examine commitment to other targets. For example, Morin, Morizot, Boudrias, and Madore (2010) examined profiles of affective commitment toward the employees' organization, workgroup, supervisor, customers, job, work, and career. Similarly, McNally and Irving (2010) extended the idea of AC and CC profiles to college students' relationships with their schools.

These examples illustrate the diversity of commitment profiles literature as well as a central problem with any profile-based literature. Studies vary considerably in how they choose commitment measures from the theoretical set of possible combinations of targets and mind sets. These differences are highly likely to influence the number and nature of the profiles obtained. In such circumstances, there are relatively few instances where one study can reasonably be characterized as a replication and extension of another—hampering generalizability efforts.

Including every conceivable commitment measure in a study often is not feasible, and even when it is, inclusion of large lists of measures may create more confusion rather than clarity. The decision to include a particular variable in a profile study should be guided by its potential impact on the profiles obtained, with the outcome of interest being an important criterion. For example, some researchers have begun to acknowledge the multidimensionality of continuance commitment and include separate measures of two CC concepts: perceived costs of leaving and quality of alternatives (cf. Sears, 2010; Stanley, Vandenberg, Vendenberghe, & Bentein, 2011). In a retention-focused study, more sophisticated measures of CC may be useful for deriving profiles; in a performance-focused study, more sophisticated measurement of CC may be unnecessary. Similarly, in a study of occupational health outcomes, commitment to other targets, such as supervisors or work teams, might be advisable given the importance of social influences on health (e.g., social support, incivility).

The commitment profiles literature suggests some general strategies to guide variable choices. For example, it would be difficult to defend a profile-based commitment study that did not include AC to the targets of interest, because of its centrality to the literature and fact that it is typically the form most strongly related to outcomes of interest. The relevance of CC and NC might vary depending on the nature of the study. Extending this logic to work stress, it would be difficult to envision a study of job stress profiles that did not include measures of job demands and control, given their centrality to the stress literature. However, inclusion of work–family concerns might be more pertinent to some kinds of person-centered research than others.

What Hypothesized Profiles Best Capture the Variables Chosen?

Strong person-centered research is no different than strong variable centered research in the sense that it relies on theoretically justified hypotheses. In addition to having a strong rationale for the variables chosen, researchers need to develop a rationale about the expected patterns of variables (i.e., the specific profiles). Indeed, modern analytical methods enable researchers to test a priori models comparing different profile solutions. This is particularly important given that past person-centered research has been frequently criticized as highly exploratory in nature.

Although strong conceptual models are desirable, person-centered analyses allow a middle ground between completely exploratory approaches and completely confirmatory approaches, such as by comparing a hypothesized profile model to models with more or fewer profiles. For example, in a study of retention among part-time workers, Martin and Sinclair (2007) hypothesized seven profiles of part-time workers, based on a set of demographic characteristics. However, they found eight profiles. They had expected a group they called married supplementers, who they described as married individuals whose spouse provided the majority of their household income. They found two distinct profiles that fit their general description of married supplementers. Older married supplementers averaged 56 years in age and typically did not have any children at home; younger married supplementers averaged 35 years old and with 1.9 children at home. Thus, their findings confirmed their general set of expectations, but with some slight modifications, illustrating how researchers can start with a set of expected profiles but, based on their findings, make minor modifications, rather than accepting or rejecting the total model.

Researchers face the challenge of striking a balance between conducting atheoretical research in which limited theoretical rationale is offered for predicted profiles and what Kerr (1998) referred to as HARKing (Hypothesizing After the Results Are Known). In our view, neither of these approaches is likely to yield profiles that replicate across multiple studies as the clusters obtained are influenced by idiosyncrasies of the sample studied, a problem that is particularly likely when researchers study relatively small convenience samples and do not replicate their findings. Although HARKing is problematic in any research literature, the highly exploratory nature of profile research makes it particularly susceptible to researchers' abilities to develop inventive explanations for any particular set of findings they obtain. On the other hand, given the relative paucity of empirically derived profile research, commitment profiles scholarship could be argued to be in an early enough stage that even largely exploratory research is useful for establishing an empirical foundation for future theorizing.

Why Might Some Profiles Not Be Observed?

Hypotheses about the number of profiles resemble hypothesized factor models in confirmatory factor analysis. In each case, the researcher makes a prediction about an expected number of latent variables that account for variance in a set of observed indicators. In each case, a researcher might compare the hypothesized model to other plausible alternatives such as more parsimonious models (with fewer factors/profiles) or models suggested by alternate theories (different sets of profiles). However, in person-centered studies, researchers often need to go one step further and offer at least some explanation for why a particular set of profiles would be expected *not* to occur. In other words, they should carefully consider the full matrix of possible profiles and give some thought to why certain profiles would not be expected to occur—perhaps using a process of elimination to help arrive at a final set of profiles. This sort of theorizing can provide additional insight into the nature of the relationships among the constructs under investigation, as it highlights the idea that certain combinations of a set of predictors may not occur in reality.

For example, Sinclair, Tucker, Cullen, and Wright (2005) described nine possible combinations of high, medium, and low affective and continuance organizational commitment. In constructing their hypotheses about which profiles were most likely to occur, they included justifications about why some profiles would be unlikely to occur. They discussed the idea that researchers should not expect to find profiles based on two opposing mind-sets, such as a worker with very strong affective commitment and very weak continuance commitment. This profile is theoretically implausible as it describes an employee with very strong affective ties to the target but who simultaneously feels they would not incur any particular costs from leaving. Other researchers have drawn on concepts such as retrospective rationality (J. P. Meyer & Allen, 1997) and cognitive consistency (e.g., Sinclair et al., 2005; Somers, 1995) to argue that individuals may adjust their perceptions of their commitments so they are cognitively consistent. For example, a worker who has very strong continuance commitment might increase his or her feelings of affective commitment to avoid the internal tension that might arise from feeling stuck in one's current position and feeling no particular affective ties to that target.

Sinclair et al. (2005) also argued that it would be unlikely to find profiles of people with low levels of both affective and continuance commitment, simply because they would, by definition, have no particular reason to remain members of the organization. Although Sinclair et al. did not find an uncommitted profile, other researchers have done so (e.g., Carson et al. 1999; Gellatly, Hunter, Currie, & Irving, 2009; Somers, 2009, 2010; Stanley et al., 2011). Such differences across studies highlight a challenge for profile research—explaining why a profile might occur in some research contexts but not others. For example, one might expect not to find an uncommitted profile of nurses in a thriving economy, because they might have multiple desirable employment options, and be more likely to leave. Alternately, one might expect to find profiles of uncommitted retail workers in poor economic circumstances, because those workers might lack viable alternatives. Studies also may differ in their definitions of high and low scores for classification, which are driven by the distribution of variables in specific samples. We discuss this issue next.

How Do Definitions of High and Low Scores Affect Profile Interpretation?

Profile researchers often differentiate profiles as groups of people who fall into various combinations of high, medium, and/or low scores on a set of variables. To test their models, they often sort samples into high and low commitment groups based on whether employees fall above a sample median or mean. The meaning of terms such as *high* and *low* is critical to testing profiles. Operationalizing these terms involves some subtle concerns that are particularly important for person-centered/typological research.

Some researchers simply define high and low based on the distance of a particular score from the mean. For example, Carson et al. (1999) split a sample of librarians into four commitment profiles based on whether they were above or below sample means on measures of career and organizational affective commitment. Similarly, Gellately et al. (2009) divided their sample into high and low AC and CC by splitting the sample at the median. Other researchers use clustering strategies to derive profiles but rely on within sample descriptive statistics to interpret them. For example, Sinclair et al. (2005) derived nine profiles from cluster analysis but then interpreted the meaning of each profile based on within sample Z-scores, defining high and low scores as Z scores of +/- .80 and medium as scores between +.79 and –.79. However, their definitions of high, medium, and low were based on within sample distributions. In other words, high (low) was defined as higher (lower) than most of the other people in their sample rather than high in an absolute sense.

The primary advantage of using within-sample definitions of high, medium, and low is ease; the primary problem is that interpretation of profiles heavily depends on the distribution of commitment scores in the sample. For example, Carson et al. (1999) divided their sample into high and low groups based on sample means of 3.5 and 3.9 on organizational and career AC, respectively. Given that they used a 5-point agreement scale with neutral as the midpoint, this means that a librarian with a 3.3 on organizational AC and a 3.6 on career AC fell into their uncommitted profile, even though both scores are in the positive range of the response scale. This person might be less committed than others in the sample, but his or her scores may still suggest a person who has positive attachments. As this example illustrates, while within-sample definitions of high and low are easy to use, they create profiles that are easy to misinterpret.

Although profile research focusing on medium scores is uncommon, it may be theoretically significant. In the case of commitment research, measures typically use 5- or 7-point agreement scales ranging from strong agreement to strong disagreement and midpoint scores defined as neutral/no opinion/unsure. In this case, scores toward the middle of the scale refer to people who are neither strongly committed nor strongly negatively committed.

Most commitment profiles literature side-steps the issue of defining "medium" by focusing only on profiles of high or low scores. However, profile studies frequently obtain groups with scores suggesting that they are at least somewhat ambivalent about their relationship with the commitment target. This ambivalence may be theoretically significant but lost by studies that dichotomize groups into high or low scores or that ignore the underlying meaning of responses relative to the response scale used to gather them.

Ideally, OHP researchers should reserve the terms *high*, *medium*, and *low* for cases when measures have well-established norms or ample validation evidence supporting a particular definition. Clinical psychology research provides a good model for this as many clinical assessment tools provide empirically based distinctions between people who have severe enough symptoms to be recommended for clinical treatment and those who have subclinical but still potentially important levels of symptoms. The use of measures with well-established norms is uncommon in OHP and OHP researchers often modify measures to fit the demands of their particular context, meaning that general norms might be less applicable. Thus, we would strongly encourage researchers to carefully consider whether raw scores on their profiles truly fit terms such as high, medium, and low before using those terms to describe profiles. In the absence of norms, two simple ways to do this are to compare sample means to those from prior research and to carefully consider the meaning of a particular score relative to the definition of the response scale used to capture that score.

Hypothesizing Differences between Profiles on Other Variables

A final challenge in person-centered research concerns hypothesizing differences between groups on other variables of interest. The primary focus in past commitment research has been on profile differences in retention and performance related outcomes; health and well-being related outcomes have generally not been studied. However, in prior commitment literature, there are a couple of ideas about profile differences that could suggest useful directions for OHP research. J. P. Meyer and Herscovitch (2001) discussed several reasons to expect that AC, CC, and NC have different outcomes, because of the nature of the mindsets involved. For example, AC and CC differ in their conceptual breadth, such that CC is focused largely on retention-related behavior whereas AC concerns a much broader set of behavioral outcomes. They suggested that profiles defined by high levels of AC should predict a wider range of outcomes. This idea could be applied, for example, to make different predictions about profiles of coping styles (which would presumably affect a wide range of outcomes) as compared with profiles of safety climate dimensions (which probably would affect a narrower range of safety-related outcomes).

The commitment literature has increasingly used the term *dominant* to characterize the strongest commitment influence for a group, where the strength of the influence is typically viewed as the highest score (cf. J. P. Meyer et al., 2012). For example, members of an *AC dominant* profile would have higher AC scores and lower CC and NC scores. Similarly, a CC-NC dominant profile would include people who had relatively higher levels of CC and NC than they do AC. The basic rationale is that the dominant mind-sets should be the most informative for predicting behavior and should set the context for the effects of other commitments. More recently, Meyer and colleagues have been integrating commitment profiles research with self-determination theory to describe profile differences in motivation stemming from particular combinations of commitments. A full discussion of this topic is beyond the scope of this chapter, but for interested readers, J. P. Meyer et al. (2012) provides a useful discussion of this issue.

We believe the dominance idea is quite useful for thinking about differences between any particular set of profiles, albeit with a couple of caveats. Although dominance is generally used to refer to relatively higher scores in the measures used to build the profile, as noted above, dominance is usually defined within sample. Thus, an AC-dominant profile typically refers to people who have relatively higher AC scores than their counterparts in the same sample. Strictly speaking, low should refer to people who give low absolute responses to commitment questions, rather than to people who are simply lower than others in the same sample. But at the very least, researchers should be explicit about their definitions of high and low as being relative.

The lowest scores in a commitment profile also could be important for profile research. For example, Sinclair et al. (2005) identified a group possessing moderate levels of CC and low levels of AC as Free Agents. It could be argued that the low AC scores are the most important aspect of this group's attachment; the low AC score suggests that the group has no particular emotional attachment to their organization. Indeed, this proposition would be consistent with the idea that negative experiences, emotional states, etc. have more powerful effects on people than do positive states (cf. Baumeister, Bratslavsky, Finkenauer, & Vohs, 2001).

Other Interesting Questions and Directions

The vast majority of person-centered commitment research follows a familiar formula: (a) select input variables, (b) create profiles, and (c) examine profile differences on other variables. However, contemporary analytical methods enable researchers to address several other interesting questions. For example, using a two-wave study of acute care nurses gathered four months apart, Sears (2010) studied differences among profiles of occupational AC and CC. In addition to deriving commitment profiles and studying retention-related outcomes, she investigated patterns of stability and change in profile membership over time, addressing questions such as the stability of the profiles over time and which kinds of changes from membership in one profile to another were more likely. One particularly interesting finding was that although she found the same set of five profiles at each time point, there were interesting patterns of profile membership change over time. For example, 82% of the Time 1 free agents (moderate CC, low AC) were members of the same profile at Time 2; but only 32% of the devoted (high AC; high CC) and attached (moderate AC; moderate CC) were in the same profile at Time 2. Moreover, the patterns of change in membership from one profile to another were apparently nonrandom, with some changes being much more likely than others.

Beginning to consider profile stability and change over time raises interesting theoretical and practical possibilities. For example, it is possible that people have dispositional tendencies to form certain patterns of attachments over time, profiles that could conceivably be similar irrespective of the target involved. Alternately, profile change may be thought of as a developmental process such that people who share common experiences may form similar patterns of change over time. Consistent with this idea, Sears (2010) found that nurses who reported more successes at work at Time 1 (e.g., positive patient outcomes, learning new work skills) were more likely to be devoted (high occupational AC and CC) or complacent (moderate occupational AC, low CC) and less likely to be free agents (low AC, moderate CC), or allied (moderate AC and CC) several months later.

Finally, multiple profile models can incorporate the idea of equifinality from open systems theory (Katz & Kahn, 1978)—essentially the idea that the same end state may be reached by multiple paths. For example, Sinclair et al. (2005) found that most combinations of affective and normative commitment showed similar levels of job performance, but one group in particular, those who were moderate on continuance commitment and low on affective commitment were lower than any other group on the performance measures. They argued that this finding offered some practical value, as it implied greater performance gains from interventions aimed at moving workers from the free agent profile to any other profile. The general concept of equifinality has not received much attention in occupational health psychology research, but it may have wide applicability to various kinds of models.

Summary and Conclusions

In this chapter, we discussed the methods, applications, and implications of person-centered analysis for OHP research. Four specific research methods were introduced (i.e., latent profile analysis, latent class analysis, growth mixture modeling, and mixture latent Markov modeling). We would like to end this chapter by discussing several important issues in using these methods in empirical research.

Substantive theories on the phenomena of interest should guide the application of the statistical methods in person-centered analysis. First, the specification of person-centered analysis models should be based on theory. For example, when using GMM, the growth curves of each latent class should be specified based on the change trajectory expected from the theories (e.g., Wang, 2007). Second, when selecting the optimal model in person-centered analysis, for example, when determining the appropriate number of latent classes, it is meaningful to consider which model is more interpretable in theories when there is no additional diagnostic information provided by statistical indices (e.g., Lubke & Muthén, 2005). For example, it is possible that from a statistical sense two models fit the data equally well. Under this circumstance, only the model that can be interpreted according to theories is useful in terms of testing and developing theories. It is also possible that a misspecified model fit the data equally well as the true underlying model and these two models cannot be distinguished by statistical means (e.g., heterogeneous mixtures of normal distributions and homogeneous nonnormal distributions; see Bauer & Curran, 2003a, 2003b; Cudeck &

Henly, 2003; Muthén, 2003; Rindskopf, 2003 for a detailed discussion on this issue).

Third, mathematically, the same set of data can be analyzed by different statistical models parameterized with different statistical approaches but fit the data equally well (Lubke & Muthén, 2007; Wang & Hanges, 2011). A famous example is that the same covariance matrix can be fit equally well by a factor analysis model or a latent class model (Bartholomew & Knott, 1999; Molenaar & von Eye, 1994). Researchers should use theoretical criteria to decide which approach is most appropriate in any particular study. Person-centered analysis are more suitable for answering questions about classifying and comparing qualitatively different subpopulations; whereas variable-centered methods are more suitable for answering questions about interrelatedness generated by processes applicable for all individuals. Therefore, whether to use the person-centered analysis in research should be decided by the substantive questions of concern. In sum, we recommend using person-centered analysis in a confirmatory rather than exploratory manner.

To use person-centered analysis appropriately also has several requirements for research design. Classification results usually depend on the information available from the data of a particular sample (e.g., covariates measured from the sample; Lubke & Muthén, 2007). Thus, classification results from representative samples are more informative. However, when random sampling is not possible, we strongly recommend cross-validation of the modeling results in another comparable sample (e.g., Wang, 2007).

Person-centered/profile approaches have been of interest throughout the history of psychology, but variable centered research has dominated the last several decades of applied psychology. Recent methodological developments have increased the sophistication of profile research as researchers have begun to move from forming groups by simply splitting scales at the midpoint, to using empirically derived methods such as cluster analyses, to contemporary applications of latent profile models. As the commitment literature illustrates, relatively few studies have taken advantage of some of these newer methods. We hope this chapter will provide some encouragement and direction to researchers considering these approaches.

We focused on the commitment literature because it includes many studies with varying degrees of sophistication. However, the lessons learned from the commitment literature apply to just about any topic researchers might choose to investigate with a person centered approach. They include, (a) choosing variables carefully, including a consideration of why certain variables might not be chosen; (b) developing strong a priori theories that predict numbers of profiles, the nature of the particular profiles, and their antecedents and outcomes; (c) developing theoretical explanations for why other profiles might not exist; (d) carefully considering the meaning of terms such as high, medium, and low in the formation of profiles and in the interpretation of profile data; and (e) developing a strong rationale for predictions about differences between profiles. Moreover, we encourage researchers to consider some of the exciting new possibilities offered by contemporary analytical methods.

References

Allen, N. J., & Meyer, J. P. (1990). The measurement and antecedents of affective, continuance, and normative commitment to the organization. *Journal of Occupational Psychology, 63*, 1–18.

Bartholomew, D. J., & Knott, M. (1999). *Latent variable models and factor analysis* (2nd ed.). London: Arnold.

Bauer, D. J., & Curran, P. J. (2003a). Distributional assumptions of growth mixture models: Implications for overextraction of latent trajectory classes. *Psychological Methods, 8*, 338–363.

Bauer, D. J., & Curran, P. J. (2003b). Overextraction of latent trajectory classes: Much ado about nothing? Reply to Rindskopf (2003), Muthén (2003), and Cudeck and Henly (2003). *Psychological Methods, 8*, 384–393.

Baumeister, R. F., Bratslavsky, E., Finkenauer, C., & Vohs, K. D. (2001). Bad is stronger than good. *Review of General Psychology, 5*(4), 323–370.

Becker, T. E., & Billings, R. S. (1992). Profiles of commitment: An empirical test. *Journal of Organizational Behavior, 14*, 177–190.

Carson, K. D., Carson, P. P., Roe, C. W., Birkenmeir, B. J., & Phillips, J. S. (1999). Four commitment profiles and their relationships to empowerment, service recovery, and work attitudes. *Public Personnel Management, 2*, 1–13.

Chan, D., & Schmitt, N. (2000). Interindividual differences in intraindividual changes in proactivity during organizational entry: A latent growth modeling approach to understanding newcomer adaptation. *Journal of Applied Psychology, 85*, 190–210.

Cohen, A. (2003). *Multiple commitments in the workplace: An integrative approach.* Mahwah, NJ: Erlbaum.

Cooper-Hakim, A., & Viswesvaran, C. (2005). The construct of work commitment: Testing an integrative framework. *Psychological Bulletin, 131, 241–259.*

Cudeck, R., & Henly, S. J. (2003). A realistic perspective on pattern representation in growth data: Comment on Bauer and Curran (2003). *Psychological Methods, 8*, 378–383.

Gellatly, I. R., Hunter, K. H., Currie, L. G., & Irving, P. G. (2009). HRM practices and organizational commitment profiles. *The International Journal of Human Resource Management, 20*, 869–884.

Gorgievski-Duijvesteijn, M., Bakker, A., Schaufeli, W., & van der Heijden, P. (2005). Finances and well-being: A dynamic equilibrium model of resources. *Journal of Occupational Health Psychology, 10*, 210–224.

Gouldner, A. W. (1958). Cosmopolitans and locals: Toward an analysis of latent social roles. *Administrative Science Quarterly, 2*, 281–306.

Katz, D., & Kahn, R. (1978). *The social psychology of organizations* (2nd ed.). New York: Wiley.

Kerr, N. L. (1998). HARKing (Hypothesizing after the results are known). *Personality and Social Psychology Review, 2*, 196–217.

Langeheine, R., & van de Pol, F. (2002). Latent Markov chains. In J. A. Hagenaars & A. L. McCutcheon (Eds.), *Applied latent class analysis* (pp. 304–341). New York: Cambridge University Press.

Little, R. J., & Rubin, D. B. (2002). *Statistical analysis with missing data* (2nd ed.). New York: Wiley.

Lo, Y., Mendell, N. R., & Rubin, D. B. (2001). Testing the number of components in a normal mixture. *Biometrika, 88*, 767–778.

Lubke, G. H., & Muthén, B. O. (2005). Investigating population heterogeneity with factor mixture models. *Psychological Methods, 10*, 21–39.

Lubke, G. H., & Muthén, B. O. (2007). Performance of factor mixture models as a function of model size, covariate effects, and class-specific parameters. *Structural Equation Modeling, 14*, 26–47.

Martin, J. E., & Sinclair, R. R. (2007). A typology of the part-time workforce: Differences on job attitudes and turnover. *Journal of Occupational and Organizational Psychology, 80*, 301–319.

McNally, J. J., & Irving, P. G. (2010). The relationship between university student commitment profiles and behavior: Exploring the nature of context effects. *Journal of Leadership and Organizational Studies, 17*, 201–215.

Meyer, A. D., Tusi, A. S., & Hinnings, C. R. (1993). Configural approaches to organizational analysis. *Academy of Management Journal*, 36, 1175–1195.

Meyer, J. P. (2009). Commitment in a changing world of work. In H. J. Klein, T. E. Becker, & J. P. Meyer (Eds.), *Commitment in organizations: Accumulated wisdom and new directions* (pp. 37–68). New York: Routledge.

Meyer, J. P., & Allen, N. J. (1997). *Commitment in the workplace: Theory, research, and application*. Thousand Oaks, CA: Sage.

Meyer, J. P., & Herscovitch, L. (2001). Commitment in the workplace: Toward a general model. *Human Resource Management Review, 11*, 299–326.

Meyer, J. P., & Maltin, E. R. (2010). Employee commitment and well-being: A critical review, theoretical framework and research agenda. *Journal of Vocational Behavior, 77*, 323–337.

Meyer, J. P., Stanley, D. J., Herscovitch, L., & Topolnytsky, L. (2002). Affective, continuance, and normative commitment to the organization: A meta-analysis of antecedents, correlates, and consequences. *Journal of Vocational Behavior, 61*, 20–52.

Meyer, J. P., Stanley, L. J., & Parfyonova, N. M. (2012). Employee commitment in context: The nature and implication of commitment profiles. *Journal of Vocational Behavior, 80*, 1–16.

Mintzberg, H. (1979). *The structuring of organizations.* Englewood Cliffs, NJ: Prentice-Hall.

Molenaar, P. C., & von Eye, A. (1994). On the arbitrary nature of latent variables. In A. von Eye & C. C. Clogg (Eds.), *Latent variable analysis: Applications for developmental research* (pp. 226–242). Thousand Oaks, CA: Sage.

Morin, A. J. S., Morizot, J., Boudrias, J., & Madore, I. (2010). A multifoci person-centered perspective on workplace affective commitment: A latent profile/factor mixture analysis. *Organizational Research Methods, 14,* 58–90.

Muthén, B. (2001). Second-generation structural equation modeling with a combination of categorical and continuous latent variables: New opportunities for latent class-latent growth modeling. In L. Collins & A. Sayer (Eds.), *New methods for the analysis of change* (pp. 291–322). Washington, DC: American Psychological Association.

Muthén, B. (2003). Statistical and substantive checking in growth mixture modeling: Comment on Bauer and Curran (2003). *Psychological Methods, 8,* 369–377.

Muthén, B. (2004). Latent variable analysis: Growth mixture modeling and related techniques for longitudinal data. In D. Kaplan (Eds.), *Handbook of quantitative methodology for the social sciences* (pp. 345–368). Thousand Oaks, CA: Sage.

Muthén, B. (2006). Should substance use disorders be considered as categorical or dimensional? *Addiction, 101,* 6–16.

Muthén, B., & Shedden, K. (1999). Finite mixture modeling with mixture outcomes using the EM algorithm. *Biometrics, 55,* 463–469.

Ployhart, R. E., & Hakel, M. D. (1998). The substantive nature of performance variability: Predicting interindividual differences in intraindividual performance. *Personnel Psychology, 51,* 859–901.

Rindskopf, D. (2003). Mixture or homogeneous? Comment on Bauer and Curran (2003). *Psychological Methods, 8,* 364-368.

Sears, L. E. (2010). *Predictors and outcomes of occupational commitment profiles among nurses* (Unpublished doctoral dissertation). Clemson University, Clemson, SC.

Sinclair, R. R., Tucker, J. S., Cullen, J. C., & Wright, C. (2005). Performance differences among four organizational commitment profiles. *Journal of Applied Psychology, 90,* 1280-1287.

Somers, M. J. (1995). Organizational commitment, turnover, and absenteeism: An examination of direct and interaction effects. *Journal of Organizational Behavior, 16,* 49–58.

Somers, M. J. (2009). The combined influence of affective, continuance and normative commitment on employee withdrawal. *Journal of Vocational Behavior, 74,* 75–81.

Somers, M. J. (2010). Patterns of attachment to organizations: Commitment profiles and work outcomes. *Journal of Occupational and Organizational Psychology, 83,* 443–453.

Stanley, L. J., Vandenberg, R. J., Vendenberghe, C., & Bentein, K. (2011). *Commitment profiles and employee turnover.* Unpublished manuscript.

Tsoumbris, P., & Xenikou, A. (2010). Commitment profiles: The configural effect of the forms and foci of commitment on work outcomes. *Journal of Vocational Behavior, 77,* 401–411

Vandenberghe, C. (2009). Organizational commitments. In H. J. Klein, T. E. Becker, & J. P. Meyer (Eds.), *Commitment in organizations: Accumulated wisdom and new directions* (SIOP Organizational Frontiers Series) (pp. 69-95). New York: Routledge.

Wang, M. (2007). Profiling retirees in the retirement transition and adjustment process: Examining the longitudinal change patterns of retirees' psychological well-being. *Journal of Applied Psychology, 92,* 455–474.

Wang, M., & Bodner, T. E. (2007). Growth mixture modeling: Identifying and predicting unobserved subpopulations with longitudinal data. *Organizational Research Methods, 10,* 635–656.

Wang, M., & Chan, D. (2011). Mixture latent Markov modeling: Identifying and predicting unobserved heterogeneity in longitudinal qualitative status change. *Organizational Research Methods, 14,* 411–431.

Wang, M., & Hanges, P. J. (2011). Latent class procedures: Applications to organizational research. *Organizational Research Methods, 14,* 24–31.

Wiggins, L. M. (1973). *Panel analysis.* Amsterdam, the Netherlands: Elsevier.

20

Longitudinal Research and Data Analysis

E. Kevin Kelloway and Lori Francis

Researchers in occupational health psychology (OHP), as in other areas of organizational research, are continually exhorted to use longitudinal research designs rather than relying on cross-sectional data, which does not allow for causal inference (Taris, 2003) and may result in misleading or biased parameter estimates (Maxwell & Cole, 2008). Despite consistent agreement on the need for longitudinal approaches, cross-sectional data continue to predominate in OHP (Zapf, Dorman, & Frese, 1996) and, more generally in organizational research (Austin, Scherbaum, & Mahlman, 2002; Stone-Romero, 2011). Moreover, our understanding of (a) what constitutes "longitudinal research," (b) what these techniques might tell us, and (c) how to analyze these data properly, is continuing to evolve (e.g., Chan, 2011).

Although we believe that cross-sectional designs still have a role to play in OHP research, we agree with the increased emphasis on collecting and analyzing longitudinal research data. Moreover, we assert that consideration of the three issues identified above will lead not only to better research on questions of traditional interest in OHP, but also in the identification of new, or at least underresearched aspects of the phenomena we study. Accordingly, the goal of this chapter is to review what we see as the three major evolving issues: the definition of longitudinal research, understanding what longitudinal research can and cannot do, and understanding how to analyze longitudinal data. In doing so, we both reference the use of and discuss the implications of these techniques in conducting research in OHP.

A brief note on presentation: We recognize that there are multiple ways of presenting longitudinal analyses, and we opt throughout the chapter to develop presentations based on commonly accepted practices for depicting structural equation models. We do so in the belief that most researchers in OHP are familiar with such diagrams. Unless explicitly noted, we do not mean to imply that structural equation models are preferred over other techniques such as random coefficient modeling in the analysis of longitudinal data.

Defining Longitudinal Research

Perhaps the most commonly cited rationale for using a longitudinal research design is to enhance causal inference. Cook and Campbell (1979) defined the conditions necessary for such an inference as comprising (a) covariation between the hypothesized predictor and outcome, (b) temporal ordering such that the predictor precedes the outcome, and (c) the exclusion of competing explanations (e.g., third variable effects). Cross-sectional data can provide, at best, evidence for covariation, although the difficulty in ruling out other explanations such as common method variance (e.g., Lance, Dawson, Birkelbach, & Hoffman, 2010; Spector & Brannick, 2010) makes even this modest contribution questionable (Maxwell & Cole, 2007). Longitudinal research goes beyond cross-sectional data and allows one to establish temporal order. Third variable effects remain a competing explanation even in the case of longitudinal research (e.g., Zapf et al., 1996) and, as a result, do not absolutely establish a causal relationship or allow causal claims (Taris, 2003); however, our ability to make a causal inference can be enhanced by the appropriate use of longitudinal methods.

In its simplest sense, establishing temporal order simply means that the predictor must occur before the outcome. In the case of specific events such as injuries resulting from accidents, it is clear that the accident preceded the injury and it is not plausible to suggest that the injury caused the accident in question (although an injury may be implicated as a causal factor in a subsequent accident). Establishing temporal order is less clear when one considers psychological variables such as stress or safety climate. These variables are not events and there is a danger in confusing the time of measurement with the time of occurrence. A research design that, for example, assesses role conflict at Time 1 and strain symptoms 6 months later at Time 2 (see Figure 20.1a) does not allow one to establish temporal order because the strain symptoms measured at Time 2 may have been evident, but unmeasured, at Time 1 and there is no way of telling whether these have changed. Similarly, role conflict may not have changed over the 6 month interval and the measurement at Time 1 is simply a proxy for role conflict at Time 2. Although this design may reduce common method effects by separating the collection of predictor and outcome measures, in a more substantive sense this design tells us no more, and indeed may introduce more ambiguity, than would a purely cross-sectional approach. As a result, this design is rarely used in OHP research (Zapf et al., 1996).

A simple improvement on this design is to measure the outcome at both time periods and to covary the stability and subsequently test the effect of the predictor (e.g., predict strain at Time 2 by strain at Time 1 and then predict strain at Time 2 using role conflict at Time 1; see Figure 20.1b). Zapf et al. (1996) refer to this as the incomplete two wave panel design. This analysis has the principal advantage of explicitly

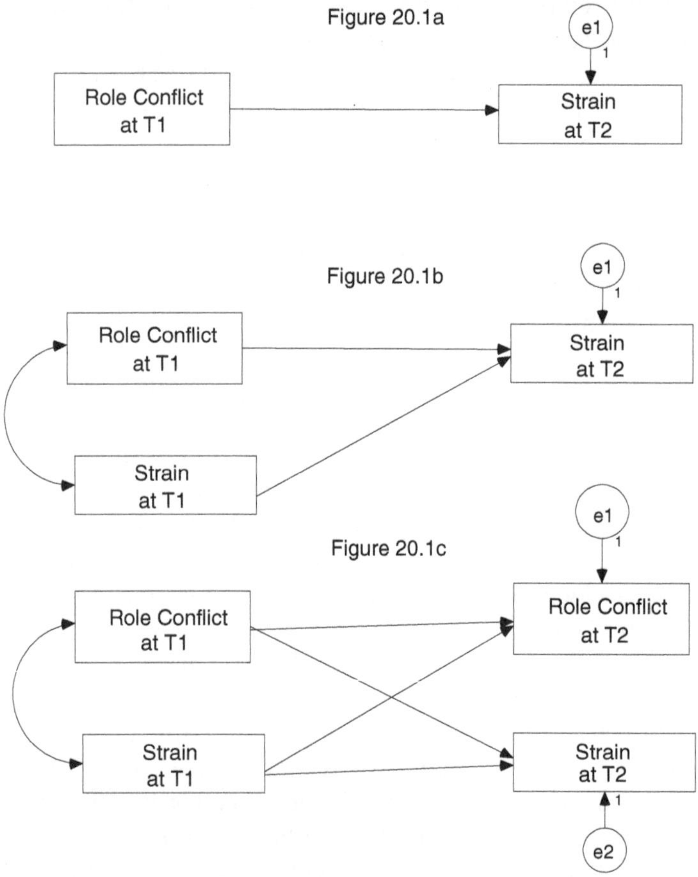

Figure 20.1 Two wave designs.

estimating the change in the outcome (i.e., the outcome at Time 2 after controlling the same variable at Time 1) with the disadvantage of leaving the stability in the predictor uncontrolled and unknown. Thus, this design is also unable to rule out reverse causality.

Perhaps the most popular resolution of this issue and by extension the most widely used "longitudinal" research design in OHP is the two wave panel design in which both the predictor and the outcome are measured at both time periods (see Figure 20.1c). These designs can be analyzed through cross-lagged correlations, cross-lagged regression analyzes, or structural equation modeling techniques. Cross-lagged correlation analysis, where one compares the magnitude of the correlation between role conflict at Time 1 and strain at Time 2 with the correlation between strain at Time 1 and role conflict at Time 2, is now generally discredited (e.g., Williams & Podsakoff, 1989; Zapf et al., 1996) in favor of the latter two approaches.

Cross-lagged regressions are based on examining predictor–outcome relationships over time while covarying out the stability in the variables (e.g., Kelloway & Barling, 1994). Thus, strain at Time 2 is regressed on strain at Time 1 (the stability) and role conflict (predictor). Reverse causality is tested by regressing role conflict at Time 2 on role conflict Time 1 (the stability) and strain (the predictor). A similar analysis can be implemented using structural equation modeling techniques (e.g., Kelloway, Gottlieb, & Barham, 2000). Structural equation modeling offers some advantage over regressions in that the former allow one to incorporate measurement error, to estimate several causal relationships simultaneously (Zapf et al., 1996), and to incorporate correlated errors (Kelloway et al., 2000)

One of the most striking changes in our understanding of longitudinal research is the recognition of the limitations imposed by two-wave data. Most notably, two-wave studies are limited in their conceptualization of change as an increment and are unable to describe the process of change (Singer & Willett, 2003). In other words, two-wave studies limit our conceptualization of change to a linear function. Additionally, two-wave studies may confound measurement error with substantive change (Ployhart & Vandenberg, 2010; Singer & Willett, 2003). As a result of these concerns, many reviewers do not see much value in two-wave studies. Taris (2003), for example, agrees with Rogosa's (1995) conclusion that "Two waves of data are better than one, but maybe not much better" (p. 174).

In response to these concerns, longitudinal research is increasingly defined in terms of several (i.e., more than two) waves of data collection.

Singer and Willett (2003), for example, suggest that longitudinal studies amenable to the measurement of change include at least three waves of data collection. Ployhart and Vandenberg (2010) define longitudinal research as "research emphasizing the study of change and containing at minimum three repeated observations (although more than three is better) on at least one of the substantive constructs of interest" (p. 97). In the remainder of this chapter, we adopt a similar view, advocating that longitudinal research comprises at least three (and preferably more) repeated observations on a variable of interest.

The Focus of Longitudinal Research

Aside from the number of observations, it is important to recognize that longitudinal research is primarily focused on understanding change in the variables of interest (e.g., Chan, 2011; Ployhart & Vandenberg, 2010). One important implication of this observation is that the variables of interest must in fact change over time. This seemingly trivial observation has important implications for the design and analysis of longitudinal methods.

A focus on change means that researchers need to develop theories about the nature of change. Such a theory would encompass a specification of the nature (or form) of the change, the duration and timing of the change, and potential causes of the change (Ployhart & Vandenberg, 2010). In discussing dynamic relationships (see below), Pitariu and Ployhart (2010) similarly suggested that such a theory should consider the time (when a specific relationship should occur), duration (how long a relationship should exist), and shape of a relationship; that is, its form over time.

We suggest that theoretical development in OHP could benefit from incorporating a conceptualization of change (e.g., Garst, Frese, & Molenaar, 2000). The typical predictions of the form 'x is associated with y' fall far short of the requirements expressed above and at best provide a weak test of the adequacy of our theories. In contrast, conceptualizing change within the context of our theories is likely to lead to hypotheses that are more open to falsification (e.g., Maxwell & Cole, 2007) and, ultimately, to better theory and better understanding of OHP phenomena.

There are also considerable practical benefits to theories focused on change. In light of continued calls for longitudinal research, we find considerable irony in the observation that most theories in OHP offer little practical guidance to the researcher who is planning such research. Key design considerations such as the number of observations and the timing of (or lag between) observations (Ployhart & Vandenberg, 2010) should be based on theoretical considerations. Conclusions drawn about the validity of the theory in question will be dramatically influenced by the researchers' choice of the number and timing of observations. Simply put, if we get these specifications wrong then we are likely to get the wrong answer from our research. If we do not understand the nature of change in the constructs we are studying, we may be reaching false conclusions even with rigorously conducted longitudinal studies. We suggest that this observation leads to at least two practical recommendations for occupational health research. First, as discussed below, we need to conduct more descriptive studies of change in order to understand the nature and timing of change in the variables we study. Second, theories in OHP should explicitly consider the role of time (see Garst et al., 2000) specifying, for example, the latency between experiencing a condition (e.g., increased role overload) and the outcomes (e.g., strain, absenteeism, burnout).

A concern particularly relevant to psychological research is the need to distinguish between "true" change and measurement change in longitudinal research. Chan (1998) notes that even when we observe change in the variables of interest, our observation can reflect "true" changes in the construct of interest, changes in the calibration of our measurement instruments, or changes in the conceptualization of the constructs we are studying. These different types of changes correspond to Golembiewski, Billingsley, and Yeager's (1976) definitions of alpha, beta, and gamma change. We would suggest that researchers in OHP are most interested in alpha change; that is, a change in the construct of interest. It follows that analysis of change should begin by establishing measurement invariance over time (i.e., the absence of beta and gamma change, Chan, 1998). Vandenberg and Lance (2000) present a comprehensive overview of tests for measurement invariance and we limit our consideration of this issue to noting the need to establish measurement invariance in longitudinal research as a basis for understanding change in the construct (i.e., alpha change) over time.

It is clear, then, that the focus of longitudinal research is on change in constructs at a descriptive level and how change in one construct might influence change in other constructs. We now turn our attention to identifying analytic techniques that are appropriate to answer these questions about change in longitudinal research.

What Longitudinal Research Can Tell Us: Questions and Analysis

In their review, Ployhart and Vandenberg (2010) offer a useful distinction between descriptive and explanatory longitudinal research. Descriptive longitudinal research is focused on "how a phenomenon changes over time," whereas "explanatory longitudinal research seeks to identify the cause of the change process by the use of one or more substantive predictor variables" (Ployhart & Vandenberg, 2010, p. 99).

Descriptive Longitudinal Techniques

Our understanding of the field of OHP suggests that its near exclusive concern has been with explanatory research to the virtual exclusion of descriptive analyzes. Indeed, researchers often find it difficult to publish descriptive data in OHP outlets even when they stem from large, nationally representative sources (Frone, personal communication, 2009). It is not putting it too strongly to observe that as a result of this state of affairs researchers in OHP have very little descriptive understanding of the phenomena, including the time courses of important relationships, which we study.

Consider for instance, the lack of descriptive understanding of workplace stressors. Decades of research have resulted in taxonomies of workplace stressors (e.g., Kelloway & Day, 2005; Sauter, Murphy, & Hurrell, 1990; Warr, 1987); however, even researchers who are highly familiar with this literature would likely be unable to describe the temporal development or progression of chronic stressors in the workplace. For example, we recognize the role overload—having too much to do in the time available—is a common stressor in the modern workplace. However, we suspect that researchers would be hard pressed to describe the manifestation of role overload. Is it a constant such that individuals who are affected experience a constant state of overload? Or does overload vary on a daily or weekly basis? (e.g., MacEwen, Barling, & Kelloway, 1992). Is the resultant experience of stress cumulative (i.e., increasing from time period to time period)? If so, is this increase linear in nature or is there a tipping point beyond which individual health becomes impaired (e.g., Karanika-Murray, 2010; Warr, 1987)? Longitudinal designs offer us the potential to ask, and answer, such questions.

In particular, we suggest that there is a potential for researchers in OHP to learn much from the analysis of univariate models in which the same persons are measured on a single variable on multiple occasions (e.g., Rosel & Plewis, 2008). There are at least three useful approaches to analyzing such data; application of the general linear model; application of time series analyses, and modeling growth curves.

Application of the general linear model to perform repeated measures analyses is, perhaps, the most familiar technique for researchers in OHP. Given a construct of interest measured on multiple occasions, it is reasonable to do a within-groups analysis of variance to ask if the variable has changed over time.[1] By coding the representation of time in the analysis, one can test for specific trends, contrast, and so on.

Another approach to dealing with the same data would be to model the observations as a time series (Rosel & Plewis, 2008). With their emphasis on description and forecasting, the substantive questions underlying times series analyses are not common in OHP research. They are included here because we believe that such questions are of interest in that they offer the potential to aid researchers to engage in descriptive research and increase our understanding of how change manifests over time.

Perhaps the simplest model for time series data is the first order autoregressive model (see Figure 20.2a), also known as a simplex or Markov model. The substantive hypothesis underlying this model is that each observation is a function of the immediately preceding observation. Once the first order relationships have been accounted for, second order or higher autoregressive effects (Figure 20.2b) can be introduced. In general, with K observations over time, the highest order relationship that can be tested is K-1. One can modify these models to constrain the paths between variables to equality (i.e., hypothesizing a constant autoregressive effect over time) or a variety of other constraints (Rosel & Plewis, 2008). Moreover, if one views the error terms associated with each measurement as an innovation (i.e., the difference between the forecasted and actual value of a variable; Box, Jenkens, & Reinsel, 1994), then one can model a moving average as well. Figure 20.2c shows a first order autoregressive model with a first order moving average in which each variable is hypothesized to be a function of the same variable at the preceding time period as well as the preceding error. Time series approaches typically focus on observed variables but can also be implemented using latent variables (Rosel & Plewis, 2008).

---

1. We note that one can introduce predictors other than time into the analysis, but deal with this case in subsequent discussion.

Figure 20.2a First Order Autoregressive Model

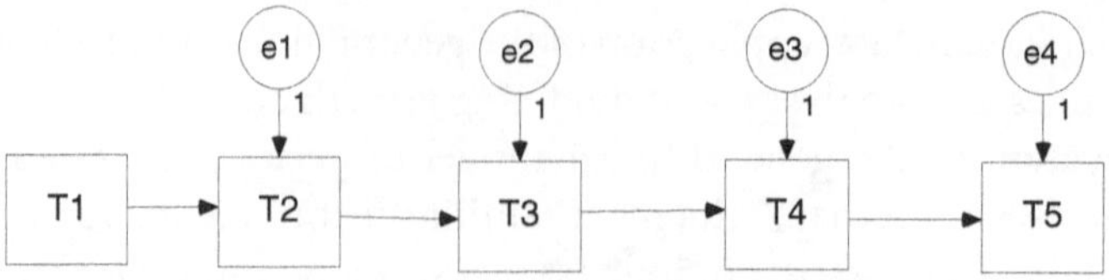

Figure 20.2b Second Order Autoregressive Model

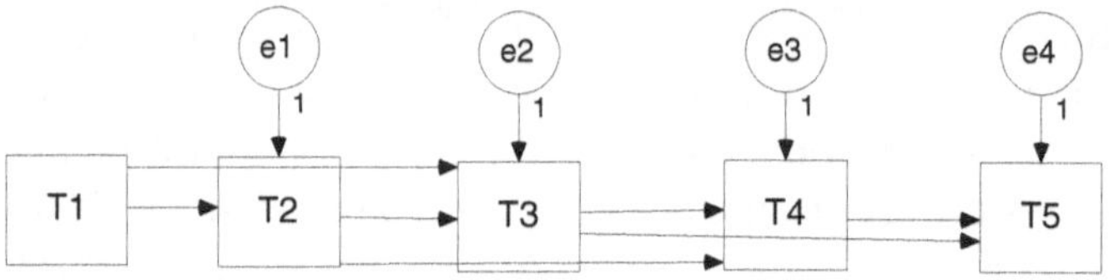

Figure 20.2c First Order Autoregressive Model with First Order Moving Average

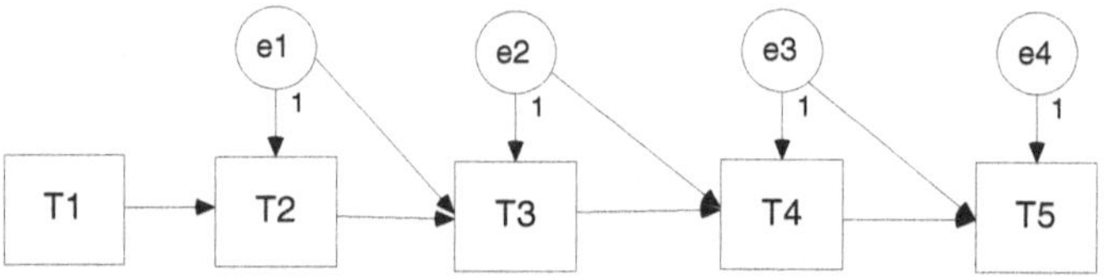

Figure 20.2 Time Series designs.

However, given multiple measurements over time, one can use techniques such as growth curve modeling to model the change occurring in a particular variable (McArdle, 1988). Growth curve models model change as a function of intraindividual change and interindividual change and can be modeled using either random coefficient or structural equation modeling techniques. As noted previously, we focus on the latter mode of presentation.

A sample growth curve model for a single variable is shown in Figure 20.3. The two latent variables representing the intercept and slope predict each of the three indicators. The paths from the intercept to the indicators are set at 1.0 as the intercept is by definition a constant. In contrast, the paths linking the slope latent variable to the indicators are set at 0, 1, 2 respectively. In doing so, we have specified a linear change over the three time periods. By manipulating the values of these path coefficients we could specify a different form of change (e.g., a quadratic change or a plateau). We could also add new latent variables to incorporate a quadratic slope in addition to a linear slope. As in time series analyses, the number of observations dictates the order of change that can be tested (e.g., linear, quadratic, cubic). Within this constraint, there is considerable flexibility to test hypotheses about the form of change in an individual variable over time.

As in most applications of structural equation modeling, questions of model fit assess how well the model, in particular, the form of change specified, describes the data. However, tests of individual parameters also offer considerable information about the nature of change in the data. The model in Figure 20.3 comprises two latent variables: the intercept (i.e., the starting point or score at the beginning of the study), and the slope of the curve (i.e., the mean rate of change of the outcome variable over time). Both of these latent variables have means and variances. The mean of

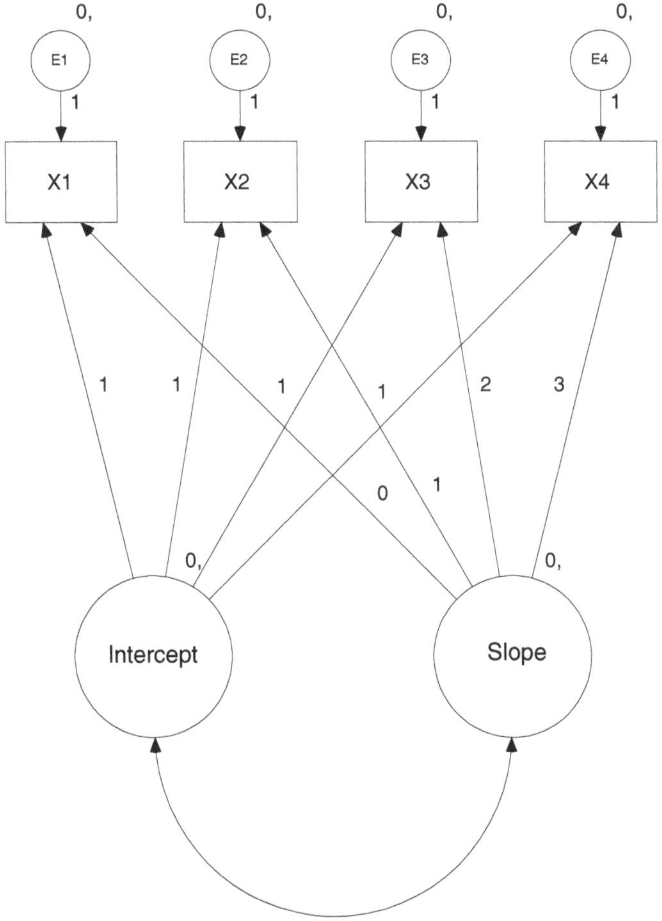

Figure 20.3 Latent growth curve model.

the variable represents the average (i.e., the mean slope is the average rate of change across all participants), whereas the variance represents the "random" coefficient. A significant variance in slope, for example, means that individuals changed at different rates. Such models capture both the intraindividual change over time as well as the existence of interindividual changes that may become the focus of prediction studies.

An elaboration of the unconditional latent growth curve is presented by Ployhart and Vandenberg (2010) using turnover intention as the construct of interest. In their formulation, the two latent variables for the intercept and slope are presented as predictors of latent variables representing the construct of turnover intention (i.e., one latent variable for each time period). In turn, these latent variables are hypothesized as the manifest indicators of turnover intention at each time period. This model incorporates the measurement model for the construct at each time period and allows one first to establish the validity of the measurement model (i.e., measurement invariance, Vandenberg & Lance, 2000) at each time period. In particular, Ployhart and Vandenberg describe testing the hypotheses of configural invariance (i.e., the same number of factors over time) and metric invariance (i.e., invariant factor loadings over time).

McArdle (2001) has developed the latent difference score model as a means of assessing change over time. Similar to latent growth curve models, the latent difference score model incorporates a constant (intercept) and a slope (overall change). However, the latent difference score model also incorporates first order latent variables that represent the true score for each construct, which in turn are represented by the manifest indicators. Second order latent variables at each observation period represent the change or gain in latent scores. An example of such a model is shown in Figure 20.4.

Latent growth curves largely assume that change is unitary (Chan, 1998) in that the observed trajectory applies to all respondents. Although observed variance in the parameters (i.e., intercepts and slopes) provides evidence of individual variance in the growth trajectory, latent growth curves do not allow the easy identification of subgroups for whom different trajectories might apply. If one knows that there are distinct identifiable groups within the population then operationalizing a growth curve model as a multigroup analysis is easily accomplished in most structural equation modeling programs (Wang & Bodner, 2007). Thus, one can easily model separate growth curves for men and women, or for managers as opposed to front line staff. However, the key requirement for doing so is that the groups are identifiable in the data.

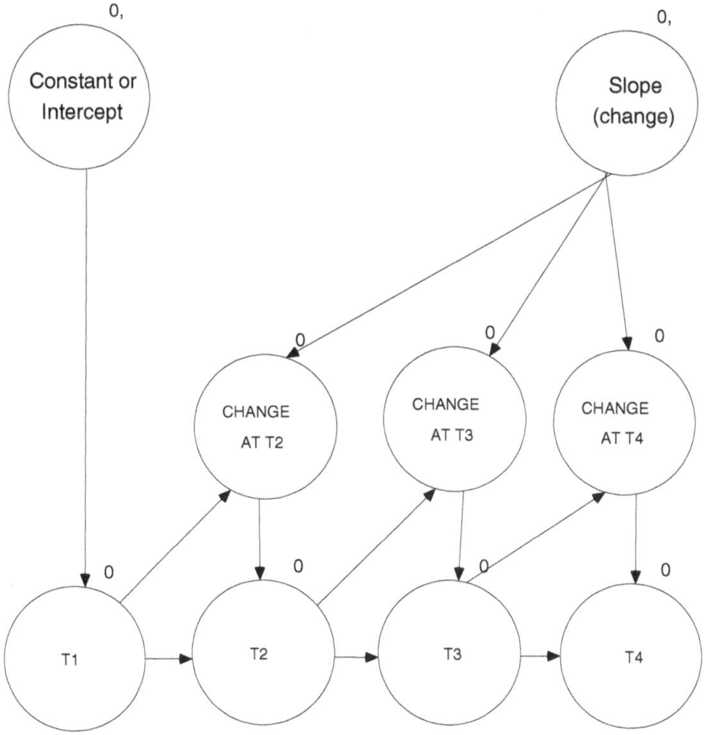

Figure 20.4 Latent difference score model. *Note:* Manifest variables have been omited for T1-T4.

Growth mixture models (Muthén, 2001, 2004) have been developed as a means of identifying unobserved subgroups in such data. In essence such models incorporate "categorical latent variables that represent mixtures of subpopulations in which population membership is not known but is inferred from the data" (Wang & Bodner, 2007, p. 638) along with a latent growth curve model (see Figure 20.5). Analytically, fitting a mixture model involves estimating a series of models in order to determine the optimal number of classes. One would start by fitting the conventional latent growth curve model (i.e., a model with one class) and then proceed to fitting a model with two latent classes, three latent classes, and so on until the optimal number of latent classes is determined. Conventional fit statistics in structural equation modeling are not available for mixture models[2] but comparisons using Akaike's

2. Currently many structural equation models do not allow for the estimation of latent classes or mixture models. Accordingly, most applications of mixture modeling utilize the Mplus (Muthén & Muthén, 1998–2007) software to conduct the analysis. Wang and Bodner (2007) provide Mplus source code for their example of growth mixture modeling.

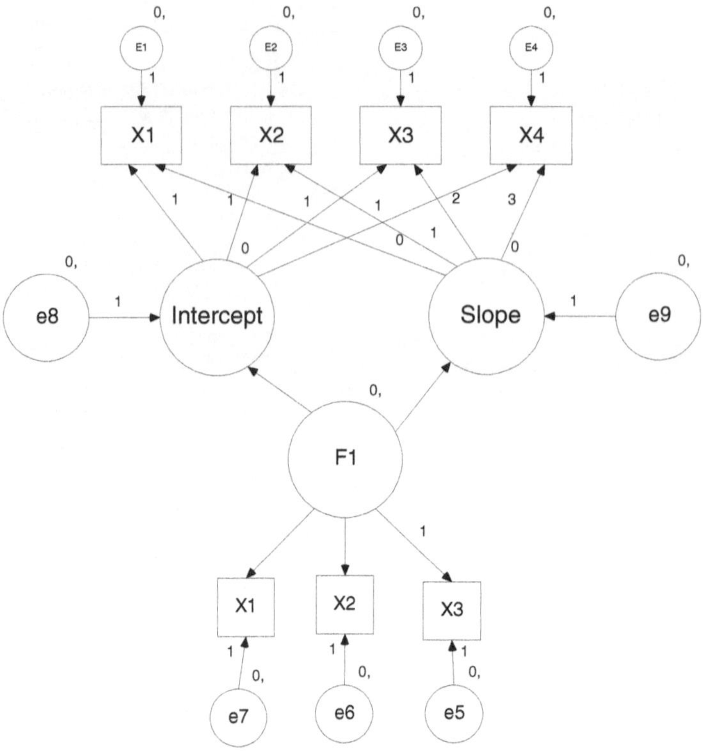

Figure 20.5 A growth mixture model. *Note:* The latent class variable can be thought of as a categrocial variable with the number of categories corresponding to the number of latent classes.

and Bayesian information criteria as well as differences in log likelihood and a measure of latent classification accuracy known as entropy are available (see Wang & Bodner, 2007 for a discussion of these measures). Once the optimal number of classes have been determined, the focus of analysis switches to defining the subpopulations thus identified (i.e., predicting membership in the classes).

Explanatory Longitudinal Techniques

As noted earlier, explanatory longitudinal techniques introduce a predictor to explain why change has occurred rather than simply modeling the form of change. Introducing a predictor is not the same as focusing on dynamic relationships: as Pitariu and Ployhart (2010) note, predictors can be static or time invariant. For example, one might hypothesize that some stable personal characteristic such as a personality trait or demographic characteristic, predicts change in an outcome variable. Utilizing the familiar general linear model, one would estimate a between and within groups analysis of variance that incorporates both change in the dependent variable over time as well as a static predictor that is hypothesized to predict the outcome variable.

Such predictors are also easily incorporated into latent growth models. A relatively minor modification to the model presented in Figure 20.3, for example, would replace the curved arrow linking intercept and slope with a unidirectional arrow leading from intercept to slope. This change, which would also require an error term to be added to the slope latent variable, would correspond to the substantive hypothesis that the starting value on a particular variable determines the rate of change in that variable. Alternatively, one could introduce a new variable as a predictor. For example, Figure 20.6 introduces a static predictor that is hypothesized to predict both the initial level (i.e., the intercept) and the rate of change (i.e., the slope) in a longitudinally measured construct. Although these elaborations are attempts to study change, they do not reflect the modeling of dynamic relationships as defined by Pitariu and Ployhart (2010). Such models would require both the predictor and outcome variable to be measured repeatedly over time.

Rosel and Plewis (2008) point to the possibility of studying dynamic relationships over time using a variant of the time series analyses discussed earlier. As shown in Figure 20.7, one could simply estimate two parallel autoregressive models and incorporate crossed effects into the model. Moving beyond this basic model, one could test for different orders (i.e., first order, second order, etc.) of both the autoregressive effects as well as the hypothesized lagged effects. One would suspect this might be a useful form of modeling when the available theory does not permit a precise specification of the appropriate lag between predictor and outcome. Within the constraints imposed by the number and timing of measurement occasions, researchers could test hypothesized longitudinal relationships across various lags. Again, a wide variety of parameter constraints can be imposed on such models depending on the hypotheses of interest. Moreover, it is straightforward to move from the focus on observed variables to incorporate latent variables into the model.

Cross-domain latent growth curves (McArdle & Hamagami, 1996) implement another approach to studying dynamic relationships. The essence of this technique is to estimate two or more growth curves for

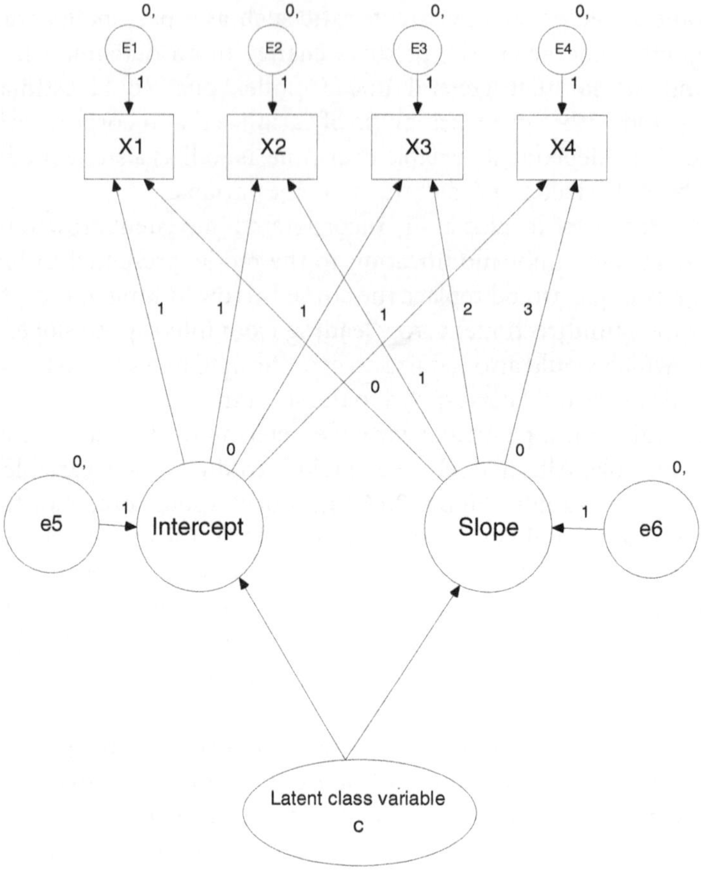

Figure 20.6 Latent growth curve model with a time invariant predictor.

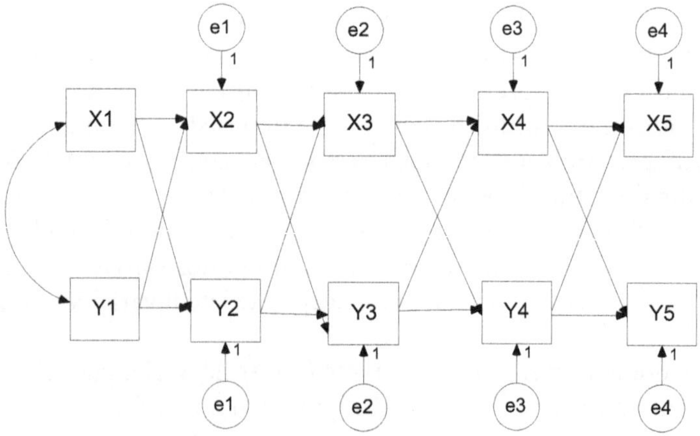

Figure 20.7 Cross effects in Time Series designs.

different variables and to test for relationships between these different variables. Figure 20.8 provides an example of a two variable cross-domain latent growth curve. As shown, growth curves are estimated for both the predictor and the outcome. A directional path is estimated from the slope of the predictor to the slope of the outcome variable corresponding to the hypothesis that a change in the predictor results in a change in the outcome variable. A variety of other hypotheses (e.g., the starting value or intercept of the predictor predicts the rate of change in the outcome) can be operationalized by incorporating the relevant paths.

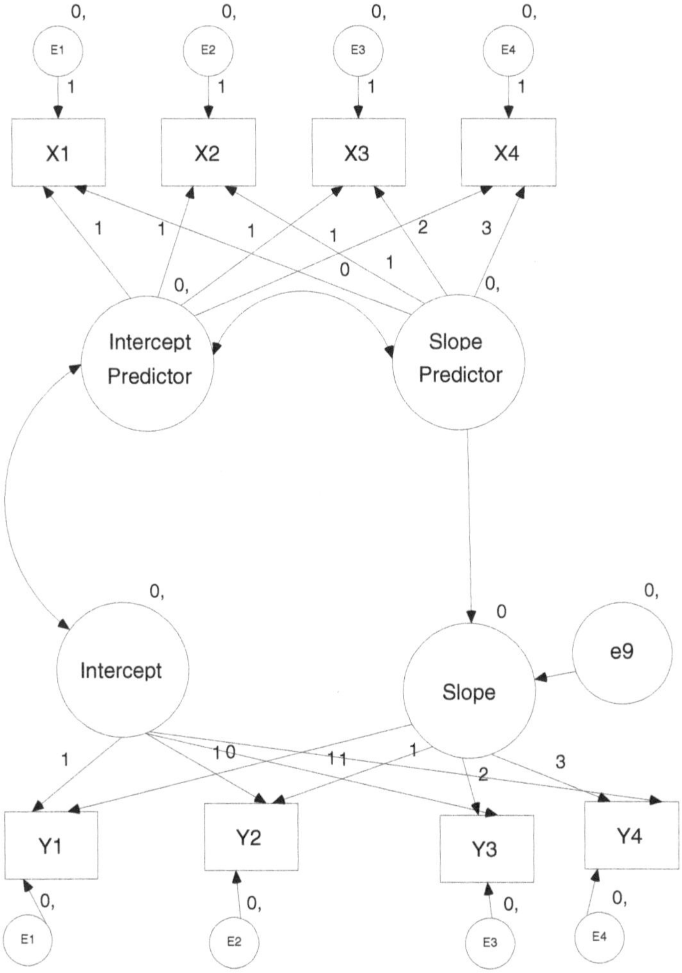

Figure 20.8 Cross domain latent growth curve model.

Pitariu and Ployhart (2010) present a model of dynamic mediation that incorporates a growth curve for each of the predictor, mediator, and outcome. Paths are hypothesized between the slope of the predictor and the slope of the mediator (i.e., hypothesizing that a change in the predictor results in a change in the mediator) and between the mediator and the outcome (i.e., hypothesizing that a change in the mediator results in a change in the outcome). One could contrast this meditational hypothesis against a model that hypothesized partial mediation (i.e., adding the path from the predictor directly to the outcome) or common cause (i.e., the predictor is hypothesized to cause both the mediator and the outcome).

Latent difference score models can also model dynamic predictive relationships (e.g., Ferrer & McArdle, 2010; McArdle, 2009; McArdle & Hamagami, 2001). As in the examples discussed above, the process involves simultaneous estimation of latent difference score models for each of two (or more) variables and incorporating the latent variable for X at time T with the change in Y at the same time period as well as paths linking the latent variable for Y at time T with the change in X at the same time period (see Ferrer & McArdle, 2010 for a discussion of this model).

Modeling Discrete Events Over Time

The models discussed thus far have been based on the implicit definition of the constructs of interest as continuous variables. In contrast, we also wish to briefly consider the longitudinal prediction of specific events and suggest that the prediction of events has considerable applications in the field of OHP. For example, one might use measures of workplace features (e.g., the perception of injustice) to predict the occurrence of cardiac events or death from such events (Elovainio, Leino-Arjas, Vahtera, & Kivimaki, 2006). Recognizing the correctness of Glomb`s (2002) critique of the common practice of aggregating measures of workplace violence, one might suggest that research could be profitably directed toward predicting the occurrence of violence acts, or the time course of the progression from a given perception or experience to such an occurrence (i.e., from an experience of injustice to the perpetration of an act of aggression/violence).

The broad class of techniques used to predict events is known as survival analysis. Singer and Willett (2003) suggest that there are three elements common to all studies of event prediction "(a) a well-defined "event" the occurrence of which is being explored; (b) a clearly defined "beginning of time"; and (c) "a substantively meaningful metric for clocking time" (p. 306). Event occurrence is understood as a transition of states—thus individuals go from being a nonvictim to being a victim of workplace violence, or from being present at to absent from work and the moment of transition is the "event" of interest. The "beginning of time" is understood to refer to a moment in time in which everyone in the population occupies only one of these states. Although strictly speaking studies should begin at the moment when all respondents are theoretically able to experience the event, but have not yet done so, in practice researchers often choose an arbitrary start time with the proviso that the start time is unrelated to event occurrence. Finally, Singer and Willett (2003) note that time should be measured in the smallest possible unit of time relevant to the study. Moreover they propose a simple test for researchers to determine when survival analysis is appropriate—suggesting that the techniques should be used for any research question that asks whether a specific event occurs or when a specific event occurs.

Finally, the application of survival analysis is dependent on the metric used for time. Most applications are based first on the determination of whether time is measured discretely (i.e., the event occurs within a certain time interval) or continuously (i.e., the timing is measured precisely). Different analytic techniques apply for these two situations (e.g., Singer & Willett, 2003; Willett & Singer, 1993). Although space does not permit an elaboration on survival analysis methods, we note that both Luke (1993) and Landau (2002) provide introductions to the technique specifically for psychologists. Most statistical packages provide routines for survival analysis (e.g., Cox regressions, Kaplan-Meier survival curves). Techniques for discrete and continuous survival analysis can also be conceptualized and operationalized within a structural equation modeling framework (see for example, Muthén & Muthén, 1998–2007).

Conclusion

It is clear that longitudinal methods are increasingly necessary to explore and explain pertinent constructs and relationships in OHP. Certain elements contributing to successful longitudinal investigations are well established, one being that at least three, ideally more time periods are necessary to model change effectively. Occupational health psychologists have several techniques available to them, such as time series analysis and latent growth curve modeling that allow detailed exploration of the nature of change over time. These models can easily be extended to allow occupational health psychologists to focus on questions pertaining to the prediction of health and safety events. That said, further theoretical development and descriptive research are needed to guide researchers in determining the appropriate time lags to incorporate in their longitudinal data collection efforts.

References

Austin, J. T., Sherbaum, C. A., & Mahlman, R. A. (2002). History of research methods in industrial and organizational psychology: Measurement, design, analysis. In S. G. Rogelberg (Eds.), *Handbook of research methods in industrial and organizational psychology* (pp. 1–33). Malden, MA: Blackwell.

Box, G. E. P., Jenkins, P. M., & Reinsel, G. C. (1994). *Time series analysis: Forecasting and control.* Englewood Cliffs, NJ: Prentice Hall.

Chan, D. (1998). The conceptualization and analysis of change over time: An integrative approach incorporating longitudinal mean and covariance structures analysis (LMACS) and multiple indicator latent growth modeling (MLGM). *Organizational Research Methods, 1,* 421–483.

Chan, D. (2011). Advances in analytical strategies. In S. Zedeck (Ed.), *APA handbook of industrial and organizational psychology: Vol. 1. Developing and building the organization.* Washington, DC: APA Books.

Cook, T. D., & Campbell, D. T. (1979). *Quasi-experimentation: Design and analysis for field settings.* Boston, MA: Houghton Mifflin

Elovainio, M., Leino-Arjas, P., Vahterra, J., & Kivimaki, M. (2006). Justice at work and cardiovascular mortality: A prospective cohort study. *Journal of Psychosomatic Research, 61,* 271–274.

Ferrer, E., & McArdle, J. J. (2010). Longitudinal modeling of developmental changes in psychological research. *Current Directions in Psychological Science, 19,* 149–154.

Garst, H., Frese, M., & Molenaar, P. C. M. (2000). The temporal factor of change in stressor-strain relationships: A growth curve model on a longitudinal study in East Germany. *Journal of Applied Psychology, 85,* 417–438.

Glomb, T. M. (2002). Workplace aggression: Confirming and informing conceptual models with data from specific encounters. *Journal of Occupational Health Psychology, 7,* 20–36.

Golembiewski, R. T., Billingsley, K., & Yeager, S. (1976). Measuring change and persistence in human affairs: Types of change generated by OD designs. *Journal of Applied Behavioral Science, 12,* 133–157.

Karanika-Murray, M. (2010). Work and health: Curvilinearity matters. In J. Houdamont & S. Leka (Eds.), *Global perspectives on research and practice: Vol.1. Con-*

temporary occupational health psychology (pp. 151–168). Chichester, England: Wiley Blackwell.

Kelloway, E. K., & Barling, J. (1994). Stress, control, wellbeing, and marital satisfaction: A causal correlational analysis. In G. P. Keita & J. J. Hurrell (Eds.), *Job stress in a changing workforce: Investigating gender, diversity, and family issues* (pp. 241–252). Washington, DC: American Psychological Association.

Kelloway, E. K., & Day, A. L. (2005). Building healthy organizations: What we know so far. *Canadian Journal of Behavioural Science, 37,* 223–236.

Kelloway, E. K., Gottlieb, B. H., & Barham, L. (1999). The source, nature, and direction of work and family conflict: A longitudinal investigation. *Journal of Occupational Health Psychology, 4,* 337–346.

Lance, C. E., Dawson, B., Birkelbach, B., & Hoffman, B. J. (2010). Method effects, measurement error, and substantive conclusions. *Organizational Research Methods, 13,* 435–455.

Landau, S. (2002). Using survival analysis in psychology. *Understanding Statistics, 1,* 233–270.

Luke, D. A. (1993). Charting the process of change: A primer on survival analysis. *American Journal of Community Psychology, 21,* 203–246.

MacEwen, K. E., Barling, J., & Kelloway, E. K. (1992). Effects of acute role overload on individual well-being and marital interaction. *Work & Stress, 6,* 117–126.

Maxwell, S. E., & Cole, D. A. (2007). Bias in the cross-sectional analysis of longitudinal mediation. *Psychological Methods, 12,* 23–44.

McArdle, J. J. (1988). Dynamic but structural equation modeling of repeated measures data. In J. R. Nesselroade & R. B. Cattell (Eds.), *Handbook of multivariate experimental psychology* (2nd ed., pp. 561–614). New York: Plenum Press.

McArdle, J. J. (2001). A latent difference score approach to longitudinal dynamic structureal analysis. In R. Cudeck, S. DuToit, & D.Sorbom (Eds.), *Structural equation modeling: present and future* (pp. 341–380). Lincolnwood, IL: Scientific Software International.

McArdle, J. J., & Hamagami., F. (1996). Multilevel models from a multiple group structural equation perspective. In G. Marcoulides & R. Schumaker (Eds.), *Advanced structural equation modeling techniques* (pp. 89–124). Hillsdale, NJ: Erlbaum.

Muthén, B. O. (2001). Second-generation structural equation modeling with a combination of categorical and continuous latent variables: New opportunities for latent class-latent growth modeling. In L. Collins & A. Sayer (Eds.), *New methods for the analysis of change* (pp. 291–322). Washington, DC: American Psychological Association.

Muthén, B. O. (2004). Latent variable analysis: Growth mixture modeling and related techniques for longitudinal data. In D. Kaplan (Ed.), *Handbook of quantitative methodology for the social sciences* (pp. 345–368). Thousand Oaks, CA: Sage.

Muthén, L. K., & Muthén, B. O. (1998–2007). *Mplus user's guide* (5th ed.). Los Angeles, CA: Muthén & Muthén.

Pitariu, A. H., & Ployhart, R. E. (2010). Explaining change: Theorizing and testing dynamic mediated longitudinal relationships. *Journal of Management, 36,* 405–429.

Ployhart, R. E., & Vandenberg, R. K. (2010). Longitudinal research: The theory, design, and analysis of change. *Journal of Management, 36,* 94–120.

Rogosa, D. (1988). Myths about longitudinal research. In K. W. Schaie, R. T. Campbell, & M. E. Rawlings (Eds.), *Methodological issues in aging research* (pp. 171-209). New York: Springer.

Rosel, J., & Plewis, I. (2008). Longitudinal data analysis with structural equations. *Methodology, 4,* 37–50.

Sauter, S. L., Murphy, L. R., & Hurrell, J. J. (1990). Prevention of work-related psychological disorders: A national strategy proposed by the National Institute for Occupational Safety and Health (NIOSH). *American Psychologist, 45,* 1146–1158.

Singer, J. D., & Willett, J. B. (2003). *Applied longitudinal data analysis: Modeling change and event occurrence.* New York: Oxford University Press.

Spector, P. E., & Brannick, M. T. (2010). Common method issues: An introduction to the feature topic in organizational research methods. *Organizational Research Methods, 13,* 403–406.

Stone-Romero, E. (2011). Research strategies in industrial and organizational psychology: Non-experimental, quasi-experimental, and randomized experimental research in special purpose and nonspecial purpose settings. In S. Zedeck (Ed.), *APA handbook of industrial and organizational psychology: Vol. 1. Building and developing the organization* (pp. 37–72). Washington, DC: APA Books.

Taris, T. (2003). Challenges in longitudinal designs in OHP. *Scandinavian Journal of Work and Environmental Health, 29,* 1–4.

Vandenberg, R. J., & Lance, C. E. (2000). A review and synthesis of the measurement invariance literature: Suggestions, practices and recommendations for organizational research. *Organizational Research Methods, 3,* 4–70.

Wang, M., & Bodner, T. E. (2007). Growth mixture modeling: Identifying and predicting unobserved subpopulations with longitudinal data. *Organizational Research Methods, 10,* 635–656.

Warr, P. B. (1987). *Work employment and mental health.* Oxford, England: Oxford University Press.

Willett, J. B., & Singer, J. D. (1993). Investigating onset, cessation, relapse and recovery: Why you should, and how you can, use discrete time survival analysis to examine event occurrence. *Journal of Consulting and Clinical Psychology, 61,* 952–965.

Williams, L. J., & Podsakoff, P. M. (1989). Longitudinal field methods for studying reciprocal relationships in organizational behaviour research: Towards improved causal analysis. In L. L. Cummings & B. M. Staw (Eds.), *Research in organizational behaviour* (Vol. 11, pp. 47–292). Greenwich CT: JAI Press.

Zapf, D., Dormann, C., & Frese, M. (1996). Longitudinal studies in organizational stress research: A review of the literature with reference to methodological issues. *Journal of Occupational Health Psychology, 1,* 145–169.

21

Looking Toward the Future of OHP Research

Robert R. Sinclair, Mo Wang, and Lois E. Tetrick

As we noted at the outset of this volume, occupational health psychology (OHP) has arrived as an established occupational health discipline with a thriving scientific community. OHP scholars have made tremendous progress in understanding the nature of occupational health problems and great strides in understanding the kinds of interventions necessary to address those problems. With this great progress in mind, what are the important next steps in the scientific progress of OHP? One answer to this question is to simply continue to do more and better science. However, in applied fields, such as OHP, there is no single clear answer as to what constitutes better science. For some, better science might mean using ever more sophisticated designs and analytic methods to capture the complexity of occupational health phenomena. For others, better science might mean choosing to analyze important occupational health problems, such as problems that are widespread in frequency, socially significant in their consequences, or both. Still others might argue that the positive psychology revolution challenges OHP scholars to reframe the orientation of the field toward creating healthy workplaces, rather than focusing solely on "problems" associated with undesirable health outcomes.

In our view there is no single best answer as to what constitutes good science, and thus, no easy answer to the question of how to improve occupational health science. We are reminded of Weick's (1979, 1999) discussion of Thorngate's (1976) trade-offs inherent in research, in which he points out that no theory can simultaneously be general (widely applicable to a range of phenomena), accurate (provide a precise explanation of phenomena), and simple (testable, verifiable).

Recommendations for best practices in research methods face similar kinds of potential trade-offs between measurement, design, and analysis, as well as between scientific rigor and scientific relevance. Namely, it is difficult to conduct studies that are simultaneously "strong" in all of the desired attributes of research because resources invested in one area typically result in a lack of resources in others. For example, researchers using multiwave longitudinal questionnaire-based research design typically use shorter measures in order to sustain participation over time. Similarly, researchers who seek to capture the causal relationships among occupational health constructs rely on experimental designs with a limited set of measures or with a sample that might be small or nonrepresentative of the workers they wish to understand. We suspect that most people reading this chapter have struggled with versions of these trade-offs in their own research programs. Better science requires simultaneous expansion across all dimensions of scientific quality, but in doing so introduces other challenges with integrating and disseminating knowledge. Despite these challenges, we promised our publisher we would discuss issues that, in our view, need to continue to be addressed by OHP scholars. And so, we shall.

In Order to Progress as a Scientific Discipline, OHP Researchers Need to …

Our goal with this volume was to raise the bar for future OHP research, both by encouraging scholars to take advantage of best practices recommendations in the measurement of important occupational health constructs, and in the design and analysis of OHP research. The chapter authors clearly responded to this challenge with an informative and engaging collection of chapters. They have helped us push the methodological boundaries that limit some OHP studies. But, we have unfinished business.

So, what is left on the methodological to-do list? To address this question, we chose to discuss six challenges/recommendations/observations related to continuing improvement in OHP scholarship. Although some of the issues were inspired by the chapters in this volume, our comments are not intended to integrate the chapters. Rather, our choices reflect topics that are common or important concerns in OHP research or that represent opportunities for better research in the future. As readers will note, several of these issues are not purely methodological. Rather, they reflect the intertwining of theoretical or practical considerations with design, measurement, and analytic choices.

Talk about Our Values

In the conventional view of the scientific process, the party-line position might be that OHP researchers should focus on accumulating reliable empirical generalizations about occupational health and that "better" methodology will produce ever better sets of generalizations as well as ever clearer boundary conditions to those generalizations. Yet, in applied sciences such as OHP, researchers often study research questions where the outcomes may be of considerable social significance (e.g., physical and mental health, quality of family life) and where the same organizational stakeholders control both the proximate causes of OHP related problems (i.e., the organization of work) and the resources necessary to address those problems (through hiring practices, family supportive work life policies, etc.). This situation leads to potential tensions in the relationship between OHP and private enterprise. Where there are such tensions, resulting from uncertainty, complexity, and differences of opinion, values creep into the conversation. Researchers' scientific, political, or ideological value systems shape not only their own research (what they find interesting or important), but also their perspectives on others' studies, something that can have important implications for a wide range of issues, such as the nature of the peer review process, advocacy about federal funding, debates about defining acceptable OHP content for conferences, and issues related to graduate student training.

We have observed at least three value orientations to "good" OHP science. Those advocating a *pure applied science* approach focus on theoretically or methodologically interesting research questions with less regard for their implications for organizations, or even their social significance. We distinguish this perspective from a basic science view in that the context is still organizational and workers are still the population of interest. In the pure applied science approach, good science is defined in terms of the extent to which it includes attributes such as reliable and valid measures, sufficient and generalizable samples, theoretical sophistication, and research design that promotes causal generalizations. Whether the research actually leads to healthier workplaces is a secondary consideration, or perhaps something assumed to happen down the road, but is not the scholar's primary interest.

In what might be called a "*productive engagement*" view, OHP scholars partner with organizations to solve problems that are linked to desired organizational outcomes such as performance, turnover, and absenteeism. Many of the attributes of the pure applied science view are still present, but the goodness of the research is defined more in terms of its ability to help organizations accomplish their goals, which reflects the ability of the study to explain variance in something the organization cares about. The productive engagement view can be seen in efforts to advocate the "business case" for occupational health. This effort seems to reflect the assumption that if researchers can gather enough evidence to demonstrate the added value of focusing on OHP concerns, businesses will start implementing OHP interventions.

Finally, in what might be viewed as a more *adversarial* perspective, organizations are viewed as the primary barrier to occupational health. In this view, OHP research should focus on health (or other socially valued outcomes) outcomes with business concerns as secondary, at best. Those who hold this view might support the position that labor unions and public policy interventions are required to create healthy workplaces, and that most organizations will ignore or resist efforts to create a healthy workplace.

One example of where ideological differences surface is in how people view research focused on individual attributes (e.g., personality and stress management) or individual behaviors (e.g., behavior based safety programs) in the role of safety and health. We have witnessed many discussions among scholars with widely divergent perspectives on the merits of behavior-based safety and stress-management interventions largely because they focus on changing qualities of workers (actions, belief systems, etc.) rather than changing characteristics of workplaces that produce exposures.

OHP scholars have widely differing and strongly held views about such individually focused interventions: those with a pure applied science perspective might view this research as important to isolate the components of variance related to safety/health outcomes for the purpose of building theories of occupational health. Those with a productive engagement view focus on how person and situation factors contribute to organizationally valued outcomes, perhaps asking questions about whether there is greater gain from efforts to hire for safety or from changes in safety climate/working conditions. Those in the adversarial camp might believe it is fundamentally unethical to focus on worker characteristics (e.g., hiring workers based on traits that might make them more resilient to stressors) without first exhausting all possible options to change the workplace (e.g., eliminating or controlling workplace hazards). These views generally reflect differing values about science, different ideological perspectives on the relationship between science and social action, and differing theoretical orientations. They rarely are discussed in journals.

As should be evident, most people probably do not fit neatly into one of these three categories and the discussion above oversimplifies a complex problem space. Indeed, any particular scholar might conduct one study that fits the pure science view, and another that is more focused on productive engagement. However, we believe it is fair to say that certain individuals, certain research literatures, and certain methodological traditions reflect a tendency toward one or the other of these value orientations. Because they reflect passionately held positions, people often have trouble discussing them in public; and, researchers who share the same views tend to publish in the same journals and read the same research. Social psychologists have discussed how such group dynamics lead to polarization of views over time (Isenberg, 1986). As a result, there is an ideological undercurrent to OHP scholarship that is not explicitly discussed in published articles, but that has theoretical, methodological, and practical implications for the field.

But what is the problem? Isn't it desirable to have a diverse set of values? Of course, the answer is probably yes, but such issues are never openly discussed. A practical case of where the issue comes up is in how different people view the mission of the *Journal of Occupational Health Psychology*; pure applied scientists view it as a venue for publishing the best empirical research in OHP; productive engagement scholars might raise questions about studies addressing issues of limited practical significance; scholars in the adversarial camp might see the journal as missing opportunities to be part of the vanguard leading efforts at social change. With this example in mind, we see several possible directions. One is to simply accept the status quo and leave these issues alone. They are difficult to address and it is unclear whether they can be addressed constructively. A second direction would be to articulate one or another set of values as the dominant ideology—to say that OHP is about science, or about helping organizations, or about stimulating social change, and let the dominant ideology guide the field. A third possible direction, which we recommend, is open discussion and debate about issues such as the relationship between OHP and private enterprise. We could ask legitimate questions about whether theoretical and methodologically oriented research does enough to help workers. We could ask whether OHP should be doing more in the way of advocating social policy. We could ask whether the values we espouse about safer and healthier workplaces are sufficiently consistent with our actions. None of these questions are purely methodological in nature, but they have methodological implications, both in terms of defining methodological quality in OHP research and in terms of the relationship between methodology, theory, and practical issues in OHP scholarship. Moreover, answering each of these questions requires clear and consistent terminology, a problem that still plagues many areas of the OHP literature. We deal with this issue next.

Find a Cure for Terminological Diversity

Like us, we suspect that most readers have encountered certain bodies of scientific literature with what we might call terminological diversity—many different names for the same essential ideas. For example, leadership literature has produced many different concepts all reflecting the basic idea that followers benefit when leaders treat them well: examples of this concept include literature on leader–member exchange (Thomas & Lankau, 2009), perceived organizational support (Casper, Martin, Buffardi, & Erdwins, 2002), transformational leadership (Arnold, Turner, Barling, Kelloway, & McKee, 2007), supervisor social support (Baruch-Feldman, Brondolo, Ben-Dayan, & Schwartz, 2002), and interactional justice (Kausto, Elo, Lipponen, & Elovainio, 2005). Each of these concepts can be captured with self-report items that, either directly or indirectly reflect followers' perceptions about how they are treated by their supervisor, and studies that include multiple measures of these constructs are quite likely to find high correlations among them. Although fine grained theoretical distinctions may be made among these concepts, in effect, they all address the basic idea that leaders should treat their employees well. This is an idea that had been clearly identified in the behavioral models of leadership dating back to the mid-20th century with the description of consideration as a form of leader behavior in the Ohio State studies (e.g., Hemphill & Coons, 1957) and Likert's democratic model of leadership (e.g., Likert, 1961).

Problems with terminological diversity are not unique to leadership literature; Block (1995) discussed similar issues in personality literature and pointed out that problems associated with having multiple terms for the same basic concept were recognized among psychologists over 80 years ago (e.g., Kelly, 1927 as cited in Block, 1995). Terminological diversity reflects a commendable desire to bring fresh perspectives to the literature, and innovating thinking certainly may generate new antecedents, outcomes, interventions, and processes. However, terminological diversity comes with costs. As Block bluntly states, useless concept redundancy is a waste of time; scholars often wind up rediscovering the same basic phenomena over and over.

Block characterized problems with terminological diversity as stemming in part from a lack of historical knowledge. However, in the age of electronic literature searches, terminology arguably becomes even more important because choosing the wrong search term (or even the wrong search database) can lead scholars to draw incorrect conclusions about a topic even before they read any of the relevant literature. These problems present significant challenges to most scientific disciplines, but are particularly problematic in fields involving applied research conducted by scholars in multiple disciplines (OHP being a prototypical example). The result is that scientific advancement occurs at a much slower pace than what might otherwise be possible and workers do not benefit as much from OHP research as researchers might prefer. Moreover, inability for a discipline to settle on clear labels for constructs creates communication challenges both among researchers in different disciplines and between researchers and practitioners.

Ultimately, a stronger emphasis on construct validity is the best way to address terminological diversity problems. This includes evidence about both incremental predictive validity (i.e., of any "new" constructs beyond other conceptually similar constructs) and structural distinctiveness (i.e., whether measures of purportedly different constructs actually load on different factors). As one example, Judge and colleagues (e.g., Judge, Locke, & Durham, 1997) as well as numerous other scholars, have provided substantial evidence that personality traits such as self-esteem, neuroticism, locus of control, and self-efficacy all reflect so-called core self-evaluations (CSE): fundamental judgments people make about themselves. The CSE literature proposes that a single global factor may explain much of the variance between people on these traits, as well as their correlates (cf., Chang, Ferris, Johnson, Rosen, & Tan, 2012). Thus, as compared with the more fragmented literature on individual personality traits, the CSE literature provides a more parsimonious model of self-assessments and a foundation upon which to investigate individual differences related to OHP. We can envision how similar investigations might be useful in a wide variety of OHP-related topics.

In addition to considering empirical evidence about the relationships among constructs, researchers might address terminological diversity problems by constructing stronger conceptual and operational definitions (cf., Shadish, Cook, & Campbell, 2002). Indeed, researchers still express concerns about the seeming inability of OHP scholarship to arrive at a clear and consistent usage of even fundamental terms such as stress, stressor, and strain. Efforts to clarify conceptual and operational definitions should also create opportunities for new scholarship.

For example, Hershcovis (2011) expressed concerns about construct proliferation in studies of bad behavior in organizations (e.g., incivility, social undermining, bullying). She posed two questions: "First, is construct differentiation and proliferation yielding new insights? Second, is it feasible to reconfigure these constructs to enable their examination in a more concise and informative manner?" (p. 500). She addressed these issues in part by conducting an integrative review of the literature around several similar constructs, concluding that many of them fit under the general rubric of aggression, with differences between them more appropriately viewed as moderators of the relationship between aggression and outcomes. Such reviews help to identify common themes and core concepts in literature as well as providing stronger conceptual standards for the introduction of new constructs. Thus, we would encourage researchers to carefully consider the questions Hershcovis raised when considering the introduction of a new concept into the OHP literature.

Researchers also might consider building models that integrate similar concepts across multiple levels of aggregation, as appropriate to the phenomena in question. For example, broad measures of the five-factor model (FFM) of personality emphasize the idea that each of the FFM traits may be further divided into multiple facets (cf. Costa & McCrae, 1992), even as others have found evidence that the FFM traits themselves may be organized into smaller numbers of global personality factors (e.g., Digman, 1997; Ones & Visweswaran, 2001; van der Linden, Nijenhuis, & Bakker, 2010).

Similarly, much of the literature on organizational justice might be viewed as fitting a hierarchical model with the broad idea of fair treatment in organizations divided into distributive, procedural, and interpersonal, and informational dimensions (cf. Colquitt, Conlon, Wesson, Porter, & Ng, 2001). Each justice type can be further subdivided into more specific types of justice. For example, there are at least three distinct distributive justice rules: equity, equality, and need, and multiple forms of procedural justice. As Colquit, Greenberg, and Zapata-Phelan (2005) point out in a historical review of the justice literature, justice scholars have wrestled with the issue of the relationship among different types of justice over the years, and that the struggle has resulted in both greater understanding of existing constructs, and "frame-breaking" progress in understanding of justice.

The idea of multilevel concepts may be extended by building models that explicitly address conditions under which broader or more specific concepts might be more useful. This might be viewed as building a theory of construct operationalization that stipulates the conditions under which narrower or broader measures are appropriate, as well as the conditions under which broader measures are sufficient. Such an approach could draw on discussions about bandwidth and fidelity in personality literature, where researchers have discussed whether broad or narrow personality measures predict broad or narrow criteria (cf. Hogan & Roberts, 1996; Ones & Viswesvaran, 1996).

Although research streams typically flow toward greater precision and specificity of ideas over time, and thus, larger sets of narrow (and potentially more fragmented) relationships among constructs, it is possible that more global models will provide useful insights about occupational health. For example, Harrison, Newman, and Roth (2006) found support for what they termed an "attitude-engagement" model in which overall job attitudes (a combination of job satisfaction and organizational commitment) predicted a general behavioral criterion construct consisting of measures of job performance, citizenship behavior, and withdrawal (lateness, absenteeism, turnover). It certainly seems possible that similar ideas could be investigated in occupational health literature, with general measures of organizational climate, subjective well-being, or physical health as examples.

Understand the Role of Time

As the preceding section discusses, the challenge of terminological diversity may create lack of precision in our understanding of the phenomena within OHP and the relations of these phenomena to the related fields of industrial and organizational science, management and organizational science, public health, and health and well-being.

Perhaps this is also a result of incomplete development of the theories which guide the design and conduct of OHP research. We would raise an additional concern in the OHP literature, acknowledging that this is not a challenge exclusively for OHP as it is an issue in many of the OHP-related fields, and this is the lack of explicit consideration of time.

Major theories in the area of occupational stress, workplace violence, work–family interactions, and safety are relatively silent about how quickly or slowly an effect emerges. For example, if one takes a physiological approach to stress such as the general adaptation syndrome (Selye, 1955) then one might expect the stress response to be much quicker, and perhaps to dissipate more quickly, than if one takes a more sociocognitive approach as in the job demands–control model of stress (Karasek & Theorell, 1990) or the effort–reward imbalance model (Siegrist, 1996). These latter theories seem to imply that there is a cumulative or chronic effect of exposure to stressors. These implied differences in time perspective may account for the general disconnect between physiological measures of experienced stress and self-reports of stress (Fried, 1994; Fried, Rowland, & Ferris, 1984). However, these differences have not been specifically discussed in our theoretical development efforts, considered in the design of our research, or incorporated into the development of our measures.

There are very few studies that have addressed the distinction between acute versus chronic stressors and most of our measures consider the phenomena to be relatively chronic. As discussed in the chapter on work–family concepts measurement, most measures of work–family conflict have relied on Likert-type response scales. This potentially confounds the occurrence of the stressor and the experience of the stressor. As Belavia and Frone (2005) and others have argued (see Chapter 3), it is important for us to be able to distinguish the occurrence of the stress in order to better determine the prevalence of a particular stressor and the effects of the stressor. Does it matter whether an individual experiences an incident of work interfering with family once a week or once a month? Can we develop dose–response curves relating the occurrence of certain stressors and specific indicators of positive health and ill-health?

The need to more explicitly incorporate time into our theoretical perspectives is not new (McGrath & Rotchford, 1983), but it is interesting and disappointing that we have made so little progress. There is some movement forward with new designs that take a more refined look at the progression of health effects. For example, Frone (2008) carefully considered the theoretical perspective on substance use and work stress to incorporate temporal context. By doing so he was able to clarify the inconsistencies in the literature about the relation between substance use and work stress. Similarly, Sonnentag and her colleagues (Binnewies, Sonnentag, & Mojza, 2010; Fritz, Sonnentag, Spector, & McInroe, 2010; Kühnel & Sonnentag, 2011; Sonnentag, 2011; Sonnentag & Niessen, 2008) have enhanced our understanding of the time perspective in recovery from work. This work has examined daily, weekly, and short-duration absence (e.g., vacations and weekends) to determine the duration of effects. This extends the relative few longitudinal studies which often select a causal interval of one year without necessarily any theoretical framework on which to determine the most appropriate causal interval other than perhaps what the participating organization will support (e.g., DeJoy, Wilson, Vandenberg, McGrath-Higgins, & Griffin-Blake, 2010). More theoretical and empirical research needs to be undertaken to guide the selection of appropriate timing of assessing the effects of exposure to specific events and interventions.

We know little based on our theories about whether the interval between assessments should be nanoseconds, minutes, hours, days, weeks, months, years or longer. The answer to these and similar questions about the role of time in OHP events, be they exposures or interventions, lies in better developed theories with respect to timing of events and responses to these events and incorporation of these theoretical refinements in our designs and measures.

Develop (and Use) Norms for Measures

Another consideration in the OHP literature is that many, if not most, of the measures that we employ do not have established norms. Without such norms, it isn't really possible to know what is high or low. For example, in many studies examining the job demands–control model, the values of job demands are determined to be "high" or "low" based on the distribution of responses in the specific study. Taking this approach, however, does not enable us to examine results across studies since "high scores" in one sample might be lower than "high scores" in another study (see chapter 19).

As expressed elsewhere, there are no generally agreed upon measures of many of the constructs in OHP. Wiegand et al. (2012) identified a set of OHP-relevant measures which might serve as a basis of standardized measures. If researchers agreed to consistently use the same measures, then it would be possible to develop appropriate norms that might hold in different contexts and for different people. Perhaps we would need norms for women versus men or different norms for people in manufacturing positions versus those in sales positions. Similarly, it is likely that we would need norms for people and situations in North American and European countries versus those for people and situations in Eastern countries. The literature is basically silent at this point in time and there is little consensus as to which measures might be the best to adopt as the standard.

The press of applied field research to keep measures short may further delay the acceptance of certain measures in favor of others. However, we would argue that without appropriate norms for the measures that we use, the most we can hope for is to express the relations we find among variables in relative terms and we do not have the ability to make a determination that in a specific situation people may be experiencing unsafe or unhealthy environments or that the effects of a specific intervention enhanced employees' health status from "at risk" to "optimal functioning."

Clarify Some External Validity Related Issues

Although several chapters in this volume have discussed issues related to external validity (e.g., Chapters 12, 14, 21), their scopes are more specific and technical. Given that OHP research often involves various types of samples, especially those that are from different countries and cultures, we feel it is important for us to conceptually address some common confusion related to external validity here. In particular, we hope to caution OHP researchers about the bias in judging a sample's representativeness, as well as recommend distinguishing theoretical generalizability and statistical generalizability as two different research design goals.

External validity refers to the extent to which the results from a scientific study can generalize to other populations, settings, or contexts. Specifically, inferences based on research findings are said to possess external validity if they may be generalized from the unique and idiosyncratic settings, procedures, and participants to the populations and conditions that are of interest (Shadish et al., 2002). One important factor that often influences a study's external validity is the representativeness of the research sample. The representativeness of the sample is critical when the study aims to answer research questions about a specific, well-defined population of interest (Highhouse & Gillespie, 2008). For example, a researcher who is specifically interested in studying the safety process among firefighters could not sample from any population other than the population of firefighters to answer his or her research question. To achieve the representativeness of the sample, researchers often use random sampling, which involves selecting people by chance from a clearly defined population. Following this procedure, a sample that matches the population on all attributes (e.g., mean and variance) is generated, thus eliminating the possibility that some members of the population may be oversampled or undersampled (Shadish et al., 2002). This way, the data from this sample can be used to draw statistical conclusions about the target population, thus establishing the generalizability of the statistical findings to the target population. Unfortunately, random sampling is often too costly and is less often used in conducting OHP research (for some exceptions, see Frone, 2008; Wang, 2007).

Highhouse (2009) cautioned about judging a sample's representativeness merely based on the superficial similarity between the sample and the target population. For example, researchers often prefer using field samples to using college student samples in studying OHP phenomena. A typical argument for this preference is that workers from field samples are more generalizable to "people" in organizations than college student samples. However, scrutinizing this argument, it certainly does not apply universally. The superior generalizability of field samples depends on whether the specific work experiences of the research sample would influence the phenomena being studied in a way that would confound the results of the study (Campbell, 1986). If there is no good reason to expect the confounding effect of full-time work experience, college student samples would be just as generalizable as field samples in answering the research question (Highhouse, 2009). In other words, if the constructs and processes are essentially the same across field and student samples (i.e., maintaining the psychological fidelity), then using student samples should render same generalizability as using field samples.

It is also important to note that social science researchers, including OHP researchers, are typically more interested in theoretical generalizability than statistical generalizability (Highhouse & Gillespie, 2008; Sackett & Larson, 1990). In other words, we are more interested in whether a causal relationship, but not the particular effect size of that causal relationship, may hold across populations. To achieve theoretical generalizability, it is not necessary to use samples that are strictly representative of the population, but only necessary to use samples that do not systematically differ from the population in a way that would interact with the causal relationship. Oftentimes, we see researchers who used samples from non-Western cultures/countries apologetically note in the limitation sections of their papers that findings from their samples may not generalize to Western populations. However, before hastily reaching that conclusion, they should scrutinize whether their samples provide adequate theoretical generalizability or psychological fidelity. Studies using samples from Western cultures/countries should not be immune to this scrutiny either. For example, to the extent that the differences between a sample of American participants and a sample of Chinese participants (e.g., in their cultural values) do not interact with the causal effect of supervisor support on work–family conflict, using either the American sample or the Chinese sample in the study should provide the same theoretical generalizability of this causal relationship to both American and Chinese populations. Given that true random sampling is rarely used in applied behavioral research (Shadish et al., 2002), scrutinizing the theoretical generalizability of the research sample is particularly important in ensuring the external validity of a study.

Apply Dynamical Systems to OHP Research

It has become increasingly common for researchers to conceptualize psychological, organizational, and broader social science phenomena in terms of self-organizing, dynamic systems (e.g., Vallacher & Novick, 1994; Vancouver, Weinhardt, & Schmidt, 2010). Although we did not allocate a particular chapter in this volume to provide a detailed introduction of this dynamic system approach, we feel it is necessary to mention it here as a potential future direction that could be fruitful for advancing OHP research. It also presents a natural extension to Kelloway and Francis's chapter (i.e., Chapter 20) in this volume.

A dynamic system is a collection of elements whose behavior or output, at the system level, evolves over time (Byrne, 1998). This change of behavior or output can result from environmental pressure or it could result from the system itself, as the result of multiple feedback loops that either amplify or minimize fluctuations among elements within the system (DeShon, in press; Vallacher & Nowack, 1994). Therefore, dynamic systems describe a type of causal model that we rarely consider in OHP research, focusing on explicating multiple causal relationships (unidirectional or reciprocal) simultaneously as well as the collective outcome of these causal relationships at the system level as a function of time.

Here, we use the study of safety climate development as an example to illustrate how a dynamic system perspective may be applied to OHP research. When a work group is formed, members of the team may exhibit various types of safety behaviors. We can conceptualize the work group as a dynamic system, with each member of the group as an element of the system. Now, each member's safety behavior may be influenced by two causal processes. One is the interpersonal process—each group member's safety behavior may be influenced by other members' safety behaviors through mechanisms such as social learning, social exchange, and impression management. This represents a reciprocal causal process through which elements of the system influence each other. The other process is a top-down process, where the organization's formal and informal policies may also shape the work group members' safety behaviors. Of course, depending on how each element may react to this top-down influence, its effect on each work group member will likely to vary. This represents the unidirectional environmental pressure on the output of the elements of the system. Now, imagine that both processes have been allowed to happen for a sufficient amount of time, what can be expected then is that as a whole system, the work group's collective safety practice will be different from when it was started. This change in collective safety practice, or safety climate, represents how the system itself is influenced by causal processes from internal and external sources.

Based on the above illustration, it is evident that dynamic systems can offer researchers a conceptual way to connect and analyze OHP phenomena that could form collective patterns or changes that have important implications over time. The key is to recognize how elements in a system are interconnected and how the causal processes present in the system may shape the system's output as a whole. Generally, the dynamic system approach argues that a system typically is composed of multiple, richly interconnected elements that interact and influence one another over time through various feedback loops (Levy, 1992). The nerve system, social groups, organizations, and societal cultural systems are all examples of such systems. For example, neurons are connected by a rich network that communicates using electrical impulses and neurochemicals, whereas people are increasingly interconnected and can communicate through face-to-face communications, e-mails, social media, phone, and written letters. These elements influence one another and their interactions leave traces and marks on the system. For example, the interactions among the elements tend to become more hierarchically organized over time in the system (Morel & Ramanujam, 1999). This self-organization may further result in emergent behaviors that are observed at the holistic, system level (e.g., organizational human capital, leadership, team learning and development of team mental-models, culture; Bell, Kozlowski, & Blawath, in press; Vallacher & Nowak, 1994).

Given the very tangled feedback loops among system elements, the utility of applying research designs that only assume unidirectional causality in an attempt to understand how elements within a dynamic system produce emergent behavior is clearly suspect. Therefore, it is absolutely necessary to study both elements' and systems' states and outputs over time. As such, the research design has to incorporate both multilevel and longitudinal design features. However, researchers can also use computational models to study dynamic systems (Vancouver et al., 2010). With this methodology, researchers build a computational model as suggested by their verbal theories of the dynamic system. The model specifies the elements of the system and the connections as well as feedback loops among the elements. The patterns of emergent behavior produced by the computational model are compared to the behavioral patterns of the real system. The computational specification of the theory is supported if there is a match between the predicted pattern and the data. Interested readers can see Vancouver and colleagues' work (Vancouver, Tamanini, & Yoder, 2010; Vancouver et al., 2010) for excellent illustrations and discussions of this methodology. Finally, we note that qualitative research designs can collect useful information with regard to dynamic systems in addition to quantitative research designs. Given that the causal influence flows in all directions for a dynamic system, qualitative research designs are extremely useful for generating initial hypotheses regarding how elements are connected and how external factors of the system may shape the elements' behaviors.

Conclusion

Occupational health psychology is in a period of tremendous growth in the overall quality, theoretical sophistication, and social significance of OHP research. Each year, more scholars are drawn to the field, and each year we see more and better OHP research published in leading psychology journals. As noted throughout this volume and specifically in this chapter, some persistent challenges remain for improving the overall quality of OHP research, and thus, for the ability of our field to help improve the quality of workers' lives. We hope this volume will inspire scholars to redouble their efforts to conduct high quality and high impact research. As we noted at the outset, there is no one right answer to how to accomplish this goal, but we look forward to seeing how the field develops in the years ahead.

References

Arnold, K. A., Turner, N., Barling, J., Kelloway, E. K., & McKee, M. C. (2007). Transformational leadership and psychological well-being: The mediating role of meaningful work. *Journal of Occupational Health Psychology, 12*, 193–203.

Baruch-Feldman, C., Brondolo, E., Ben-Dayan, D., & Schwartz, J. (2002). Sources of social support and burnout, job satisfaction, and productivity. *Journal of Occupational Health Psychology, 7*, 84–93.

Bell, B. S., Kozlowski, S. W. J., & Blawath, S. (in press). Team learning: A theoretical integration and review. In S. W. J. Kozlowski (Ed.), *The Oxford handbook of organizational psychology*. New York: Oxford University Press.

Bellavia, G., & Frone, M. R. (2005). Work-family conflict. In J. Barling, E. K. Kelloway, & M. R. Frone (Eds.), *Handbook of work stress* (pp. 113–147). Thousand Oaks, CA: Sage.

Binnewies, C., Sonnentag, S., & Mojza, E. J. (2010). Recovery during the weekend and fluctuations in weekly job performance: A week-level study examining intra-individual relationships. *Journal of Occupational and Organizational Psychology, 83*(2), 419–441. doi:10.1348/096317909X418049

Block, J. (1995). A contrarian view of the five-factor approach to personality description. *Psychological Bulletin, 117*, 187–215.

Byrne, D. (1998). *Complexity theory and the social sciences: An introduction*. New York: Routledge Press.

Campbell, J. P. (1986). Labs, fields, and straw issues. In E. A. Locke (Ed.), *Generalizing from laboratory to field settings* (pp. 269–279). Lexington, MA: Heath.

Casper, W. J., Martin, J. A., Buffardi, L. C., & Erdwins, C. J. (2002). Work–family conflict, perceived organizational support, and organizational commitment among employed mothers. *Journal of Occupational Health Psychology, 7*, 99–108.

Chang, C-H., Ferris, D. L., Johnson, R. E., Rosen, C. C., & Tan, J. A. (2012). Core self-evaluations: A review and evaluation of the literature. *Journal of Management, 38*, 81–128.

Colquitt, J. A., Conlon, D. E., Wesson, M. J., Porter, C. O. L. H., & Ng, K. Y. (2001). Justice at the millennium: A meta-analytic review of 25 years of organizational justice research. *Journal of Applied Psychology, 86,* 425–445.

Colquitt, J. A., Greenberg, J., & Zapata-Phelan, C. P. (2005). What is organizational justice? A historical overview. In J. Greenberg & J. A. Colquitt (Eds.), *The handbook of organizational justice* (pp. 3–56). Mahwah, NJ: Erlbaum.

Costa, P. T., & McRae, R. R. (1992). *Revised NEO personality inventory (NEO-PI-R) and NEO five-factor Inventory (NEO-FFI) professional manual.* Odessa, FL: Psychological Assessment Resources.

DeJoy, D. M., Wilson, M. G., Vandenberg, R. J., McGrath-Higgins, A. L., & Griffin-Blake, C. S. (2010). Assessing the impact of healthy work organization intervention. *Journal of Occupational and Organizational Psychology, 83,* 139–165. doi:10.1348/096317908X398773

DeShon, R. P. (in press). Multivariate dynamics in organizational science. In S. W. J. Kozlowski (Ed.), *The Oxford handbook of organizational psychology.* New York: Oxford University Press.

Digman, J. M. (1997). Higher-order factors of the Big Five. *Journal of Personality and Social Psychology, 73,* 1246–1256.

Fried, Y. (1994). The future of physiological assessments in work situations. In C. L. Cooper & R. Payne (Eds.), *Studies in occupational stress: Causes, coping and consequences of stress at work* (pp. 343–373). Oxford, England: Wiley.

Fried, Y., Rowland, K. M., & Ferris, G. R. (1984). The physiological measurement of work stress: A critique. *Personnel Psychology, 37*(4), 583–615. doi:10.1111/j.1744-6570.1984.tb00528.x

Fritz, C., Sonnentag, S., Spector, P. E., & McInroe, J. A. (2010). The weekend matters: Relationships between stress recovery and affective experiences. *Journal of Organizational Behavior, 31*(8), 1137–1162. doi:10.1002/job.672

Frone, M. R. (2008). Are stressors related to employee substance use? The importance of temporal context assessments of alcohol and illicit drug use. *Journal of Applied Psychology, 93,* 199–206.

Harrison, D. A., Newman, D. A., & Roth, P. L. (2006). How important are job attitudes? Meta-analytic comparisons of integrative behavioral outcomes and time sequences. *Academy of Management Journal, 49,* 305–325.

Hemphill J. K., & Coons A. E. (1957). Development of the leader behavior description questionnaire. In R. M. Stogdill & A. E. Coons (Eds.), *Leader behavior: Its description and measurement* (pp. 6–38). Columbus, OH: Ohio State University Press.

Hershcovis, S. (2011). "Incivility, social undermining, bullying … oh my!": A call to reconcile constructs within workplace aggression research. *Journal of Organizational Behavior, 32,* 499–519.

Highhouse, S. (2009). Designing experiments that generalize. *Organizational Research Methods, 12,* 554–566.

Highhouse, S., & Gillespie, J. Z. (2008). Do samples really matter that much? In C. E. Lance & R. J. Vandenberg (Eds.), *Statistical and methodological myths and urban legends: Doctrine, verity and fable in the organizational and social sciences* (pp. 247–266). New York: Psychology Press.

Hogan, J., & Roberts, B. W. (1996) Issues and non-issues in the fidelity-bandwidth trade-off. *Journal of Organizational Behavior, 17,* 627–637.

Isenberg, D. J. (1986). Group polarization: A critical review and meta-analysis. *Journal of Personality and Social Psychology, 50,* 1141–1151.

Judge, T. A., Locke, E. A., & Durham, C. C. (1997). The dispositional causes of job satisfaction: A core evaluations approach. *Research in Organizational Behavior, 19,* 151–188.

Karasek, R. A., & Theorell, T. (1990). *Healthy work, stress, productivity and the reconstruction of working life.* New York: Basic Books.

Kausto, J., Elo, A., Lipponen, J., & Elovainio, M. (2005). Moderating effects of job insecurity in the relationships between procedural justice and employee well-being: Gender differences. *European Journal of Work and Organizational Psychology, 14,* 431–452.

Kühnel, J., & Sonnentag, S. (2011). How long do you benefit from vacation? A closer look at the fade-out of vacation effects. *Journal of Organizational Behavior, 32*(1), 125–143. doi:10.1002/job.699

Likert, R. (1961). *New patterns of management.* New York: McGraw-Hill.

Levy, S. (1992). *Artificial life: The quest for new creation.* New York: Random House.

McGrath, J. E., & Rotchford, N. L. (1983). Time and behavior in organizations. *Research in Organizational Behavior, 5,* 57–101.

Morel, B. M., & Ramanujam, R. (1999) Through the looking glass of complexity: Organizations as adaptive and evolving systems, *Organization Science, 10,* 278–293.

Ones, D. S., & Viswesvaran, C> (1996), Bandwidth-fidelity dilemma in personality measurement for personnel selection. *Journal of Organizational Behavior, 17,* 609–626.

Ones, D. S., & Viswesvaran, C. (2001). Personality at work: Criterion-focused occupational personality scales used in personnel selection. In B. W. Roberts & R. Hogan (Eds.), *Personality psychology in the workplace* (pp. 63–92). Washington, DC: APA Books.

Sackett, P. R., & Larson, J. R., Jr. (1990). Research strategies and tactics in industrial and organizational psychology. In M. D. Dunnette & L. M. Hough (Eds.), *Handbook of industrial and organizational psychology* (2nd ed., Vol. 1, pp. 419–489). Palo Alto, CA: Consulting Psychologists Press.

Selye, H. (1955). Stress and disease. *Science, 122,* 625–631.

Siegrist, J. (1996). Adverse health effects of high-effort/low reward conditions. *Journal of Occupational Health Psychology, 1,* 27–41.

Shadish, W., Cook, T., & Campbell, D. (2002). *Experimental and quasi-experimental designs for generalized causal inference.* Boston, MA: Houghton Mifflin.

Sonnentag, S. (2011). Recovery from fatigue: The role of psychological detachment. In P. L. Ackerman (Ed.), *Cognitive fatigue: Multidisciplinary perspectives on current research and future applications* (pp. 253–272). *Decade of Behavior/Science Conference.* doi:10.1037/12343-012

Sonnentag, S., & Niessen, C. (2008). Staying vigorous until work is over: The role of trait vigour, day-specific work experiences and recovery. *Journal of Occupational and Organizational Psychology, 81*, 435–458. doi:10.1348/096317908X310256

Thomas, C. H., & Lankau, M. J. (2009). The effects of LMX and mentoring on socialization, role stress, and burnout. *Human Resource Management, 48*, 417–432.

Thorngate, W. (1976). Possible limits on a science of social behavior. In J. H. Strickland, F. E. Aboud, & K. J. Gergen (Eds.), *Social psychology in transition* (pp. 121–139). New York: Plenum.

Vallacher, R. R., & Nowak, A. (1994). *Dynamical systems in social psychology.* New York: Academic Press.

Vancouver, J. B., Tamanini, K. B., & Yoder, R. J. (2010). Using dynamic computational models to reconnect theory and research: Socialization by the proactive newcomer as example. *Journal of Management, 36*, 764–793.

Vancouver, J. B., Weinhardt, J. M., & Schmidt, A. M. (2010). A formal, computational theory of multiple-goal pursuit: Integrating goal-choice and goal-striving processes. *Journal of Applied Psychology, 95*, 985–1008.

van der Linden, D., Nijenhuis, J., & Bakker, A. B. (2010). The general factor of personality: A meta-analysis of Big Five intercorrelations and a criterion-related validity study. *Journal of Research in Personality, 44*, 315–327.

Wang, M. (2007). Profiling retirees in the retirement transition and adjustment process: Examining the longitudinal change patterns of retirees' psychological wellbeing. *Journal of Applied Psychology, 92*, 455–474.

Weick, K. E. (1979). *The social psychology of organizing* (2nd ed.). Reading, MA: Addison-Wesley.

Weick, K. E. (1999). Theory construction as disciplined reflexivity: Trade-offs in the 90s. *Academy of Management Review, 34*, 797–806.

Wiegand, D. M., Chen, P. Y., Hurrell, J. J., Jex, S., Nakata, A., Nigam, J., … Tetrick, L. E. (2012). A consensus method for updating psychosocial measures used in NIOSH health hazard evaluations. *Journal of Occupational and Environmental Medicine, 54*, 350-355.

Author Index

Subject Index

Page locators in *italics* indicate figures and tables.